Louis Bell, Joseph Wetzler, Thomas Commerford Martin

The Electric Motor and Its Applications

By T.C. Martin and Joseph Wetzler - Vol. 1

Louis Bell, Joseph Wetzler, Thomas Commerford Martin

The Electric Motor and Its Applications
By T.C. Martin and Joseph Wetzler - Vol. 1

ISBN/EAN: 9783337734060

Printed in Europe, USA, Canada, Australia, Japan

Cover: Foto ©Andreas Hilbeck / pixelio.de

More available books at **www.hansebooks.com**

The Electric Motor

AND ITS APPLICATIONS.

BY

T. C. MARTIN and JOSEPH WETZLER.

THIRD EDITION

WITH

AN APPENDIX

ON

THE DEVELOPMENT OF THE ELECTRIC MOTOR SINCE 1888.

BY

DR. LOUIS BELL,

NEW YORK:

THE W. J. JOHNSTON COMPANY, L'T'D,

TIMES BUILDING.

1892.

PREFACE TO THE THIRD EDITION.

THE two previous editions of this work have served such a useful purpose, and have so thoroughly covered the development of the practical electric motor to about the year 1888, that in issuing this third edition no attempt whatever has been made to revise the text of the work. In the last two years the theory of the electric motor has come to be better understood, and practice has undoubtedly improved, but the efforts of the earlier inventors and their explanations of their own work are now a matter of history, and the record of them should be preserved. In arranging this third edition, however, it became necessary to add to it a brief appendix, delineating, in a way necessarily somewhat sketchy, the growth of the electric motor up to the present time. The machine itself has been improved in various ways as the principles of construction of dynamo-electric machines have become more widely known, and the use of the motor in modern industry has increased to an extent little short of marvelous. Probably not less than 25,000 motors are in operation to-day in the United States alone, while the electric railways that could have been counted on the fingers in 1888, have now increased in number to something like 300, operating over more than 2,000 miles of track. It is especially this field of electric traction in which the motor has come into use since the last edition of this book was issued, and hence, in reviewing the progress that has been accomplished, it is electric traction to which attention has been especially called. The stationary motor has been given brief consideration, and it has been thought wise to describe a comparatively small number of typical forms rather than to make an attempt to catalogue the efforts of inventors in this particular line. It would, indeed, be difficult to keep track of the course of invention in producing slight modifications of existing motors, or in devising unusual details of winding and mechanical construction. The electric motor to-day, whether for stationary or traction purposes, differs from its predecessors only in more careful mechanical construction and better electrical design. The great tendency throughout has been towards simplicity and slow speed, the latter having been enforced in railway motors by hard experience.

Comparatively little has been said with reference to foreign motors and motor systems; first, because the use of the motor has not increased abroad to anything like the same extent as here; and second, because rapid improvement has not been compelled by necessity as it has been in America. An exception, however, has been made in favor of the City and South London Railway, as it exhibits a very advanced type of apparatus for electric traction; a model that might often be followed to advantage here. The lesson of that successful experiment has done much to spur on American electrical engineers, and is at least the proximate stimulus that has produced the recent gearless motors for electric railway service.

L. B.

June, 1891.

CONTENTS.

LIST OF ILLUSTRATIONS.

THE ELECTRIC MOTOR

AND ITS APPLICATIONS.

CHAPTER I.

ELEMENTARY CONSIDERATIONS.

As the functions of the electric motor are dependent upon the action of the electric current, it is necessary at the beginning of the present work to give a review, however brief, of the principal facts involved.

The time is not so very remote when the existence of an electric current was chiefly, if not wholly, made manifest by effects other than magnetic. Up to the year 1820 one of the most important facts generally known was that if the circuit from a voltaic battery was completed through acidulated water, the water was decomposed. Another interesting thing ascertained was that if a conductor were made sufficiently thin, the current would bring it to incandescence. To these, we may add the discovery of the electric arc by Sir Humphry Davy during the first decade of this century, and that of the reaction of the muscles and nerves to the passage of a current.

For many years the idea had prevailed, as Prof. Forbes puts it, that there was some "hidden connection between the power that impelled the compass to point to the north and the lightning in the sky. It had been believed that when lightning had disarranged the compass-needle and reversed its polarity, it showed that there was some connection between electricity and magnetism." What this connection was long remained a matter of mere speculation, but while carrying on some experiments with the object of solving the problem, Oersted found that when the current from a wire held over a compass-needle from the south to the north, the needle's north pole was swerved to the west; see Fig. 1.

Thus, practically, was demonstrated for the first time the correlation between electricity and magnetism. The experiment attracted universal attention, and gave to many investigators a clue to follow up. It was not long after this in the same year, 1820, that Arago and Davy discovered, independently of each other, that iron and steel could be magnetized by the passing of a current through a wire wound around them; and Sturgeon was prompt to apply the principle to the construction of powerful electro-magnets.

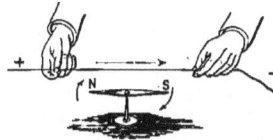

FIG. 1.—OERSTED'S EXPERIMENT.

It was also noted by Barlow, that by passing a current from the centre to the circumference of a copper disc placed between the poles of a magnet, the disc would revolve. This, familiar as "Barlow's wheel," was the first electric "motor," in the true sense of the word.

Faraday, in 1831, recognizing with the insight of genius, the relation and convertibility of the phenomena so far observed, obtained an electric current by electro-magnetic induction. This grand discovery may be summarized in the statement that when a conductor in closed circuit is made to cut magnetic " lines of force," a current is generated in that conductor. Faraday also reversed Barlow's experiment and obtained a current from a copper disc rotated between the poles of a magnet. But he left the application of the fruitful laws he had discovered to others, who invented numerous types of magneto-electric generators, into a description of which it is not needful here to enter.

Although the reversibility of the electric motor and the magneto-electric generator had already been noticed, little thought, apparently, was given to the fact; and meantime the other phenomena exhibited in the action of electro-magnets were employed in the construction of electric motors, some of which are described in subsequent chapters.

It was not until 1873, after the substitution of electro-magnets for permanent ones in electric generators, that the reversibility of the "dynamo" was fully recognized or realized, in the action of the Gramme machines exhibited at the Vienna exhibition of that year. As said above, electric motors had been built and operated many years before this, but they found no extended practical application, chiefly because they depended upon the galvanic battery for a supply of current. Now, as the energy developed by the oxidation of a quantity of zinc of a given value in a battery is far less than that which can be obtained by burning a quantity of coal of the same value under a steam boiler, it follows that electric motors could not compete with other forms of motors. Hence it remained for some cheaper source of current to be discovered, as it was in the dynamo-electric generator.

With this brief sketch, intended to be suggestive rather than exhaustive, of the facts relating to the evolution of the electric motor of to-day, we come to consider the motor in regard to the manner in which it operates as a machine for converting electrical energy into mechanical energy.

Beginning with the earlier forms of motors, we note at once several ways in which the current can be applied for mechanical purposes. The term "electric motor," it should be promised, includes all apparatus by which energy in the form of electric current is converted into mechanical energy, through whose employment work is performed, such as the driving of a fan or a lathe, the raising of an elevator, the propelling of a locomotive, and the like. Thus we

may use the attraction that an electro-magnet exerts upon an iron or steel armature; or the mutual attraction between two electro-magnets. These and analogous principles are evidently based upon the attractive force exhibited between masses of magnetic metal.

In the practical, commercial motor of to-day, however, the action, though apparently similar to the above, is quite different. It depends upon the principle that when a current passes through a wire, the latter becomes surrounded by a field of force similar in nature to that pertaining to "permanent" magnets. This is clearly shown in Fig. 2, which represents a plate through which two wires pass, the lines of force being exhibited by the positions assumed by the iron filings sprinkled upon the

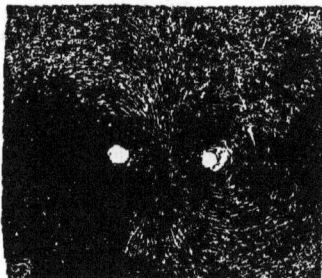

FIG. 2.—FIELD OF FORCE AROUND TWO WIRES.

surface of the plate. As will be observed, the lines of force encircle the wire, and when such a wire is brought into the vicinity of a magnet, it acts to all intents and purposes as if it were a magnet having circular lines of force at all points in planes at right angles to its length.

Now, if the current in the wire is in a certain direction, these lines of force will appear to circulate from left to right; and with the current in the contrary direction, from right to left. The wire in this condition, when brought in proximity to a magnet, is attracted or repelled in the same way as if it were a magnet, the attraction or repulsion being determined by the direction in which the lines of force circulate—in other words, according to the direction of the current.

The action, then, in the majority of electric motors of to-day is primarily that exhibited between a magnet and a wire carrying a current, and is the reverse of the action seen in the magneto-electric or dynamo-electric machine. In the magneto and dynamo, the motion of a conductor generates current; in the former, a current in the conductor produces motion. In the dynamo, it requires mechanical power to force the wires through the magnetic field, in order to generate current, and, according to Lenz's law, the currents generated have a direction such that their reaction upon the magnet tends to stop the motion which produces them. Conversely, in the motor it is the current acting upon the magnet that produces mechanical power. In this case, however, the reactions of the current being free to exert their power, the motion obtained will be the opposite of that in the dynamo. Hence we understand why a dynamo set to work as a motor, runs in the reverse direction, where the construction of the machine is such that the polarity of the field magnets remains the same in both functions of generator and motor.

It was said above that the action in nearly all the motors of to-day is due primarily to the reaction between magnet and current, in contradistinction to that exhibited in the older motors, which operated by the attraction of magnetised iron or steel. But as the modern motors have iron in their armatures, the question may be asked, What is the object in putting it there? The answer is, that by the presence of the iron the magnetic lines of force are strongly concentrated upon the wires. Dynamos will generate current and motors will perform work without the presence of an iron core in their armatures, but both are less efficient when so constructed.

Let us now see what takes place when the electric motor starts to work. We will suppose, for the sake of simplicity, that a galvanic battery is connected with a motor, and that a galvanometer is interposed in the circuit. If now we clamp the motor down so that it cannot revolve, and hence cannot do work, we get a certain strength of current. If we then release the motor so as to allow it to revolve, we find that as the speed of the armature increases, the current decreases. This fact was observed by the earliest experimenters with electric motors, and not being correctly interpreted was consid-

ered one of the greatest drawbacks to successful operation, for it was believed at that time that as the motor revolved it created a resistance which tended to diminish the current. It was argued, therefore, that the electric motor was necessarily a wasteful machine. We know to-day that this is by no means the case, and that the action observed in the electric motor is one upon which its true value as a working and useful machine depends. Jacobi was the first to point out that the diminution of current indicated by the galvanometer was not due to resistance engendered by the motor, but was due to a counter-electromotive force generated by it. This is easily understood when we consider that the motor, in revolving in a direction opposite to that of the generator, creates a current in the other direction. The opposing current, of course, lessens the original current. We may here appropriately and correctly apply the converse of Lenz's law above quoted, and say that the motion produced is always such that by virtue of the magneto-electric induction which it sets up, it tends to stop the current. From his observations, Jacobi formulated the important law that when an electric motor does its greatest possible work, it diminishes the original current one-half; and hence a loss of 50 per cent. is experienced when the motor is exerting its greatest power.

This can be shown by a simple illustration. Let us take two machines exactly alike and connect them up as generator and motor, the former being driven at a constant speed. Now, if while the generator is at work we clamp the motor so that it cannot revolve, no work is being done by it, and, at the same time, no counter-electromotive force is generated. If now we release the motor, and allow it to revolve freely, its speed will gradually increase until reaching that of the generator. But at this equal speed it will create a counter-electromotive force equal to that of the generator, so that, practically, there will be no current at all, and, evidently, the motor will again be incapable of doing work. We see, therefore, that at zero counter-electromotive force and at maximum counter-electromotive force, the motor does no work. The mean between these two limits is one-half the original electromotive force; so that, as Jacobi pointed out, the motor is doing its greatest work when the original current is reduced one-half, i. e., when the counter-electromotive force is 50 per cent. of the original electromotive force.

But this law, relating only to the maximum *work* or *activity* of the motor, has unfortunately been misinterpreted to mean that the maximum efficiency of the electric motor is 50 per cent., thus making it appear to be a comparatively inefficient and wasteful machine.

The *efficiency* of the electric motor can, theoretically considered, be made to include anything between 0 and 100 per cent. In order to make this clear, we will take the same example as before. It has been seen that when the generator is in operation and the motor is clamped down, the latter does no work, and the current is dissipated and wasted in heating the circuit. Hence the efficiency, being the ratio of the power applied in the generator to that obtained from the motor, is zero. It has also been seen that when the motor is running without a load, it generates a counter-electromotive force equal to the original electromotive force, and reduces the current in the line to zero. But, evidently, with no current existing it requires no power, excepting that for overcoming friction, to drive the generator. We thus have virtually no power required at the generator and virtually none developed at the motor—which gives an efficiency of 100 per cent. This, as we have seen, takes place when the counter-electromotive force is 100 per cent. of the original.

Now let us put a small load upon the motor. Its speed, which has hitherto been equal to that of the generator, will be diminished in consequence, and hence the counter-electromotive force will fall off. This allows a certain amount of current to flow through the circuit, and the generator requires a certain corresponding amount of power to drive it. If we go on increasing the load of the motor, we reduce, in equal degree, the speed and the counter-electromotive force, while increasing the current and the power necessary to energize the generator. Each accession of current, therefore, means more power applied to the generator. But while at 100 per cent. efficiency both machines were running at the same speed, the increase in load on the motor has checked its speed, and hence it is not doing as much work as is required to run the generator, which retains its original speed. It follows that the efficiency can no longer be 100 per cent., but must be something less; and that something

loss is clearly the ratio between the relative speeds of the generator and the motor.

Now it has been seen that the counter-electromotive force is in direct ratio to the speed of the motor. We infer, therefore, that the efficiency of an electric motor is in direct ratio to the counter-electromotive force developed. If we run a motor so as to allow it to develop a counter-electromotive force of 90 per cent., its efficiency will be 90 per cent.; and thus we can, by reducing the power developed by the motor, increase its efficiency to any desired degree. It being true also that in order to obtain the maximum *power* from a motor we reduce its efficiency to 50 per cent. it is evident that when we wish to work with higher economy we must not tax the motor to its full working capacity.

As these principles are new to many, we have sought, even at the risk of making a labored and reiterative explanation, to state them fully. A very apt analogy is encountered in the working of the steam engine, with which greater familiarity exists. By letting the valve of an engine follow full stroke, one might admit steam to the cylinder at boiler pressure during the entire stroke. This would give a mean effective pressure upon the piston equal to the boiler pressure and would cause the engine to exert its greatest power. But this method of operating is not economical, and in actual practice, the steam is cut off at different points in the stroke. That of course reduces the mean effective pressure, and hence the power of the engine below that which it could be made to exert if run uneconomically.

As just said, we have entered with careful explicitness into the description of the action involved in the operation of the electric motor, because considerable misapprehension appears to exist as to the true meaning of Jacobi's law. We have refrained, as will be noted, from any mathematical demonstration of these important facts and principles, for the reason that to many deeply interested in this subject, mathematical symbols and formulæ furnish no adequate picture of the reasoning involved.

In describing the causes of the efficiency of electric motors, we have assumed that the generator and motor are both perfect converters of energy; that is, that the generator is capable of converting into electricity all the power applied to it, and that the motor is capable of converting into mechanical power all the electricity supplied to it (friction being omitted in both cases). Now in actual practice such is not the case, because all dynamos present resistance to the current, which generates heat in them; and other causes besides friction tend to reduce the efficiency of the machine. Dynamos have nevertheless been built giving an efficiency of over 90 per cent., showing that they are excellent converters of energy.

But it has been found in the past that when efficient dynamos were employed as motors their efficiency was reduced considerably. This difference was generally sought to be explained by an assertion or a supposition

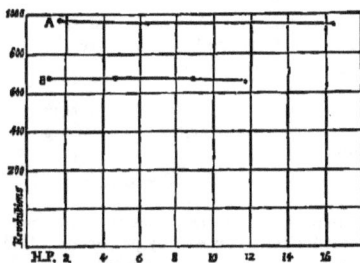

FIG. 3.—DIAGRAM SHOWING CONSTANCY OF SPEED UNDER VARYING LOADS.

that, having been designed and made for one purpose, the machine was, therefore, not suitable for the other. No satisfactory explanation was given of the cause of this lower efficiency of motors, or, in other words, of how the missing power is expended.

Taking up this question Mr. Mordey, an able English electrician, has recently shown the principal cause of this loss, which can be readily avoided, and his experiments also establish the fact that a well constructed motor approaches very closely in its action the limit that theory would assign, as indicated above. In the search for general principles, says Mr. Mordey, all those ways of considering the actions which depend on the idea of magnetic poles in the armature were abandoned, and the conclusion was arrived at that the armature should have no polar action whatever, that the iron of the armature should have only the function of a conductor of lines of force, and that the power

of the motor should be due to the simple action between the lines of force of the magnetic field and the armature wires conveying currents at right angles to those lines of force. This mode of regarding motor action is convenient on several grounds, and leads to certain conclusions, which, if correct, form substantial bases for practical construction. Thus the armature, instead of being, as hitherto, considered as a strong electro-magnet placed in the field of another electro-magnet, is to have its electro-magnetic functions reduced as much as possible, or preferably suppressed altogether. The field is to be very strong.

As with such an arrangement there is no polar effect in the armature except that due to the direct magnetic induction of the field-magnets, it follows that the maximum power is obtained with any given current when the brushes occupy an absolutely neutral position, or, in other words, when there is no "lead" and no distortion or rotation of the field. These conditions do away with the most troublesome and prolific cause of the sparking at the brushes.

But by working backward in this way, Mr. Mordey saw that the conditions which seemed to be best for a motor were precisely those which the would-be designer of a perfect dynamo would set before him as his goal.

Certain perfect analogies had been arrived at. In both dynamos and motors, according to this briefly sketched view: (1.) The field should be a very strong, the armature a very weak, electro-magnet. (2.) In both generators and motors "lead," distortion, or displacement of brushes or of magnetic field is wrong, and is to be avoided by attention to (1). Whatever "lead" there may be in either case, there is this difference, that in dynamos this "lead" is in the direction of rotation; in motors it has the opposite direction, as the course of the current through the armature is reversed, but the field is the same. (3.) In both generators and motors absence of sparking at the brushes depends mainly on the conditions of (1) being complied with. (4.) Reversal of rotation. In neither generators nor motors is movement of the brushes necessary.

But having got so far, a little consideration suggested the probable existence of another analogy. Since a dynamo, having the above theoretically perfect form and action, with a constant field, would produce a constant electromotive force if run at a constant speed, independently of the load or amount of current generated, a motor constructed on the same principles and having a constant field, if supplied with energy at a constant difference of potential, should run at a constant speed, independently of load.

If this should prove to be a true analogy, a simple means of obtaining results of great use in the practical application of electricity would be obtained.

The experiments were carried out with a "Victoria" dynamo. The results ultimately obtained, which are given in the following tabulated account of the experiments, and in the curves plotted from them, show that this fifth analogy is as true as the preceding ones, a constancy of speed being obtained that was very remarkable, even when the load was increased, until much more than that which, as a generator, was usually considered the full working current was traversing the armature.

Two sets of readings were taken, working up to about the same current in each set, but with the potential-difference of the supply different in the two cases, as stated. The field was of the same strength throughout. The load consisted of another "Victoria" dynamo driven through a modified White's transmission-dynamometer, the work being varied as required by altering the external circuit of this dynamo.

At first it appeared that the counter-electromotive force of the motor was dependent neither on speed nor strength of field, as the latter was constant and the former very nearly so, while the current rose with the work; but calculation showed that this was not the case; indeed, it could not be so.

Calling the counter-electromotive force e, and the loss of potential caused by the resistance of the armature e_1, and the difference of potential at the terminals E, then

$$e + e_1 = E.$$

Calling the current C, and the resistance of the armature R, we know that

$$e_1 = C R.$$

The resistance of the armature of the motor in question was .027 ohm, which we call .03 in order to make some allowance for the effect of heating.

Experiments with a D_2 "Victoria" Shunt Motor at Constant E. M. F.

Speed.	Current.	Difference of potential at terminals.	H. P.*	
975	36.3	1·10	1.8	Curve A. Maximum speed-variation 3 per cent.
965	60.5	1·10	6.6	
948	97.1	1·10	12.87	
915	1:30.8	1·10	16.3	
680	20	100	1	Curve B. Maximum speed-variation 3 per cent.
677	61.4	100	4.8	
675	102	100	9.14	
660	125	100	11.7	

* Including an unascertained loss in transmission.

In the first and last readings of the second set of tests (curve B) Fig. 3, we have, therefore, the following conditions:

	Speed.	e_1.	e.
	680	.87	99.13
	660	3.75	96.25

Now

$$\frac{680 \times 96.25}{660} = 99.13.$$

From which it appears that the counter-electromotive force was exactly proportional to the speed, as is to be expected where the field is constant and the magnetic distortion *nil*.

The other cases do not work out with the same accuracy. The results are, however, quite within the limits of error inseparable from the rather rough conditions of workshop tests.

One other fact may be pointed out in connection with these tables of results and curves. In the case of dynamos working with a constant field, the output with the same current is almost exactly proportional to speed, as the electromotive force is also simply proportional to speed. So with the motor, the speed is proportional to the electromotive force of supply, and the work, with the same current, is in the two cases, is simply proportional to speed, therefore, to electromotive force.

Turning now to the question of the efficiency above alluded to, in order to localize the loss which was found to occur in motors and to ascertain its cause, the several possible sources of waste were carefully considered. These are:

(*a.*) Friction at the bearing, air friction, and friction of the brushes against the commutator.

(*b.*) Loss of energy in heating the armature

and field-conductors, and a certain loss due to self-induction. (c.) Loss by the production of eddy currents in the iron.

Now it is evident especially with a generator or motor having the qualities sketched above, that at the same speed, and working with the same currents in its conductors, the losses under (a) and (b) must be identical, whether it be working as a generator or as a motor. And as with such conditions its efficiency as a motor is lower than as a generator, the cause of the loss must be sought under (c), i. e., the eddy or Foucault currents in a dynamo must be less than in a motor, all other conditions being the same. And such is the case, the explanation arrived at by Mr. Mordey being a very simple one.

To quote again from Mr. Mordey, in a dynamo the rotation of the armature causes eddy currents to be generated in the iron core, in the same direction as in the conductor proper with which the core is surrounded. Of course, as the armature is always more or less subdivided or laminated in a direction at right angles to the lines of force, any circulation of currents round the core is avoided, but local currents, which are aptly called eddies, are set up, and, taken as a whole, these eddy currents on the outside of the core are in the same direction as the current flowing in the copper conductors.

In an electric motor, however, the eddy currents and the currents in the copper conductor are in opposite directions; as, although the electromotive force set up in the conductor is in the same direction in a motor as in a dynamo, the current in the former is forced through the armature in a direction contrary to the electromotive force, or opposite to its course in a generator. According to the laws of induction, therefore, it will be seen that while in a dynamo the two sets of currents, those in the iron and those in the conductor, tend to oppose and reduce one another, in a motor they act in such a manner as to mutually assist each other. Thus, with the strength of field, the current in the conductor, and the speed, the same in the two cases, it will be seen that in a motor the eddy currents in the iron core of the armatures will be greater than in a generator, and therefore the heat lost in the former will be more than in the latter. There is little doubt that this is the cause of the efficiency of motors being lower than that of generators; and it points to the advisability of giving more attention in the former to those principles which are well understood for the reduction or elimination of eddy currents.

This precaution to be observed was first pointed out by Mr. Mordey but had been recognized by others, and is of the greatest importance. We have personally seen the results of experiments upon small motors which with solid iron cores gave an efficiency of only 20 per cent. but with laminated cores were brought up to 70 per cent. efficiency.

Regarding the difference which should exist between a dynamo and a motor, Profs. Ayrton and Perry have advanced the theory, that in the dynamo the field magnets should be large and strong and the armature small and weak magnetically, while the contrary applies in the case of the motor. Practice, however, does not bear out this assumption, for only recently Dr. John Hopkinson in some tests of identical machines of his design, run respectively as dynamo and motor, obtained almost identical efficiencies.

We may conclude, therefore, that there is no radical difference in the relative actions of motor and dynamo and that the losses which have heretofore been experienced were due to faultiness in internal construction.

CHAPTER II.

EARLY MOTORS AND EXPERIMENTS IN EUROPE.

THE first experiments with electric motors to attract general attention throughout Europe appear to have been those of Jacobi, from 1834 to 1838, although prior to that time the field had been boldly entered by other acute investigators, who sought in various and ingenious ways to utilize the principles we have outlined in the preceding chapter. Thus in 1826, Barlow showed how to employ electricity as a continuous motive power by rotating a disc of copper between the poles of a magnet. The current was sent perpendicularly through the disc from its axis to circumference, when it passed into a cup of mercury. In 1830, the Abbé Salvatore dal Negro, professor of natural philosophy at the University of Padua, made a motor in which a permanent magnet oscillated between the legs of an electro-magnet, the polarity of the limbs changing at each movement. The oscillation was converted into continuous rotation.

In 1832, before the Zurich Society of Engineers, Dr. Schulthess suggested that " a force such as we obtain by interrupting the current and establishing it again could be advantageously applied to mechanics," and in 1833 he exhibited to the society a machine in which his ideas were embodied. About this time, too, Botto is said to have invented a motor in which a lever worked like that of a metronome, by the alternate action of two fixed electro-magnetic cylinders on a third movable cylinder attached to the lower arm of the lever. The upper arm imparted a continuous circular movement to a metal fly-wheel.

Thanks to the substantial aid of the Emperor Nicholas of Russia, who contributed a sum of $12,000 to the work, Professor Jacobi, the discoverer of electro-plating, was enabled to prove in 1838 at St. Petersburg, on the Neva, that his electro-magnetic motor of 1834, as improved, could replace the oarsmen in a boat carrying a dozen passengers. Fig. 4 is a perspective of the Jacobi motor of 1834, which was composed

of two sets of electro-magnets. One set was fastened to the square frame T, disposed in a circle and with the poles projecting parallel with the axis. The other set S was similarly fastened to the disc A attached to the shaft and revolving with it. Each set comprised four magnets, and there were consequently eight magnetic poles. The current from a powerful battery passed through the commutator C to the coils of the electro-magnets, and as the magnets attracted each other the disc rotated. By means of the commutator on the shaft, the current was reversed eight times during each revolution, just as the poles of two sets of magnets arrived opposite each other. Attraction ceasing, repulsion took place, and the motion was thus accelerated. As the poles were alternately of different polarity, the reversals had the effect of causing attraction between each pole of one set and the next pole of the other. In his historic experiments of 1838, Jacobi used a modified form of this motor, so as to obtain greater power. In the new form, two sets of electro-magnets were attached to stationary vertical frames, one on each side of a rotating disc or star. Each set was composed of twelve electro-magnets. The electro-magnets on the rotating star were made in the form of bars passing entirely through the star. The axis carried a commutator formed of four wheels, regulating the direction of the current with the result that when the straight bar magnets were between two consecutive poles of the horse-shoe-magnets on the frames, they were always attracted towards the one and repelled from the other. The reversal of the current took place when the rotating poles were exactly opposite the fixed ones. The boat upon which this motor was placed, and which it propelled by means of paddles on the Neva, was 28 feet long, 7 feet wide, and 2 feet 9 inches draught. No fewer than 14 passengers were carried. The battery power was furnished by 320 Daniell cells, the weight of which was far from inconsiderable. In 1839, on a repetition of the experiment, 138 Grove cells were used. At no time was a higher speed attained than 3 miles per hour.

At this time, 1838-9, an inventive Scotchman, named Robert Davidson, had built a lathe and a small locomotive for which electricity was the driving power. The motor for the locomotive consisted of two cylinders of wood fitted to

magnet. By this arrangement, it followed that the current was interrupted in the active electro-magnet and sent into the other, its *vis-a-vis;* and thus the axle was continuously turned. Acting together, the four sets of armatures and the two axles served to propel the car. Two sets of cells were employed, one for the electro-magnets on the right and the other for those on the left. At each extreme end of the axles, in-

FIG. 4.—JACOBI MOTOR.

the axles of four wheels, and furnished with four sets of iron armatures arranged to pass between the poles of eight electro-magnets. These were placed horizontally at the bottom of the car, two and two, by their opposite poles, in two opposite rows, so that each of the cylinders carried two sets of iron bars parallel to the axles. The bars presented themselves successively, as the cylinders rotated, to the poles of the corresponding opposite electro-magnets. When one of the bars on one side was opposite its magnet, one on the other side was just within range of the attraction of its electro-

side the driving-wheels, were two small cylinders or commutators of ivory and metal upon which bore brushes leading the current from the batteries. Davidson's car was 16 feet long, six feet broad, and of five tons weight, including batteries. He drove it at a speed of four miles an hour with 40 cells composed of plates of iron and amalgamated zinc measuring 15 inches by 12.

An excellent and very early motor was that of Elias, made at Haarlem, Holland, in 1842. It consisted of two concentric rings of soft iron, the inner one being revolvable. The exterior, fixed ring supported vertically, had six enlargements dividing it into six equal parts. Between the dividing pieces, was wound insulated copper wire, and the winding was such that a current entering at one end of the horizontal diameter was divided between the upper and lower halves of the ring, and left at the other end of the same diameter. The interior ring was of like construction, its six poles being alternately north and south. The current entered by wires with each of which three parts of the commutator were in connection. The motor was worked by two batteries, one for the exterior ring and one for the interior; or by a slight change of connections, one battery only was necessary. In either case, the alternate north and south poles of the exterior ring remained the same. The poles of the inner ring changed polarity at each sixth of a turn, the commutator being so arranged that each pole of the movable ring was always repelled by one of the fixed poles and attracted by the other and next. The windings of the two rings were very close together, so that the action of

FIG. 5.—FROMENT MOTOR.

parallel currents in them served to assist in the rotation due primarily to the attraction and repulsion of the magnets.

One of the most interesting of the early motors is that of M. Froment, made in 1845 and

illustrated in Fig. 5. It may be likened to a breast-wheel, whose paddles are acted upon by magnetism instead of water. The wheels were made of brass or other non-magnetic material, but the armature bars around the circumference were of soft iron. By means of the commutating device, the current, cut off from each electro-magnet as soon as the armature

FIG. 6.—DU MONCEL MOTOR.

arrived opposite its poles, was led to another magnet until the said armature had moved on sufficiently to allow the next armature to come within range. The desired effect being obtained, the current was again sent through the first electro-magnet. The commutator consisted of spring rollers in contact with each of the magnets and the battery, and was worked by means of a small cam on the driving shaft. Froment devised other ingenious forms.

In 1851, Count du Moncel devised a motor, Fig. 6, not unlike that of Page, and of which the arrangement reminds one, as he himself said, of an oscillating steam engine. The iron cylinder, which in the position of the crank in the figure, has passed entirely through the right-hand bobbin or solenoid and passed a short distance into the other bobbin, is shown on the point of being attracted into the latter. On arriving at the end of its stroke, it was within reach of an iron ring or disc, terminating the left-hand bobbin. This gave it an extra impetus that carried it over the dead-point corresponding to the movement of the shaft in the opposite direction. Between the bobbins was a roller upon which the iron rod or piston moved, to prevent friction. The commutator was composed of two eccentrics fixed to the axis of the fly-wheel and insulated from each other. A fixed silver spring in connection with each one of the bobbins encountered at each half revolution of the fly-wheel one of the eccentrics. A third spring large enough to bear upon both of the eccentrics brought the current to the two latter successively.

The Bourbouze motor, also modeled upon that of Professor Page—described in Chapter III.—was made like a steam engine with two pistons. This early type is shown in Fig. 7. At the two extremities of the horizontal beam were two iron cylinders working like pistons inside two long magnetizing bobbins, whose lower ends were occupied by short iron cylinders joined together by a piece of iron between the

Fig. 7.—Bourbouze Motor.

bobbins; constituting, in fact, an electro-magnet. When the current passed into one of the bobbins, the corresponding iron rod or cylinder was attracted as well by the magnetic pole at its end as by the coils, and it was drawn downward until the current was cut off by the commutator. The process was repeated in the other bobbin, and the beam was depressed at the corresponding end. This to-and-fro movement was utilized as in steam engines by means of a crank and fly-wheel, in the manner indicated. The commutator was a plate that rubbed alternately on two contacts fixed horizontally on a table, and was set in motion by an eccentric rod worked like that of a steam engine. The Bourbouze motor may be compared to an ordinary working-beam engine.

Last, though not least, but on the contrary of epochal importance, comes the Pacinotti motor invented by the distinguished Italian physicist in 1861 and described by him in *Il Nuovo Cimento* in 1864. Pacinotti builded better than he knew, and it was not until 1871 when the celebrated Gramme dynamo with ring armature made its appearance, that he recognized the true value of his motor and brought it from its obscurity and oblivion in the Philosophical Museum of the University of Pisa to be seen at the exhibitions of Vienna in 1873, and of Paris in 1881. Under the title: "A description of a Small Electro-Magnetic Machine," Dr. Pacinotti said : "I took a turned iron ring furnished with sixteen equal teeth. This ring was suspended by four brass arms B B (Fig. 8), which fixed it to the axis of the machine. Between these teeth little triangular pieces of wood were let in, wound with silk-covered copper wire. This arrangement was to obtain perfect insulation of the coils or bobbins thus formed between the iron teeth. In all the bobbins the wire was wound in the same direction, and each was formed of nine turns. Each is thus separated from the other by an iron tooth and the triangular piece of wood. On leaving one bobbin to commence the next, I end the wire by fixing it to the piece of wood which separates the two bobbins. On the axle carrying the wheel thus constructed I grouped all the wires, of which one end formed the end of one bobbin and the other the commencement of the next, passing them through holes for this purpose in a wooden collar fixed on this same axle and then attaching them to a commutator also on the axle.

"This commutator consisted of a ring or small cylinder of wood, having on its circumference two rows of grooves, in which are fitted sixteen pieces of brass (eight in each row); they are placed alternately, and concentric with the wooden cylinder on which they form a spindle. Each of those pieces of brass is soldered to the two ends of wire corresponding with two consecutive bobbins; so that all the bobbins are connected, each being joined to the following by a conductor, of which one of the pieces of brass of the commutator forms a part. If we put two of these pieces of brass in communication with the poles of a battery by means of two metallic rollers, G, the current, in dividing, will go through the coil at both points where the ends of the wire fastened to the pieces of brass communicate with the rollers ; and magnetic poles will appear in the iron circle in the diameter perpendicular to *A A*. On these poles acts a fixed electro-magnet, which determines the rotation of the circular electro-magnet; the poles of the circular electro-magnet when in movement always appearing in the fixed positions corresponding to the communication with the battery."

He said further:—"It seems to me that what increases the value of this model is its faculty for being transformed from electro-magnetic into magneto-electric with continuous current.

permanent magnet; the electro-magnetic machine resulting from this will have the advantage of giving additional induced currents all in the same direction, without necessitating the use of mechanism to separate the opposite currents or make them converge." As to reversibility, he remarked with keen foresight:—"This model further shows how the electro-magnetic machine is the complement of the magneto-electric machine, for, in the first, the current obtained from any source of electricity circulating

FIG. 8.—THE PACINOTTI MACHINE.

If, instead of the electro-magnet, there was a permanent magnet, and the circular magnet was made to turn, we should have, in fact, a magneto-electric machine which would give a continuous induced current always in the same direction. To develop an induced current by the machine thus constructed, I brought to the magnetic wheel the opposite poles of two permanent magnets, or I magnetized by means of a current the fixed electro-magnet, and I made the circular electro-magnet to turn on its axis. In both cases I obtained an induced current always in the same direction. It will easily be seen that the second method is not practicable, but that an electro-magnet is easily replaced by a

in the bobbins produces movement of the wheel with its consequent mechanical work; whilst in the second, mechanical work is employed to turn the wheel, and obtain, by the action of the permanent magnet, a current which may be transmitted by conductors to any required point."

Other early European experimenters of merit might be mentioned, such as Wheatstone, McGawley, Gaiffe, Larmenjeat, Roux, and Hjorth, but the descriptions above will serve to show the state of the art as regards electric motors down to the time when the reversibility of the dynamo-electric machine gave an entirely new direction to the efforts of inventive genius in Europe.

CHAPTER III.

EARLY MOTORS AND EXPERIMENTS IN AMERICA.

As all previous works dealing in any way with electric motors have little to say about American work in the field, no apology need be offered for the effort made in this chapter to supply, in part, the deficiency.

The first electric motor patented in this country was constructed early in 1837, and was the device of Thomas Davenport, a blacksmith, of Brandon, Vt., who styled his invention an "Application of Magnetism and Electro-Magnetism to Propelling Machinery." The frame of the machine was made of a circular ring and disc, horizontally arranged, the former being supported upon the latter by vertical posts. Upon the lower disc were mounted two copper

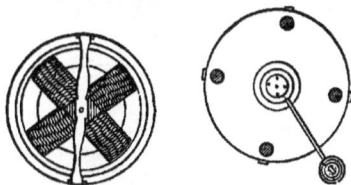

Fig. 9.—DAVENPORT MOTOR.

segments, arranged in the centre, as seen in the sectional view, which, together, constituted a circular ring pole-changer. The electro-magnets were four in number and projected horizontally in radial lines from a common centre, Fig. 9. Through this centre passed a vertical shaft having bearings in the frames so as to have a revolving motion. The conducting wires from the source of energy extended up from the copper segments parallel with the shaft of the electro-magnet. Davenport arranged within the inner periphery of the upper horizontal ring a ring of steel cut in two, forming a pair of steel segments, which he termed "artificial magnets." The description of this device in the patent specification is somewhat obscure, but the inference is that these were permanent

magnets, and being semi-circular in shape, they approximated the form of a horseshoe. The principle of operation of this machine will be apparent at a glance. The polarity of the electro-magnets was changed during their revolution by the wiping contact of their connections with the two segmental plates on the bottom disc, these segments connecting with the positive and negative poles of the battery.

As a remarkable instance of the granting of a broad claim by the Patent Office to an inventor, that of Davenport may be cited. It reads:—"Applying magnetic and electro-magnetic power as a moving principle for machinery in the manner above described, or in any other substantially the same principle."

To Davenport appears to belong the honor of first printing by electricity as well as of first building an electric railway. A paper called *The Electro-Magnet and Mechanics' Intelligencer* was published by him in 1840. It is said that he obtained the current for his machine from a battery of amalgamated zinc and sheets of platinized silver. Pieces of sheet iron platinized might, he thought, be used with advantage instead of the silver plates. The electrodes were plunged into water acidulated with sulphuric acid in the proportion of nine parts of water to one of acid. Davenport and Ransom Cook are said by Prof. Moses G. Farmer to have used in 1840, with motors, a zinc and copper battery, with a solution of blue vitriol as the exciting fluid.

Davenport was a man far ahead of his time. Having seen a magnet in use at Crown Point, on Lake Champlain, in 1833, extracting iron from pulverized ore, he jumped at once to the idea that he could apply magnetism to the propulsion of machinery. He bought the magnet, began to experiment, and by 1834 had obtained rotary motion. He then went to Washington, where he took steps to obtain the patent above mentioned, and in the autumn of 1835 he set up a small circular railway at Springfield, Mass., over which he ran an electro-magnetic engine. In December of the same year he exhibited his road in Boston for two weeks. During 1837 he showed to Prof. Benjamin Silliman, in New York, a motor in which "the exterior fixed circle is now composed entirely of electro-magnets. The conducting wires were so arranged that the same current that charged the magnets of the motive wheel charged the stationary ones placed around it, only one battery being used. It lifted sixteen pounds very rapidly, and when the weight was removed, it performed more than six hundred revolutions per minute."

In June, 1838, Nelson Walkly, of Tuscaloosa, Ala., devised an electric motor, the principal improvement being in the mode of changing the poles of the electro-magnets.

The electro-magnets employed by Walkly were semi-circular in form. Two of them were fixed to a horizontally-revolving wheel with proper insulations. The ends of the wires on the revolving magnets were connected with two segmental collars placed on the vertical shaft. These collars were placed, one above the other, on the shaft, but insulated therefrom. The currents of the two revolving magnets were taken off by wiping electrodes arranged to lie against the collars which led to the negative and positive elements of the battery; Fig. 10.

Fig. 10.—Walkly Motor.

Through a post fixed in the right-hand side of the upper platform, Fig. 11, were fixed two conductors connected with the battery. The outer ends of these spring conductors, when at rest, pressed against an insulated pin, and between their ends was interposed a lever, the end of which was just the size of the insulated pin. This lever was composed of two plates of metal with a piece of wood between them, and

was made to vibrate by means of a double elliptic cam fixed on the upper end of the vertical shaft by means of a connecting pitman. Should more revolving magnets be used than two, the cam might be fixed on a pinion, revolving more times than the main shaft, so as

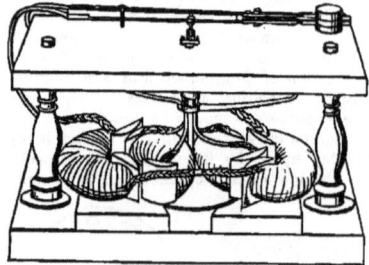

Fig. 11.—Walkly Motor.

to change the polarity every time one of the rotary magnets came opposite one of the stationary ones.

To magnetize the outer, or stationary, magnets, the current of electricity passed from the positive side of the battery to the conductor, and thence to the lower plate of the vibrating lever, and so to one of the stationary magnets. When the machine was at rest, the lever would be in contact with the spring conductors, and the ends of the rotary magnets opposite the ends of the stationary ones. By moving the rotary magnets the cam would move the end of the lever and the end of the spring away from the insulated pin, leaving the opposite spring resting against the pin. The north poles of the stationary and rotary magnets would then repel each other, causing the latter to revolve, so that the lever was vibrated back, thereby moving the spring on the opposite side and changing the polarity of the stationary magnets, and so on.

The next American patent for an electrical motor, in chronological order, appears to be that granted to Solomon Stimpson, September 12, 1838, Figs. 12 and 13. Between two vertical circular brass rings were attached the poles of a series of stationary magnets by screws. Within or between the stationary magnets were a series of revolving ones mounted upon a central shaft, the whole number of magnets—both stationary and revolving—being twelve. The wires of all the stationary electro-magnets were connected terminally with mercury-holding cells resting on the base plate. These insu-

Fig. 12.—Stimpson Motor.

lated cells, the inventor explained, were for battery communications. The electric connections of the revolving magnets passed out at one side and were connected with a pole-changer.

The galvanic current was not distributed to the revolving magnets individually, but they were charged by pairs, the magnets of each

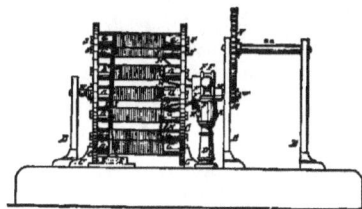

Fig. 13.—Stimpson Motor.

being charged in sequence. Wiping springs connected with the conducting wires were arranged to lie against the revolving pole-changer, which was composed of a series of metallic segments with interposing insulating material. The wiping contact was made upon the opposite sides of the pole-changer, and thus

were constituted two permanent battery poles. As the machine revolved, the two opposite extremities of the wires were presented in alternate order to the same battery pole, and thus a change of polarity was effected. Power was applied through a pinion on the shaft commencing with the cog gear.

The patent of Truman Cook, of New York, was granted in 1840. The body of the rotating armature, Fig. 14, was made of wood, brass, or any other material not affected by magnetic influence. Upon the periphery of this armature were placed six rectangular bars of soft iron, at equal distances apart, and extending from end to end parallel with the axis. The electro-magnets were of the usual horseshoe

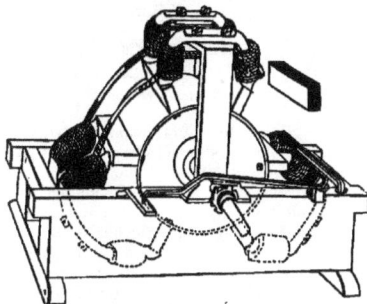

Fig 14.—Cook Motor.

form, and were placed in pairs, so that the opposite poles of each of them, at the same instant, stood immediately over the ends of the two contiguous armatures or keepers. In this machine there were three pairs of electro-magnets.

There were two mercury cups located in the frame in which the ends of the electro-magnet coils terminated, those wires which formed one termination passing into one cup, and those forming the opposite electric pole passing into the other mercury cup. A cam wheel secured to the armature shaft was made to touch the terminal wires from one electric pole, so that the ends of the wire were lifted from the mercury cup at each rotation.

The notches shown in this cam wheel corresponded with the number of the revolving armatures, and were so arranged as to suspend the transmission of the current, and, consequently, the magnetic induction at the proper moment for allowing the armatures to pass the magnets. One of the projecting teeth on the cam was insulated, and it was this one that raised the terminal wires of one electric pole by the action of the cam. In the drawings these wires are not represented as dipping into the mercury cup, but as resting upon a piece of metal which forms a conducting communication with the cup.

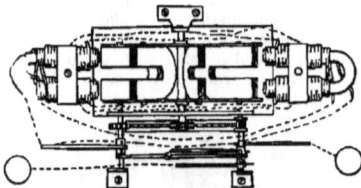

FIG. 15.—LILLIE MOTOR.

The length of the several armatures was less than that of their distance from each other, and the north and south poles of the magnets, constituting each pair, were at a distance apart corresponding to the distance of the armatures. The influence of the magnetic field was consequently exerted between the opposite poles of the magnets constituting the pair, this resulting from their proximity being greater than that of the opposite poles of each individual magnet.

FIG. 16.—LILLIE MOTOR.

It will be perceived that these magnets operated in pairs, one of them extending its influence directly to the other, thus mutually actuating the armatures as they approached. Mr. Cook showed a modified form of armature in the detail view, which consisted of several plates of soft sheet iron, placed side by side, with narrow sheets of copper interposed between them at each end.

In 1850, John H. Lillie, of Joliet, Ill., constructed an electric motor comprising a series of radially arranged permanent horseshoe magnets, revolving on a wheel in proximity to stationary electro-magnets, Figs. 15 and 16. A helix of fine wire was wound around the outside of the electro-magnet, "for the double purpose," he said, "of producing other electro-currents in my first electro-magnet."

On one end of the axis of the wheel to which the permanent magnets were attached was a large spur-wheel, which drove two pinions to which the commutator was attached. The frame of the machine received two stationary electro-magnets on a line radial from the shaft, one on each side. Around the usual coils of the electro-magnets were wound secondary coils which were connected with the electro-

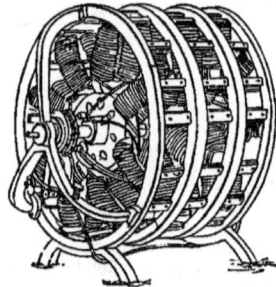

FIG. 17.—NEFF MOTOR.

magnets placed below the bed-plate of the frame, where they formed a circuit and caused the latter magnets to be energized. Secondary currents were said to be destroyed in this way. The lower magnets were so placed as to aid in the propulsion of the wheel. Mr. Lillie found it necessary to have the permanent magnets quite long, otherwise their poles would be changed by a powerful current in the electro-magnet.

Jacob Neff, of Philadelphia, devised an electric motor in 1851, Figs. 17 and 18. The metal frame of the Neff electric motor was connected by cross-bars, to which the armatures were attached in such a manner that each magnet had a separate armature. The rotating wheel of electro-magnets was secured by means of insulated nuts in the wheel-frame, which was tapped for the purpose of receiving the magnets.

The commutator was composed of three separate discs. The outside ones had flanges, by means of which they were secured to the shaft, and they were also adjusted, as circumstances might require, by means of set screws, non-conducting substances being placed between the discs. Each disc had sixteen platinum points on its periphery, corresponding in number to the armatures. Friction rollers covered with platinum were arranged to work under the commutators, they being retained in their position by set screws above their journals and spiral springs beneath. The commutators completed the circuit when the battery was connected, and the magnets were energized as they came in contact with the friction rollers and demag-

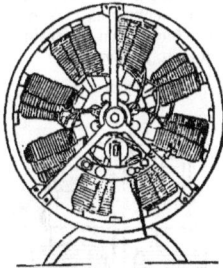

FIG. 18.—NEFF MOTOR.

netized as they left it. As will be understood, the magnets were energized when their edge was near the edge of the armatures, and continued attracting until the magnets were immediately under or opposite to the armatures. The connection was then broken and the magnets passed freely under the armatures.

Another interesting early motor patented about this time was that of Thomas C. Avery, of New York, Figs. 19 and 20. In this electromagnetic engine, Avery combined a series of electro-magnets, in pairs so as to present their poles toward a common centre, sufficient space being provided between the poles of the magnets for an intervening axis. This axis consisted of revolving sets of bars extending radially outward and passing between the poles of the magnets. At the points where the ends of the magnets approached the axis, pieces of brass or other non-magnetic material were interposed to prevent the ends of the magnets

3

acting upon the axis. On each side of these collars of brass, at a sufficient distance apart, were arranged on the axis a series of arms between each two of which the legs of the magnet were allowed to pass, and on which they exercised their attractive force alternately as the

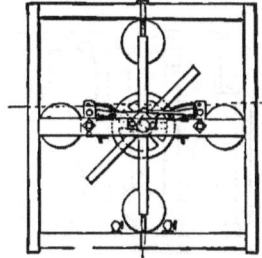

FIG. 19.—AVERY MOTOR.

circuit was made and broken through the opposite pairs of electro-magnets. Two cams were attached to the rotating axis for the purpose of breaking and changing the direction of the electric current from the vertical to the horizontal magnets continuously. Two circuit closers acted at their outer ends on these cams by means of compression springs, and were

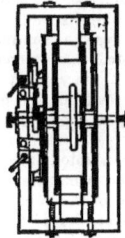

FIG. 20.—AVERY MOTOR.

attached at their back ends by a screw to support pieces fixed to the side of the frame. These support pieces were attached to a cross-bar, and were in contact with conducting material secured to the under side of the cross-bar for the purpose of reversing the direction of the electric currents.

In 1852, John S. Gustin, of Trenton, N. J., constructed an electric motor arranged for operating a pump. A side and an end elevation, Figs. 21 and 22, are here given. In connection with the oscillating beam shown was a pendulum arm extending downward and carry-

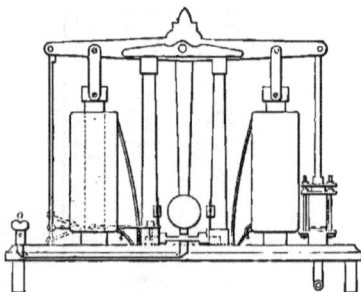

Fig. 21.—Gustin Motor.

ing a weighted ball. At the lower extremity of this pendulum was a projection which moved the valve, alternating the battery current on the magnets by the vibrating motion of the pendulum. The pendulum was intended to move between two spring-buffers at either end of its throw, which were designed to relieve the force of the blow of the pendulum, and also to assist in reversing its motion. On the valve or break-piece was a conducting plate, its length so adjusted that it could form a connection with but one side at the same time. The negative wires from the helices were both led to a like strip of copper on the opposite side of the bed-plate. At either end of the oscillating beam were pivoted links which connected or suspended the armatures of the magnets. The arrangement of the pump is shown at one side of the device, its piston-rod being connected with one end of the working-beam. At the opposite end of the beam was a long depending link which was pivoted at its lower end to a regulating spring. One end of this spring was securely fixed in the frame.

A thin piece of rubber cloth was placed on the magnet poles to prevent the adhesion of the armature to the magnets when the battery current was broken, and also to prevent violent concussion of contact. The regulating

spring was adjusted in its tension in this machine so as to require twenty pounds force to move it one inch to the point of extension (shown by dotted lines) with the rod from the working-beam, and so set as to be at rest when the pendulum was central. The object of the regulating spring was to receive the excess of power of the electro-magnets when they were closing and to give it off when they were too far extended for the attractive force to be available. With the assistance of the pendulum and this spring, nearly an equal force was said to be exerted throughout the stroke of the pump. Gustin stated that the tension of this regulating spring should be fully equal to the power required to move the pump when the spring was at its extreme point of action. At that point the electro-magnets of the Gustin motor were so feeble that the spring had nearly the whole work to perform.

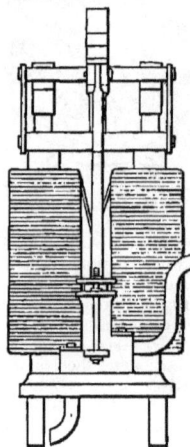

Fig. 22.—Gustin Motor.

In the same year Gustin devised another electro-magnetic motor. The legs of the magnets of this motor were nine inches long and cylindrical; the armatures were of the same length and form, and were adapted to move longitudinally in the helices which projected beyond the poles of the magnet forming a hollow core. Five magnets were employed and ten armatures, there being an armature to each pole of the magnets. Adjusting nuts were employed to secure the armatures in different relative positions on their supporting rods.

The armatures of one magnet of the series were arranged to allow of a play of one inch, the armatures of the next magnet having a play of two inches, and so on. At the energization of the series, the magnet whose armatures had the play of one inch was the most powerfully attracted, whereupon the current was broken on that magnet and and closed upon the next magnet of the series. A like result was produced successively upon all the magnets of the series until they had all performed their work. The stroke of the working-beam was not yet complete, however, when the fifth pair of magnets closed. One series of magnets

a piston-rod connected to the crank of a shaft carrying a fly-wheel. The core moved downward by its weight until its upper end was just leaving the solenoid, and thus one movement of the piston was accomplished. On passing the current the core or piston was attracted upward, and thus the second movement was completed. A commutating device was attached to the shaft which automatically admitted the current into the coil and cut it off at the right moment.

Professor Page soon improved on this single-acting electric engine by adding another solenoid, which could pull the piston in the other direction without the assistance of gravity. Fig. 23 shows this form of engine which takes electricity at both ends of the "cylinder," to borrow the expression of steam engineers. This

· FIG. 23.—PAGE MOTOR.

having performed their work, the other series, working in connection with the opposite end of the working-beam, were in readiness to perform their work in like manner. The motion of the working-beams was communicated through their bearing-rods to the crank and fly-wheel, thereby producing rotary motion.

The most celebrated early motor next to that of Jacobi was undoubtedly that of Prof. C. G. Page, of the Smithsonian Institute. This depended upon a different principle from that of the others. When the end of a bar of iron was held near a hollow electro-magnetic coil or solenoid, the iron bar was attracted into the coil by a kind of a sucking action until the bar had passed half way through the coil, after which no further motion took place. Professor Page constructed an electric engine on this principle about 1850. The solenoid was placed vertically, like the cylinder of an upright engine. A rod of iron, by way of armature, was fastened to

arrangement will be readily understood. There are two solenoids and each has its iron rod passing through it, though they are joined into one piston by a piece of non-magnetic material. The piston is attached to a frame f f' which slides through supports, and in this way it is free to move inside the solenoids. The current is sent alternately through each coil by an eccentric disc on the axle (which suggests a further resemblance of this motor to a steam-engine). This eccentric touches first one and then the other of two springs e e, connected to the solenoids.

A large motor of this description was constructed by Professor Page, in 1850, which developed over ten horse-power. Professor Page sought to apply his motor to locomotion, and he actually constructed an electric locomotive to demonstrate the practicality of his scheme. But he never achieved much success, as might have been foreseen. Among the improvements which Professor Page introduced was that of making each solenoid double, so that the arms of a U magnet could slip into them, instead of one single bar. As the solenoids attracted most strongly when the cores were almost out of them, he wound his solenoids in short sections, and a sliding commutator worked by the motion of the cores successively cut out the sections of coil which the cores had entered and transferred the current to others ahead of them, and thus the range of attraction was greatly increased.

Professor Page, it is interesting now to recall, made the trial trip with his electro-magnetic locomotive on Tuesday, April 29, 1851, starting from Washington, along the track of the Washington & Baltimore Railroad. His locomotive was of sixteen horse-power, employing 100 cells of Grove nitric acid battery, each having platinum plates eleven inches square. The progress of the locomotive was at first so slow that a boy was enabled to keep pace with it for several hundred feet. But the speed was soon increased, and Bladensburg, a distance of about five miles and a quarter, was reached, it is said, in thirty-nine minutes. When within two miles of that place, the locomotive began to run, on nearly a level plane, at the rate of nineteen miles an hour, or seven miles faster than the greatest speed theretofore attained. This velocity was continued for a mile, when one of the cells cracked entirely open, which caused the acids to intermix, and, as a consequence, the propelling power was partially weakened. Two of the other cells subsequently met with a similar disaster. The professor proceeded cautiously, fearing obstructions on the way, such as the coming of cars in the opposite direction, and cattle on the road. Seven halts were made, occupying in all forty minutes. But, notwithstanding these hindrances and delays, the trip to and from Bladensburg was accomplished in one minute less than two hours. The cells were made of light earthenware, for the purpose of experiment merely, without reference to durability. This part of the apparatus could therefore easily be guarded against mishap. The great point established was that a locomotive, on the principle of Professor Page, could be made to travel nineteen miles an hour. But it was found on subsequent trials that the least jolt, such as that caused by the end of a rail a little above the level, threw the batteries out of work-

ing order, and the result was a halt. This defect could not be overcome, and Professor Page reluctantly abandoned his experiments in this special direction.

It is interesting here to note that in 1847, the versatile and unwearying investigator, Professor Moses G. Farmer, constructed and exhibited in public an electro-magnetic locomotive, drawing a little car that carried two passengers on a track a foot and a half wide. He used forty-eight pint cup cells of Grove nitric acid battery. In 1851, Mr. Thomas Hall, of Boston, then at work for Mr. Daniel Davis, constructed and exhibited at the Charitable Mechanics Fair in Boston, the little locomotive, Fig. 24. Our illustration is taken direct from the original woodcut of the locomotive. The block was made nearly thirty-seven years ago, and first appeared in Palmer & Hall's catalogue of 1850. The engine which it represented was on

FIG. 24.—HALL LOCOMOTIVE OF 1850–1.

the principle of an electro-magnet revolving between the poles of a permanent magnet. The armature had a worm on its shaft which matched into a gear attached to the driving wheels, the latter being insulated by ivory. The track was laid in five-foot sections, and was about forty feet long and five inches wide. Under the platform of the car was a pole-changer attached to a lever; when the engine reached the end of the track it ran against an inclined plane which reversed the pole-changer and sent the engine to the other end of the track, where the same thing was repeated: thus the engine was sent automatically from one end to the other. The current, produced by two Grove cells, was, it is well to note, conveyed to the engine by the rails. We have seen, also, a photograph of the "Volta," a finely-constructed model, which was made on the same principle as the above, but so as to resemble very closely a locomotive actuated by steam. Mr. Hall says that in 1852 he made, for Dr. A. L. Henderson, of Buffalo, a model line of railroad with electric engine, with depots, telegraph line, and electric railroad signals, together with a figure operating the signals at each end of the line automatically. This, he states, was the first model of railroad signals or trains worked by telegraph signals.

Professor Page, in 1854, patented a modification of his early ideas. Figs. 25 and 26. This later motor resembled in external appearance, to

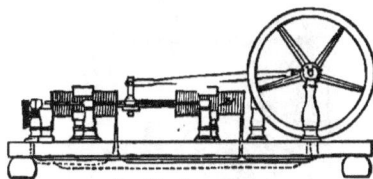

FIG. 25.—PAGE MOTOR.

some extent, a double-action, slide-valve steam pump. This Page motor comprised two parallel axial bars working through two pairs of helices, and two fixed armatures arranged at either extremity of the parallel bars. The pitman-rod connected the crank of the fly-wheel to the cross-head of the axial bars. The two pairs of helices were each connected by wires with the two conducting springs shown in the detail view, each bearing alternately against the cut-off on the fly-wheel shaft. This connection was made by means of the wires passing down under the base-board and up through

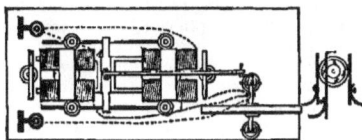

FIG. 26.—PAGE MOTOR.

to their respective connections, as shown by the dotted lines. This fly-wheel cut-off or commutator consisted of two semi-cylindrical metallic segments insulated from each other and secured to a cylinder of wood upon the shaft. An entire metallic ring was fixed upon a part of the wooden cylinder of less diameter than that to which the insulated segments were attached. This ring was connected by a strip

of metal with one of the metallic segments. The three conducting springs are shown in position in the detail view.

The spring in contact with the smaller ring connected with the positive pole of the source of electrical energy, and the current, therefore, passed through the metallic connections to the spring at the left-hand side of the detail figure. This latter spring was connected with one termination of the helices to the left of the drawing, the other being connected with the negative pole. The commutator revolved in the direction of the arrow. The axial bars are shown with thin poles passed entirely through the helices and within the influence of the armature. The instant the dead point was reached, the other pair of helices was charged to propel the frame of axial bars in the opposite direction. This was effected by the revo-

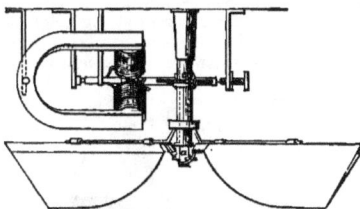

FIG. 27.—STEIN MOTOR WITH FAN.

lution of the commutator in the direction of the arrow, the metallic segments being reversed. The very short distance through which the magnets acted with power, and the rapid diminution of power as the magnets receded from each other, presented serious practical difficulties in this as in other electro-magnetic engines, whether in the reciprocating or rotary form. Dr. Page asserted that by the employment of a reciprocating core arranged to move in the line of its length through an arrangement of helices, the magnetic power could be made to act with more uniformity through a considerable distance, as some portion of the magnetic core would be always in close proximity to the helix.

In the latter part of 1854, Louis Stein devised an electric motor for operating a revolving fan, Fig. 27. The device was intended to be attached to the ceiling in the manner now familiar. The main pendent vertical shaft carried the wings of the fan. This shaft had a worm-wheel keyed to it, and to the armature shaft of the electric motor was affixed a worm which meshed with the worm-wheel and revolved the fan. The electro-magnets were arranged at equal distances apart around the horizontal shaft. Armatures were arranged at suitable distances between the series of electromagnets, so that when the battery was in action the shaft was·kept in motion and the fan revolved. This patent of Stein was more on an application of the electric motor than an improvement in the motor itself.

The electric motor of Maurice Vergnes, Figs. 28, 29 and 30, comprised four wheels or discs composed of wood and revolving upon a common axle. Each disc included an electro-magnet arranged thereon diametrically on both sides of the disc, so that the magnets were parallel. Each pair of the magnets communicated with a separate battery and revolved in a peculiarly arranged "multiplying coil," the coils being side by side and parallel to each other. There were two of these multiplying

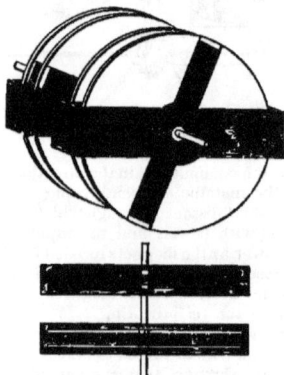

FIGS. 28 AND 29.—VERGNES MOTOR.

coils, which communicated with separate batteries, which, together with the separate batteries requisite for the electro-magnets, made in all four distinct batteries of equal intensity. By means of pole-changers the direction of the electric current was reversed in the multiplying coils at every half revolution of the wheels, and in each pair at the moment when the other

pair was exerting its greatest force. The conducting power of the multiplying coil was said to be equal to the conducting power of the electro-magnet revolving therein.

By referring to the drawings it will be seen that the magnets were straight bar magnets

FIG. 30.—VERGNES MOTOR.

and that the multiplying coils or helices formed an inclosing horizontal band through which the discs and their magnets revolved. The currents in the two multiplying coils were reversed alternately so as to produce a continuous revolution of the electro-magnets without any change in their polarity. The commutator was arranged on the shaft between the two pairs of rotating, magnet-carrying discs.

Maurice Vergnes in 1860 again appeared in the field of electricity, with an improvement upon the electric motor just described.

Instead of employing two distinct sets of electro-magnets revolving in double stationary helices, he now used a single wheel, the spokes of which were electro-magnets turning within a single set of helices. The distinguishing feature of the later construction was the disposition of the series of electro-magnets on a common axle and revolving within stationary helices, so that all the electro-magnets had, when passing through one end of the helices, a like polarity, and vice versa. Vergnes asserted that by this arrangement he obtained a continuous rotary motion without any dead point, and could develop considerable power. His second device is illustrated in Fig. 31. The two rectangularly arranged helices, within which the magnet-carrying wheel revolved, were supported in a horizontal position upon a table or frame, as shown in the figure. The magneti-

wheel revolved within these helices, the axle of which passed between them. The wheel itself was composed of two flat electro-magnets placed at right angles with each other, and on a common centre on the shaft. The disposition of the elements of this apparatus was such that when one electro-magnet approached an inclination of forty-five degrees with the helices in

Fig. 31.—Modified Vergnes Motor.

its rotation, the current passed into the magnet. The other bar magnet was energized in the same manner. The commutator was provided with anti-frictional contact electrodes.

The patent of Yeiser, granted in 1858, employed what was at that date a novel mechanical arrangement for obtaining the full measure of the attractive power of an electro-magnet upon its armature. The operation of this device will be readily apparent from an inspection of Fig. 32. A series of balance beams arranged one above the other was employed. To both ends of these beams were attached armatures of equal weight, which came into the magnetic field successively and were attracted to the magnets so that each one in turn became momentarily in effect an elongation and a part of the electro-magnet. At each end of the machine was arranged a series of upright electro-magnets side by side, so that all their poles were in the same horizontal plane. The circuits of the two series of magnets, were, however, independent of each other. The length of the series of armatures was sufficient to cover the poles of all the magnets at each end.

The commutator for closing the circuits alternately through the two series of magnets was mounted upon the driving shaft above the

armature beams. The commutator consisted of a wheel, one-half of the periphery of which was insulated in the usual way. The circuit was changed from one to the other of the series of magnets twice in every revolution of the shaft.

The distance between the horizontally arranged armature beams was such that the beams had a limited amount of movement independent of each other, the lower beam being pivoted so that it could have no more vibratory movement than that of the several beams pivoted above it.

In the electric motor patented to Lewis H. McCullough, February 26, 1867, a vertically arranged vibratory armature-carrying beam was the main feature. Fig. 33 is an illustration of the McCullough motor. Two pairs of electro-magnets, one above the other, were arranged on both sides of the vertical oscillating beam, and so that the double armatures lay in the same horizontal plane as the magnets.

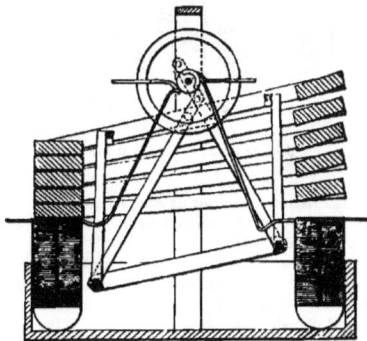

Fig. 32.—Yeiser Motor.

Each pair of armature plates was equidistant from the pivotal point of the axis of the vibrating beam. Attached to the upper end of the beam by means of a wrist-pin was a pitman through which rotary motion was communicated. An endwise disposition of the magnets for the purpose of increasing or diminishing their attractive force was accomplished by adjusting screws at the rear of the magnets. The lower oscillating end of the vertical beam was arranged to make and break the electrical connection between the oppositely arranged magnets. McCullough stated that in the operation of his motor there was no positive breaking of the current at any point in the vibration of the

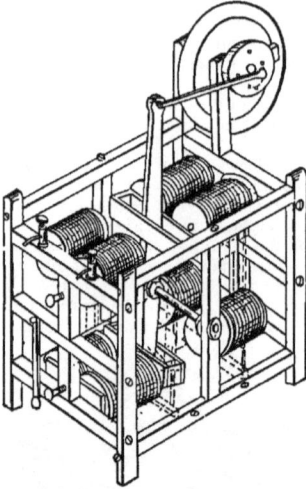

FIG. 33.—McCullough Motor.

central beam, and that as a consequence, there was no loss of the electrical or exciting force upon the magnets.

On April 2, 1867, Chas. J. B. Gaume, of Iowa, patented an electro-magnetic engine of which a side elevation in Fig. 34, and a plan view in Fig. 35 are shown. In the Gaume construction a series of electro-magnets were placed on the periphery of a wheel, and journaled to the same axis another wheel revolving between the adjacent magnets, carrying a series of armature plates attracted successively. The battery wires were so connected through the motor that a reserve power might be attached or detached by the motion of a governor upon the engine, the speed of which determined the battery connection.

By an inspection of the figures it will be seen that the electro-magnets were mounted upon the horizontal shaft, the wheel carrying the armatures being mounted upon the same shaft,

but revolving in an opposite direction. Each of these wheels carried a bevel pinion, and both meshed with a third bevel gear, mounted upon a vertical shaft, to which the governor was attached. The wires of the electro-magnets were led to the commutator in the usual manner.

Below the armature beams and between the magnets was a supplementary oscillating arm, having pivoted to its outer ends two upright rods, the upper ends of which were attached to the beam which carried the topmost pair of armatures. To the ends of the lower oscillating beam were also pivoted two crank arms or pitmen, the upper ends of which were coupled to the driving shaft by means of crank arms. As all of the series of bars which were operated upon came down as close as possible together within the magnetic field of each pole, the commutator broke the circuit of that series of magnets and closed the circuit of the other series, whereby the other ends of the series of bars were brought into action. In this way an

FIG. 34.—Gaume Motor.

oscillating motion of the beams was produced, and the upper beam served through its connections to produce a rotary motion of the driving shaft. When the circuit was first closed through the series of magnets the lowest of the corresponding series of armature bars was attracted directly to the magnets, and by its movement all the other armatures opposite, whose ends rested upon each other, were caused to move a corresponding distance, upon which the lowest bar became magnetic, attracted the second one and drew it down in contact with it, thus giving all the beams a further move-

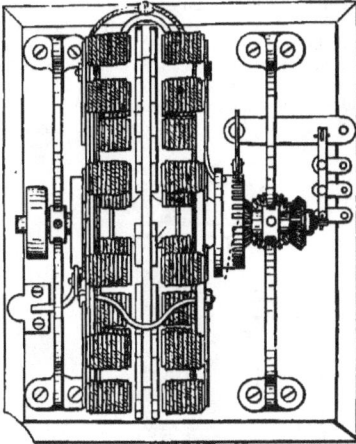

Fig. 35.—Gaume Motor.

ment. The second bar, as it came in contact with the first, became magnetic and attracted the third, and so on through the series till all the bars were in contact, as shown in the figure.

The electric governor was of the usual pivoted ball construction, and revolved upon a sliding collar on a vertical shaft rotated by an arrangement of bevel wheels, as before indicated. When the balls rose under increase of speed, a central rod was depressed, raising by an arrangement of levers the horizontal pivoted circuit breaker shown at the bottom of the side-elevation.

This circuit breaker or switch had three keys, which, when the switch was in a horizontal position were in contact with three corresponding plates to which were attached wires from auxiliary batteries. When the governor reached a certain high speed it disconnected one of the keys and consequently one of the sources of electrical power. If the speed still increased, the electrical connection between the second or

central key was broken, and so on. Thus it will be seen, the amount of electrical power was graduated to the speed, the successive connections being severed as the speed increased, and, conversely, being restored when the speed decreased.

As is usual with this type of machine, a determinate impulse in a given direction having been communicated to the wheels, their impetus carried them in the intervals of time when the electric circuit was broken, and the electric impulse being imparted at a certain period, the armatures were individually attracted toward the electro-magnet next in series, and an additional impulse was obtained, producing an increment of speed.

The principal feature of novelty in the electric motor which William Wickersham patented June 2, 1868, and is shown in Fig. 36, was the employment of an endless electro-magnetic chain, the alternate links of which were magnetic bars, the remaining connecting links being non-magnetic. The magnetic links were surrounded by helices through which an electric current passed.

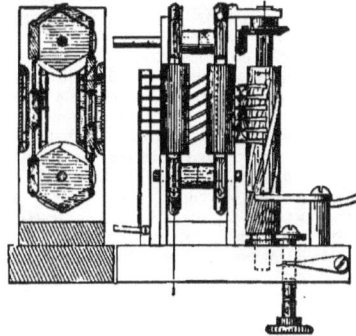

Fig. 36.—Wickersham Motor.

The machine itself had two of these endless magnetic chains, arranged vertically on parallel shafts so as to revolve thereon. The bars of the chain, at fixed periods in their revolution, passed through helices having hollow cores.

The strips of metal of which the stationary helices were formed, extended at one end of the motor beyond the helices, and were arranged in parallel lines. A commutator.

consisting of a vertically arranged revolving cylinder, had metal conductors in the shape of strips of metal placed at .intervals around it. These conductors were wound spirally, and extended from one end of the cylinder to the other. The extended ends of the helices before referred to were arranged so as to wipe this revolving commutator, and thus close the circuit successively in the different helices. The closing of the circuit in the independent series of coils which constituted the completed helices, was made successively from one end to the other of each helix, and at the same speed that the magnetic endless chain moved in passing through the helices. This was effected by the spiral form of the conductors in the commutator.

The motor was adapted to be stopped or reversed, the commutator being vertically adjustable upon its shaft for this purpose. This commutator, which Wickersham styled a " circuit-cylinder," had a rod arranged in a parallel position with it and passing through the base, by means of which the vertical adjustment of the commutator was accomplished. On the upper end of this rod was secured a freely revolving washer, which rotated within a groove near the lower end of the commutator. The rod itself had three grooves within it, in any one of which the spring stop bolted to the side of the machine could rest. When this rod was raised or depressed, the commutator moved with it and thus was held at any vertical elevation determined by the three grooves in the rod. When the commutator was moved to its highest position the engine ran forward; at its lowest position the commutator reversed the motor, while at its intermediate position the adjustable commutator brought the motor to a state of rest.

The independent coils which constituted the helices were wound in different directions, and each one conducted the electric current around the magnetic link of the endless chain in a different direction from the one preceding it, thereby giving to the magnetic links alternate reversed polarities. When two columns of helices were used on opposite sides of the machine (the magnet chain passing downward through the one and upward through the other), the attraction of the former would be downward and that of the latter upward.

The motor constructed by Charles T. Mason and that made by Mr. A. J. B. DeMorat com-

plete the list of patented motors, the terms of protection of which expired up to the end of 1885, so that our review of some of the early American motors may well end at this point.

Mason's motor, Fig. 37, was designed for driving a fan. It consisted of an electro-magnet, one terminal of the coil of which was connected to the binding post shown in the illustration, and the other to the spindle of the fan shaft. The armature of the electro-magnet is shown pivotally secured to the fan shaft above the the electro-magnet. The fan spindle also carried a cam which, as it revolved, broke and

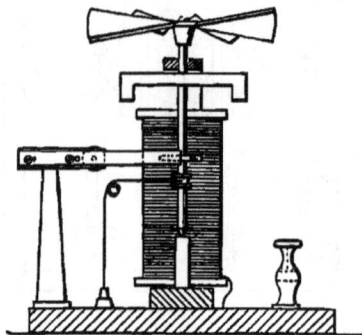

FIG. 37.—MASON MOTOR.

made connection with the horizontal wiping spring secured to a standard at the left of the figure. The cam and wiping spring formed the commutator of the motor.

De Morat's motor is shown in side elevation and in vertical section in Fig. 38. De Morat asserted that there was no interruption or breaking of the current in the use of this motor, such a result never having been practically accomplished before, and that greater velocity, more regular and constant motion, with greater power, could be obtained from his construction than from any other similar machine patented before that date.

By referring to the drawing it will be seen that the lower wheel represents a circular magnet of two or more poles. The central disc of this wheel was of iron, with contiguous coils of wire on either side, the whole being clamped together with wooden discs. The commutator was also fixed upon the shaft of this circular magnet. It consisted of two metal bands insulated from each other and electrically connected with two wiping springs which completed the circuit. Above this circular magnet was a wheel of many armatures. This wheel consisted of a number of radial arms which had flanges to receive the separate armatures composing the wheel. Each of these armatures had an independent radial movement within the flanges of the wheel, but without touching the circumference of the same. Each, in addition, had

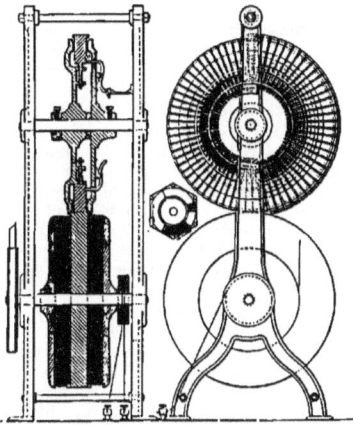

Fig. 38.—De Morat Motor.

an inward extension stem or shank, which was held in operative position by lugs upon the wheel frame. Coiled springs surrounded these stems between the lugs, and had a constant tendency to force the armatures outward when not held inward by the latches. A projecting arm was bolted to the inner side of one of the standards, on the end of which the outer ends of the latches of the armatures struck. This movement lifted the latches out of the notches in the armatures, and the springs forced them outward, as shown.

As soon as the circular magnet was energized, the armatures were attracted angularly, producing a motion by the tendency to make contact. This was not possible without producing two motions, one causing the system to revolve, the other sending the armatures in-

ward by their contact with the magnet, and fastening them there by the latches dropping into the notches on the armatures. In that position they passed beyond the magnetic field until released by the projection on the frame.

De Morat contemplated reversing the relative arrangement of the motor by converting the armatures into electro-magnets and causing them to exert an attraction on different curves or on a number of planes tangent to the circle in the form of a polygon, as shown in the small sketch representing a hexagon. The attraction would then be effected so as to form an endless chain or elastic band.

It deserves notice that between 1860 and 1867—the period of the civil war—not a single patent was issued in America on electric motors. A war to-day would probably be highly stimulative of inventiveness in this direction. But down to 1860, the interest that had begun to manifest itself twenty years earlier, continued in almost undiminished measure. A lively sketch of the condition of affairs during that period was given by Dr. Vander Weyde in May, 1886, before the New York Electrical Society. Dr. Vander Weyde was commissioned by the late Mr. Peter Cooper to examine the various motors that were submitted by inventors who desired to obtain capital for the furtherance of their work; and it was well for the distinguished philanthropist that he could enjoy the services of one so competent, and of one, too, who by continuous experiments between 1843 and 1848 had already satisfied himself that the electric motor could never be substituted to any extent for other motors so long as the main dependence was upon chemical batteries.

"Invariably I felt obliged to advise adversely," says Dr. Vander Weyde, "and, while Mr. Cooper was very slow to invest in uncertain enterprises, in some instances a great pressure was brought to bear upon him by enthusiastic inventors and their still more enthusiastic friends, to whom he might have yielded if my convictions, in which he appeared to have much confidence, had not prevented it. Those examinations took place off and on during the whole period of the erection of the Cooper Union building, which was completed in 1859, when I was appointed one of the teachers. I believe Mr. Cooper never spent a single dollar on account of electro-motors, except on such small specimens as were required for class instruction in the regular course of lectures in physical science given in the Cooper Union building.

"The electro-motors I examined differed greatly in size, from such as occupied scarcely the space of a cubic foot to those of the size of a 50 horse-power steam engine. Among the latter I must mention the motors of three inventors who operated on a large scale, viz., Professor Page, of Washington; Professor Vergnes, of New York, and Mr. Paine, of Newark. I saw Page's engine in operation in New York in 1850. His system is well known among electricians, but deserves special mention for the large scale on which he executed it. It consisted in massive iron plungers which were attracted into coils by alternate currents, and by means of a crank they revolved a fly-wheel. Vergnes' machine was exhibited at our World's Fair in 1853, in the Crystal Palace on Reservoir Square—now called Bryant Park—and consisted in elongated loops of copper wire revolving between the poles of powerful and colossal electro-magnets. In regard to this machine, I will remark that if he had reversed the function of his machine and revolved the loops by means of steam power, he would have had one of the forms of Siemens' dynamos, and would have solved the problem reserved for the investigators of twenty years later (Pacinotti, Gramme and Siemens), who transformed power into electric currents by the inverson of the function of the motor, as Gramme did with Pacinotti's ring. In fact, one of the little motors which I constructed in 1844 would have been a small dynamo, if revolved with sufficient power and velocity.

"For driving their large motors, both Page and Vergnes used proportionally large batteries—large in size as well as in the number of cups. Their batteries were always carefully hidden from view, especially those used by Vergnes, who had in the Crystal Palace several locked rooms which were filled with them.

"Paine, of Newark, did not need any battery at all for exhibition of his motor. He had, however, a small battery connected with his motor, and pretended that this did drive it, together with the circular saw connected with the same. This saw operated with such power that it aroused my suspicions, so I surreptitiously disconnected the battery, and as the saw worked just as well I was convinced that power

was obtained from elsewhere. I then discovered that next door there was a factory where steam-power was used, and that Paine's electro-motor was only on exhibition during the working hours of the factory. The whole deception was clear, the only purpose being to sell stock in Mr. Paine's electro-motor company, which was kept up for several years, but has been put in the shade by the strong vitality of the Keely motor enterprise of the present day."

This chapter should not close without mention of Pinkus, who early conceived an ingenious method of operating an electric railway. Dr. Wellington Adams, of St. Louis, in a paper read in 1884, before the engineers' club of that city, gave some interesting details of the manner in which, when working upon the idea of a railway whose motors picked up their current from the rails, he was referred to the work of his predecessor. "Although at the time (1879) actively engaged in medical practice, and connected with the Medical College in Denver, so great were the allurements, that I was induced to give up everything in Colorado and leave there rather precipitately for Washington, in quest of a generic claim upon this fundamental principle. My case being examined, it was, however, found that the same principle had been proposed and provisionally patented, as far back as 1840, by one Henry Pinkus, a remarkably inventive genius of that period. In 1840, however, the dynamo was unknown, and the electric car motors of Pinkus, which existed only in his imagination, were supposed to be operated by galvanic batteries buried in the ground. The principle of the transmission of the current to the car while in motion for the purpose of effecting its propulsion was, however, the same. The inventor even went so far as to anticipate the future use of 'mechanical generators which should be more economical than the batteries.'" Pinkus was but another of those who were allowed to see the promised land, but were unable to enter. Mention may also be made here of the electric locomotive devised in 1847 by Mr. Lilley and Dr. Colton, of Pittsburgh. This locomotive was driven around a circular track by electricity. The rails were insulated, each connecting with a pole of the battery, and the current was taken up by the wheels, whence it passed to the magnets, upon whose alternate attraction and repulsion motion depended.

CHAPTER IV.

THE ELECTRICAL TRANSMISSION OF POWER.

WHEN Dr. Antonio Pacinotti described his "Electro-Magnetic Machine" in the Italian periodical *Il Nuovo Cimento*, in June, 1864, he mentioned the fact that the machine could be used either to generate electricity on the application of motive power to the armature, or to produce motive power on connecting it with a suitable source of current. This, so far as can be determined, was the first mention of the now so well known principle of the *reversibility* of the dynamo-electric machine, the practical utilization of which implies the development of a new electrical industry—the electrical transmission of mechanical energy.

Probably Dr. Pacinotti himself did not realize that even while he was, for the first time, disclosing the principle of construction that was destined to make the dynamo-electric machine practical and efficient, he was demonstrating this principle of reversibility which promises to multiply the application and utility of dynamo-electric machines tenfold. We mention Dr. Pacinotti's name here purposely to give him the homage due for his valuable researches in this field, especially inasmuch as it was his misfortune to be too far ahead of his time. His researches failed to attract the attention and encouragement which they deserved, and we might say that the same ground had to be travelled over again by those who came later.

The principle of the reversibility of dynamo-electric machines appears to have been perceived by Messrs. Siemens about 1867, but it was not heard of in practical application until the year 1873, when it was practically demonstrated by MM. Hippolyte Fontaine and Breguet at the Vienna Universal Exposition. In this case a Gramme machine used as a motor to work a pump was run by the current produced by a similar machine connected by more than a mile of cable, and put in motion by a gas engine. This was the first instance of electrical transmission of mechanical energy to a distance.

It is always interesting to go back to the first dawn of a new invention, but it is not always easy to determine whether it was the result of accident and necessity, or the outcome of clear, intelligent foresight in the part of an inventor working for a particular purpose. As regards the first transmission of power by electricity, opinions are divided. According to M. Figuier, accident, pure and simple, was the cause of the discovery. He relates that at the International Exhibition of Vienna in 1873, the Gramme Company exhibited two machines intended for plating purposes. One of these machines was in motion, and a workman who noticed that some cables were trailing on the ground, thinking they belonged to the second machine, placed them in its terminals. To the surprise of everybody this second machine, which had been standing still, began to turn of its own accord. Then it was discovered that the first machine was working the second.

This story is romantic, but disappointing to a true lover of science, who would prefer to believe that a great discovery was the logical outcome of the working of a powerful intellect, and not the result of accidental meddling on the part of an ignorant workman. But there is another version of the story, told by M. Hippolyte Fontaine to the Société des Anciens Elèves des Ecoles Nationales des Arts et Métiers. M. Fontaine claims to have actually invented or discovered the electrical transmission of power, as will be seen from the following short abstract of his paper read before the above mentioned society:

"On the 1st of May, 1873—that is, on the date fixed four years previously by imperial decree—the Exhibition in Vienna was formally opened. At that time the machinery hall was yet incomplete, and remained closed to the public until the 3d of June, when it was also thrown open. I was then engaged with the arrangement of a series of exhibits, shown for the first time in public, which were intended to work together, or separately, as desired. There was a dynamo machine by Gramme for electroplating, giving a current of 400 ampères at 25 volts, and a magneto machine, which I intended to work as a motor from a primary battery, or from a Planté accumulator, to demonstrate the reversibility of the Gramme dynamo. There were also a steam engine of my invention heated by coke, a domestic motor of the same type heated by gas, a centrifugal pump placed on a large reservoir, and arranged to feed an artificial cascade, and numerous other exhibits. To vary the experiments I proposed to show, I had arranged the pump in such a way that it could be worked either by the Gramme magneto machine or by the steam engines (Fontaine).

"On the 1st of June it was announced that the machinery hall would be formally opened by the Emperor at 10 A. M. on the day after the morrow. Nothing was then in readiness, but those who have been in similar situations know how much can be got into order in the space of 48 hours just before the opening of a great exhibition. In every department members of the staff with an army of workmen under their orders were busy clearing away packing cases and decorating the spaces allotted to the different nations. These gentlemen visited all the exhibits in order to determine which of them should be selected for the special notice of the Emperor, so as to detain him as long as possible among the exhibitors of their respective countries.

"M. Roullex-Duggage, who superintended the work in the French section, asked me to set in motion all the machinery on my stand, and especially the two Gramme machines. I set about at once, and on the 2d of June I had the satisfaction of getting the large Gramme dynamo, the two engines (Fontaine), and the centrifugal pump to work; but I failed to get the motor into action from the primary or secondary battery. This was a great disappointment, especially as it prevented my showing the reversibility of the Gramme machine. I was puzzled the whole of the evening and the whole of the night to find a means to accomplish my object, and it was only in the morning of the 3d of June, a few hours before the visit

of the Emperor, that the idea struck me to work the small machine by means of a derived circuit from the large machine. Since I had no leads for that purpose, I applied to the representative of Messrs. Manhis, of Lyons, who was kind enough to lend me 250 metres of cable, and when I saw that the magneto machine was not only set in motion, but developed so much power as to throw the water from the pump beyond the reservoir, I added more cable until the flow of water became normal. The total length of cable in circuit was then over two kilometres. This great length gave me the idea that by the employment of two Gramme machines it would be possible to transmit mechanical energy to great distances. I spoke of this idea to various people, and I published it in the *Revue Industrielle* in 1873, and subsequently in my book on the Vienna Exhibition. The publicity thus given to it was so great that I had neither time nor desire to protect my invention by a patent. I must also mention that M. Gramme has told me that he had already worked one dynamo by the other, and I have always held that the honor of my experiment belongs to the Gramme Company."

Electric lighting had not yet left the laboratory or the lecture room, and the Gramme machines, then about the only ones made, were all constructed for electrotyping or electroplating, and were consequently ill adapted to the purposes of electrical transmission. However, the demonstration served the purpose of M. Fontaine, its author, for it called attention to this field of study. In 1877 some officers of the French army made use of two Gramme machines to transmit power from a steam engine to a dividing machine, placed at a distance of about sixty metres. Meanwhile, other experimenters were at work in the same direction and it became a lecture experiment to work machinery by an electric motor operated by a current generated at a distance. It was only in 1879, however, that the real importance of the subject was made apparent by MM. Félix and Chrétien in their experiment at Sermaize in plowing by electricity, which was conducted on a practical scale and caused great excitement. At the same time, 1879, the electrical railway of Siemens and Halske, which made its first appearance at the Berlin Exhibition, was an interesting and perhaps still more striking instance of the possibilities of the electrical transmission of power. The electrical exhibition of 1881 at Paris also afforded to electrical engineers an excellent opportunity to demonstrate the applicability of electrical transmission in providing motive power for multifarious purposes. The currents from dynamo-electric machines were used for driving motors for sewing machines, lathes, planers, drills, hammers and other workshop machinery, rock-drills, saws, pumps, elevators; and also for electric railways, one of which, from the firm of Siemens & Halske, served to convey passengers from the Place de la Concorde to the Palais de l'Industrie.

At the Munich Exposition in 1882 the subject of the electrical transmission of mechanical energy did not fail to receive a share of attention. There were several practical examples showing the production of motive power from electricity.

Our illustration, Fig. 39, represents one of these installations, a small workshop, which derived its motive power from a Schuckert machine fed from a similar machine placed in another part of the Crystal Palace and run as a generator. M. Schuckert had also another installation for demonstrating the transmission of mechanical energy by electricity. In this case the generator was placed at Hirschau in a machine shop supplied with water power from the Isar River. The motor was in the Crystal Palace, and the distance between the two was about ten kilometres. The conductors were two wires of copper, four millimetres in diameter (= No. 6 B. and S.) and had a total resistance of 9.6 ohms. The power expended on the generator was of about nine horse, and the work done by the motor, which was bolted to a counter-shaft furnishing power to two threshing machines running empty, was equal to about three horse power. The efficiency was therefore about one-third.

In another installation, the Edison "Z" machine was used as a motor to supply motive power to a German "Melkerei," or dairy, such as is frequently to be seen in the mountains in Bavaria, where the power of water-falls is made to move the churns, skimmers and other contrivances.

But there was another example — we might say proof—of the possibilities of electrical transmission of energy to a distance, which made the Munich Exposition itself memorable, and com-

pared with which all other previous ones paled into insignificance. We refer to the celebrated feat of electrical transmission of M. Marcel Deprez. The science of electrical transmission of energy over long distances may be said to date from that time, for it was in these experiments that M. Deprez revealed to the electrical world the theory of electrical transmission deduced by himself, while he furnished a proof by ocular demonstration. It is for this reason that the name of Marcel Deprez occupies such an important space in all discussions bearing on this subject.

We may omit the consideration of M. Deprez's particular theories for the present, however, in order to point out generally what are the principles entering the problem of the distribution of power by electricity, the most of which were first demonstrated by Siemens and Deprez.

To convey energy by means of electricity from one place to another, three things are necessary: a generator, a motor, and two conductors connecting both. The generator converts energy out of its mechanical form (or chemical, caloric) into electric energy; the motor reconverts it into its mechanical (chemical, caloric) form. But not all the electric energy produced by the generator will be reconverted by the motor, as it is a well known fact that if a current pass through a circuit a certain amount of its energy will appear as heat, as no circuit can be made without resistance. If then W stands for the work expended upon the generator, w for that done by the motor, and if $H J$ be the mechanical value of the heat generated (J stands for Joule's equivalent and H for the number of heat units), then, according to the law of conservation of energy first enunciated by Helmholtz:

$$W = H J + w. \qquad (I.)$$

Now, if the electromotive force of the generator be E, and the resistance of the circuit, including generator, motor, and conductors be R, then a current C would have to pass through the circuit, and, according to Ohm's law,

$$C = \frac{E}{R}.$$

And this is also the case as long as the motor stands still; but as soon as its armature rotates, i. e., the motor does work, the current C sinks

FIG. 33.—SCHUCKERT INSTALLATION AT THE MUNICH EXPOSITION OF 1882.

to C_1. Now, as no new resistance has been added to the circuit, the cause of this falling of current can only be an electromotive force, induced by the magnets of the motor in the rotating coils of the armature, and opposing the electromotive force of the generator. C_1 can therefore be expressed by the formula:

$$C_1 = \frac{E-e}{R} \qquad \text{(II.)}$$

in which e stands for the counter electromotive force.

In the preceding and subsequent formulas Mr. H. M. Schlesinger, in following Professor S. P. Thompson, has assumed the generator and motor to be such as to convert mechanical energy into electric, and *vice versa*, without loss; the sources and effect of such loss will be considered later on.

Formula (II.) may also be written in the following manner:

$$E = e + C_1 R,$$
$$\text{or} \qquad E\,C_1 = e\,C_1 + C_1{}^2 R. \qquad \text{(III.)}$$

But the work done by a dynamo can be expressed by the product of its electromotive force and the current it generates, and, according to Joule, the heat generated in a circuit is proportional to the square of the current passing through it and to its resistance. We can therefore put

$$W = E\,C_1, \text{ and } H\,J = C_1{}^2 R.$$

Equation (III.) can then be written

$$W = e\,C_1 + H\,J.$$

On comparing this with (I.) we got:

$$e\,C_1 = w. \qquad \text{(IV.)}$$

Or substituting the value of C_1 found in equation (II.),

$$w = e\,\frac{(E-e)}{R} \qquad \text{(IVa.)}$$

or *the work done by the motor is equal to the product of the current flowing through the circuit and the counter electromotive force the motor has set up.*

Referring to (II.), we find that if E and R be constant, C_1 is a function of e, for any change of e produces a change of C_1. Now, as $e\,C_1$ is the expression for the work done by the motor,

5

the question is, for what current will $e\,C_1$ be a maximum? To find this maximum, we will write (III.):

$$e\,C_1 = E\,C_1 - C_1{}^2 R;$$

and employing the differential calculus for the sake of brevity we get by placing the first differential coefficient equal to zero:

$$\frac{d\,e\,C_1}{d\,C_1} = E - 2\,R\,C_1 = 0,$$

or $$E = 2\,R\,C_1;$$

but, according to (I.), $E = R\,C$,

therefore $$C_1 = \frac{C}{2},$$

and as $$C_1 = \frac{E-e}{R} = \frac{C}{2},$$

$$E - e = \frac{1}{2}\,R\,C = \frac{1}{2}\,E,$$

$$e = \frac{1}{2}\,E.$$

That is, if the counter electromotive force is such that the current flowing through the circuit is just one-half of the current which would flow through it if the motor did no work, the motor will be doing most work *for unit time;* for any other current larger or smaller than $\frac{C}{2}$ the amount of work done within unit time will be less.

This law, generally called the law of maximum activity, was discovered by Jacobi. Although as we have pointed out in a previous chapter it has *nothing whatever to do with the efficiency of transmission,* it has often been mistaken for the law governing the latter; and as in case of maximum activity the efficiency of transmission is 50 per cent., it has been said that the highest efficiency of the system is only 50 per cent. This is entirely wrong, as the efficiency can be made as high as one likes. This will be seen later on.

Jacobi's law can easily be proved by an example. Let $E = 100$; $R = 10$; then $C = 10$, and for $C_1 = 5$, $e = 50$; as deduced from eq. (II.)

the work done by the generator is $E\,C_1 = 500$ watts, the work done by the motor is $c\,C_1 = 250$. If $C_1 = 6$, then from (II.)

$$E - c = R\,C_1,$$

or $100 - c = 60$; therefore, $c = 40$;
    $E\,C_1 = 600$, $c\,C_1 = 240$.

If $C_1 = 4$, $c = 100 - 40 = 60$,

and $E\,C_1 = 400$, $c\,C_1 = 240$.

(This also proves what has been said about the efficiency, as in the first case $\eta = \dfrac{250}{500} = \dfrac{1}{2}$;

in the second $\eta = \dfrac{240}{600} = \dfrac{2}{5}$; and in the last case

$\eta = \dfrac{240}{400} = \dfrac{3}{5}$.)   The Greek letter η stands for the efficiency of transmission.

In his work on dynamo-electric machinery, Prof. S. P. Thompson has given diagrams to show this law graphically.   Our equation (II.) is

$$C_1 = \frac{E - c}{R}$$

and $E\,C_1 = W_1$, $c\,C_1 = w$;

we can then write

$$W = \frac{E\,(E - c)}{R} \text{ and } w = c\,\frac{(E - c)}{R},$$

and as R is a constant we get the relative values,

$$E\,(E - c) \text{ and } c\,(E - c).$$

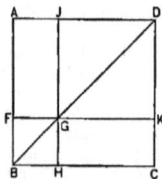

Fig. 40.

Let us now construct a square at $A\,B\,C\,D$, Fig. 40, the side of which is equal to E, and measure out from the point B the counter electromotive force $e = B\,F$, draw $F\,K$ parallel to $B\,C$ and through the point G, in which $F\,K$ intersects the diagonal $B\,D$, draw $J\,H$ parallel to $B\,A$. The rectangle $A\,F\,K\,D$ will now repre-

sent the work done by the generator, as $A\,F = E - e$ and $F\,K = E$, and the rectangle $G\,H\,C\,K$ represents the work absorbed by the motor, as $G\,H = e$ and $H\,C = E - e$.

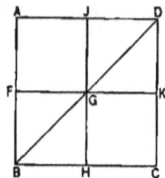

Fig. 41.

According to a well-known geometrical law the rectangle $G\,H\,C\,K$, Fig. 41, will be a maximum for $G\,B = G\,D$; then $B\,F = \dfrac{A\,B}{2}$; that is $e = \dfrac{E}{2}$, and as was shown above, if the counter electromotive force is half the electromotive force, the current passing through the circuit is one-half of the current which would pass if the motor were standing still. The square $G\,H\,C\,K$ representing w is

$$G\,H\,C\,K = w = e\,(E - e) = \left(\frac{E}{2}\right)^2 = \frac{1}{4}\,E^2.$$

The rectangle $A\,F\,K\,D$ represents W or

$$A\,F\,K\,D = W = E\,(E - e) = \frac{1}{2}\,E^2 = 2\,w.$$

According to this diagram, then, the motor will do most work in unit time when the counter electromotive force is one-half of the electromotive force, the current at the same time being one-half of the current which would flow through the circuit if the motor was standing still.

As we pointed out before, and as formulas (I.), (II.), (III.) and (IV.) show, not all the work put into the generator will be recovered through the motor. In the following, both generator and motor are still assumed to be perfect—that is, transforming energy of one form into another without loss. Of course, in practice this will not be the case, but the nature and amount of such losses are known to all good dynamo-makers, who, therefore, also know by what means they can be brought down to the lowest limit. Moreover, it will be very difficult, if not even quite impossible, to give formulas covering all these sources of loss. But, aside from these losses, part of the energy is lost in the circuit, being that part which is necessary to force the current through the circuit. The rest of the energy will appear as useful work in the motor (assumed to be perfect). Now, the efficiency of any system is the ratio of the useful work to that spent in producing it; that is, if η stand for the efficiency of transmission,

$$\eta = \frac{w}{W}, \text{ and in our case}$$

$$\eta = \frac{w}{W} = \frac{e\,C_1}{E\,C_1}; \text{ that is}$$

$$\eta = \frac{e}{E}. \tag{V.}$$

According to this equation, the efficiency of transmission is as the ratio of the electromotive forces. This again shows that Jacobi's law of maximum activity has nothing to do with the efficiency. The counter electromotive force e can range between the limits $e = 0$ and $e = E$, at the same time η will range between $\eta = 0$ and $\eta = 1$. If $e = 0$, then $\eta = 0$ also; if $e = \frac{1}{2} E$,

then $\eta = \frac{1}{2}$; if $e = E$, then $\eta = 1$. In the

first case, the motor will be doing no work, as the energy put into the generator will be lost in heating the wires of the circuit. With e the efficiency will gradually rise, and at the same time the actual work got out of the motor in

unit time till $e = \frac{E}{2}$, at this point $\eta = \frac{1}{2}$; that

is, half the energy put into the generator is lost as heat in the circuit; the other half appears as useful work, and for $e = \frac{E}{2}$ also,

the motor will be doing most work. As e continues to rise, η does the same, but at the same time the work done per unit time sinks again, till $e = E$. Now, $\eta = 1$, but the motor will be doing no work, and theoretically the generator ought to require none either; but in practice this is impossible, as e can never

rise so high that $e = E$. The cause of this is the resistance of the circuit, and, of course, the mechanical resistance of the armature, such as friction, etc. In practice, it is difficult to measure the counter electromotive force e (e must not be confounded with the difference of potential at the terminals of the motor); it is more convenient to measure the current flowing through the circuit and the difference of potentials at the terminals of the generator, and, knowing the resistance of the generator and the other part of the circuit, the electromotive force of the generator is easily found, and equation (II.) will give the counter electromotive force.

These relations between counter electromotive force, efficiency, and work per unit time, can very easily be shown graphically with the diagram used by Thompson. Let $A\,B$, Fig. 40, again be equal to E, $F\,B = e$, and the lines $F\,K\,J\,H$ drawn as before; then, R being constant:

$$A\,F\,K\,D = E\,(E - e) = W$$
$$G\,H\,C\,K = e\,(E - e) = w.$$

The efficiency of transmission will then be as the ratio of these rectangles, and the work lost in heating the circuit will be represented by $J\,G\,K\,D$, as $A\,F\,G\,J$ is equal to $G\,H\,C\,K$.

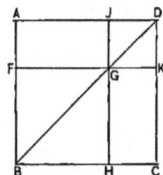

Fig. 42.

This diagram, therefore, represents a case in which either the load put on the motor is too large or the armature not properly geared to the rest of the working parts (in an electric locomotive, to the wheels). The result is that the armature of the motor cannot move with the necessary speed, and therefore the counter electromotive force is very low. Fig. 42 is the diagram of another case; $A\,B$ again is equal to E, $B\,F = e$. The work spent in the generator is, therefore, again represented by the rectangle $A\,F\,D\,K$, and the useful work by $G\,H\,C\,K$. It is easy to see that in this case the efficiency represented by the ratio of the two rectangles is far superior to that represented by Fig. 40. Whereas in Fig. 40 $F B$ is $\frac{1}{2}$ of $A B$ (i. e., $E = 3e$), and in consequence rectangle $G H C K$ is one-third of $A F K D$, the efficiency being therefore $\eta = \frac{1}{3}$, in Fig. 42 $F B$ is two-thirds of $A B$ (that is, $2 E = 3e$), and, therefore, rectangle $G H C K$ is $\frac{2}{3}$ of rectangle $A F K D$, making the efficiency $\eta = \frac{2}{3}$. In this case, therefore, although only *half the work* has been put on the generator, the motor is doing *exactly the same amount* of work as in the case represented by Fig. 40.

In the expression for the efficiency

$$\eta = \frac{e}{E},$$

there is no term representing the resistance R. This proves clearly that theoretically there is no limit for this system of transmission, and that the resistance of the circuit, or, what amounts to the same thing, the distance between the two stations, may be ever so large; so long as $\dfrac{e}{E}$ is kept constant, the efficiency will always be the same. In practice, of course, a limit exists, because if we make R larger, keeping E and (in order to have the same efficiency) e constant, a smaller current would flow through the circuit, as will be seen by referring to equation (II.)

This equation may be written·

$$C_1 R = E - e;$$

E and e being the same, $E - e$ is constant; altering R, therefore, will also alter C_1. The result is, that $E C_1$ would be smaller than before, and, consequently, if the resistance is made larger, the electromotive forces will also have to be increased.

Again, if we keep the resistance constant, but alter the electromotive forces, keeping $E - e$ constant, then, according to equation (II.) the current C_1 will be constant too. In equation (III.) the term $R C_1^2$ represents the amount of energy lost as heat in the circuit, and as it only involves R and C_1, it follows that if these remain the same, the former will not vary with the varying electromotive forces. But, on the other hand, making E and e larger also makes the amount of energy transmitted and the amount transformed into useful work larger. Fig. 43 shows this graphically.

On comparing it with Fig. 42, it will be found that $J G K D$, representing the heat wasted, is the same in both cases, but the rectangles $A F K D$ and $H G K C$ are larger in Fig. 43 than in Fig. 42. More energy has, therefore, been transmitted; but as the losses are the same, the efficiency in the case represented by Fig. 43 is larger than in that represented by

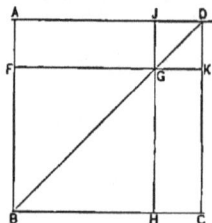

Fig. 43.

Fig. 42. The amount of energy, therefore, that can be transmitted over a circuit having a given resistance is again theoretically without limit, and the larger the amount of energy transmitted the larger will be the efficiency of transmission, if $E - e$ is kept constant.

Let E, for example, be 200 volts, $e = 150$ volts, and $R = 10$ ohms;

then $E - e = 50$

and $C_1 = \dfrac{200-150}{10} = 5$ ampères.

The amount of energy put into the generator is

$$W = E C_1 = 1000 \text{ watts};$$

the useful work is

$$w = e C_1 = 750 \text{ watts};$$

energy lost as heat,

$$H J = R C_1^2 = 250 \text{ watts};$$

the efficiency is

$$\eta = \frac{e}{E} = \frac{150}{200} = \frac{3}{4}.$$

Now let $E = 400$ volts, $e = 350$ volts, and R again $= 10$ ohms; then

$$E - e = 50.$$

$$C_1 = \frac{400-350}{10} = 5 \text{ ampères.}$$

$W = E\,C_1 = 2000$ watts.

$w = e\,C_1 = 1750$ watts.

$H\,J = R\,C_1{}^2 = 250$ watts,

and

$$\eta = \frac{e}{E} = \frac{350}{400} = \frac{7}{8}.$$

We have now put into the generator double the amount of energy; of this the same amount is lost in heating the circuit, and the efficiency has risen from $\frac{3}{4}$ to $\frac{7}{8}$. *This clearly shows that it is always more economical to use small currents with high electromotive forces*, the more so since the higher the electromotive force the greater the resistance can be made without altering the efficiency, and the less will be the cost of the circuit.

Up to the present we have assumed that only that amount of energy is lost which appears as heat in the circuit; but there is another source of loss due to the leakage of current between the two conductors. With currents of low tension this is not perceptible, as the insulation can easily be made perfect; but on long lines and with high electromotive forces this loss may be so large as to lower the efficiency somewhat.

It will now be necessary to say a few words about the resistance and efficiency of the different parts of the circuit. The circuit consists of three different parts, and they are: the generator, the conductors, and the motor. Letting r_1 stand for the resistance of the generator, r_2 for that of the conductors, and r_3 for that of the motor, then the resistance of the circuit will be:

$$R = r_1 + r_2 + r_3.$$

Let l_1, l_2, l_3 stand for the losses of energy in these different parts, then we will have:

$$l_1 = C_1{}^2\,r_1,$$
$$l_2 = C_1{}^2\,r_2,$$
$$l_3 = C_1{}^2\,r_3,$$

and the efficiency will be: $\eta = \eta_1 \times \eta_2 \times \eta_3$. and

$$\eta_1 = \frac{E\,C_1 - C_1{}^2\,r_1}{E\,C_1} = \frac{E - C_1\,r_1}{E};$$

$$\eta_2 = \frac{E\,C_1 - C_1{}^2\,r_1 - C_1{}^2\,r_2}{E\,C_1 - C_1{}^2\,r_1} = \frac{E - C_1\,r_1 - C_1\,r_2}{E - C_1\,r_1}$$

$$\eta_3 = \frac{E\,C_1 - C_1{}^2\,r_1 - C_1{}^2\,r_2 - C_1{}^2\,r_3}{E\,C_1 - C_1{}^2\,r_1 - C_1{}^2\,r_2} = \frac{E - C_1\,r_1 - C_1\,r_2 - C_1\,r_3}{E - C_1\,r_1 - C_1\,r_2}$$

But $C_1\,r_1$ is the electromotive force necessary to force a current C_1 through a resistance r_1; $E - C_1\,r_1$ will therefore be the difference of potential at the terminals of the generator, and for similar reasons $E - C_1\,r_1 - C_1\,r_2$ will be that at the end of the conductors or terminals of the motor. On the other hand, $C_1\,r_3$ gives the loss of electromotive force due to the resistance of the wire on the motor, but the electromotive force at this end of the circuit has already been shown to be equal to e, and if e_1 and e_2, respectively, stand for the other two, then we have:

$$r_1 = \frac{e_1}{E}, \quad r_2 = \frac{e_2}{e_1}, \quad r_3 = \frac{e}{e_2},$$

and

$$\eta = \frac{e_1}{E} \times \frac{e_2}{e_1} \times \frac{e}{e_2} = \frac{e}{E}.$$

We have seen from equation (V.) that the efficiency of transmission is as the ratio of the electromotive forces of the generator and receiver *i. e.*, $\dfrac{e}{E}$. As this expression does not contain the factor of resistance R, of the line or machines, M. Deprez was led to promulgate the theory that *in the electrical transmission of energy the efficiency is independent of the distance of transmission*. This theory, first started in March, 1880, has provoked considerable discussion. It means, in effect, that we can make the distance, and hence the resistance of the line, whatever we please without loss of efficiency.

We cannot here enter into a full discussion of this view, but we may say briefly, that as a *theory*, pure and simple, the proposition is correct, as was developed in the preceding pages. But there are practical difficulties which prevent its realization, and it would be dangerous to apply it in the calculations of a working enterprise. It is evident that the quantity of heat generated in a conductor by the electric current ought to increase with the length of the wire, and hence the loss would increase. On the other hand, M. Deprez argues that the production of heat varies as the square of the intensity of the current, and that the latter is diminished by an increase of the length of the line.

The question suggests itself: How can the evil effects of increased distance be obviated?

We have seen by equation (IV. *a*) that the useful work of the motor is equal

$$\frac{e\,(E-e)}{R}.$$

This equation shows that there are two ways of overcoming the difficulties of long distance transmission of a given power. Thus we can either diminish R, the resistance of the conductor, by increasing its diameter; or we can increase the relative electromotive forces of the machines.

The first method, that of increasing the size of the conductor, was first proposed by M. Maurice Levy, in February, 1882, in a note to the French Academy of Sciences, wherein he says that "the resistance of the exterior circuit can be made very small, even for long distances, by employing a very large wire." Evidently, this would be an easy way out of the difficulty if it were not for the fact that any increase in the size of the conductor means increased cost,' and hence, starting from an economical standpoint, we soon reach a limit in the size of conductor that can be used.

The second method of overcoming the difficulties of long distance transmission, as we have seen above, consists in increasing the electromotive forces of the machines. This is the method developed by M. Marcel Deprez and consists in employing *high tension* currents. According to M. Deprez, in order to obtain the same useful work, whatever be the length of the line, it suffices simply to vary the electromotive forces of the machine proportionally to the square root of the resistance of the circuit. In other words, if, as before, R represents the resistance of the circuit, and E and e, respectively, the electromotive forces of the machines, and in such a circuit we obtain useful work at the motor w, then, in order to obtain the same amount of work with other values, R^1, E^1, e^1, it is necessary to make the new values E^1 and e^1 such, that they will satisfy the following equations:

$$\frac{E^1}{E} = \sqrt{\frac{R^1}{R}}, \qquad \text{(VI.)}$$

$$\frac{e^1}{e} = \sqrt{\frac{R^1}{R}}. \qquad \text{(VII.)}$$

Or if we let w^1 represent the work produced in the second case, we have:

$$w = \frac{e\,(E-e)}{R},$$

$$w^1 = \frac{e^1\,(E^1 - e^1)}{R^1}.$$

Substituting now in the latter equation the values of E^1 and e developed from equations (VI.) and (VII.), we get:

$$w^1 = \frac{e\,\sqrt{\dfrac{R^1}{R}}\left(E\,\sqrt{\dfrac{R^1}{R}} - e\,\sqrt{\dfrac{R^1}{R}}\right) = \dfrac{R^1}{R}\,(E-e)}{R^1 = R^1}$$

$$\therefore \qquad w^1 = \frac{e\,(E-e)}{R} = w.$$

Before describing the experiments made by M. Deprez to demonstrate this theory we must allude to another form in which according to M. Deprez the efficiency of transmission can be expressed; and that is, that the efficiency is equal to the ratio of the speeds respectively of generator and motor. Calling N the former, and n the latter, the efficiency would be expressed by,

$$\frac{n}{N}.$$

This, of course, assumes that the two machines are identical, that the magnetic field and the current are the same in intensities in both, and hence that the electromotive forces developed are proportional to the speeds of the armatures. This, however, is not the case if there is any leakage along the line, for then not all the current developed by the generator passes through the motor. Moreover, as Prof. S. P. Thompson points out, when there are resistances in the line, the ratio of the electromotive forces of the machines is not the same as the ratio of the two differences of potentials, as measured between the terminals of the machines.

Further, even though the current running through the armatures and field magnets in the generator which creates the current, and in the motor which utilizes the current, be absolutely identical, the intensities of the magnetic fields of the two machines are not equal,—even though the machines be absolutely alike in build; because the reaction between the armature and the field magnet is entirely different in the dynamo used as a motor from that in the dynamo which is being used as a generator.

As we have stated above, M. Deprez, by employing high electromotive forces, seeks to diminish the current, and thus to diminish the heat which it generates in the conductor to such an extent that it shall be inappreciable.

His experiments to demonstrate the truth of his theory are very interesting, and of great value from a scientific standpoint.

The first real *long distance* transmission was undertaken by M. Deprez at the Munich Electrical Exhibition of 1882, with two Gramme machines. These were placed respectively at Munich and at Miesbach, a distance apart of 57 kilometres (37 miles). They were connected by an ordinary iron telegraph wire, 4½ mm. in diameter, and constituted a complete metallic circuit 114 kilometres (74 miles) in length. The resistance of the line measured 950.2 ohms; that of the generating machine at Miesbach 453.4 ohms, and that of the motor at Munich 453.4.

The generator was placed in the workshop of Herr Fohr, and appeared as shown in the illustration, Fig. 44. The motor was placed in the Munich Crystal Palace, and was belted to a centrifugal pump which fed a cascade nearly three metres in height, as illustrated in Fig. 45. The measurements taken by the committee were as follows:

Speed of generator at Miesbach, . . . 1611 revolutions.
Intensity of current at   "   . . . . 0.519 ampère.
Speed of motor at Munich, ... 752 revolutions.
Difference of potential at terminals of motor, 850 volts.
Work measured by brake at motor, . . 0.25 H. P.

From these data the following values were calculated:

Difference of potential at terminals of generator, ... : . . . . 1343 volts.
Total electrical energy at Miesbach, . . 1.13 H. P.
Total electrical energy at Munich, . . . 0.433 H. P.
Electrical efficiency, ... 38.9 per cent.

It will be understood here that this efficiency is not the *absolute* or commercial efficiency, but the *electrical* alone. We must explain this by an example. If, for instance, in the above case, all the power applied to the generator had been converted into electrical energy, or, in other words, if the generator were a perfect machine; and if the motor had converted all the electricity into useful work, then the efficiency would have been that given, viz., 38.9 per cent. The absolute efficiency of a system is the ratio of the power applied to the generator to that obtained from the motor. It thus includes not only the electrical efficiency but that also of the motor and generator, as con-

FIG. 44.—DEPREZ GENERATOR AT MIESBACH.

verters of energy. Thus, in the above experiment, while the electrical efficiency was 38.9 per cent., the absolute efficiency must have been less. Exactly how much less this was we have no means of telling, because the power applied to the generator at Miesbach was not measured. But if we assume the efficiencies of the motor and generator each to have been 85 per cent., we would have for the *absolute* efficiency of the transmission 0.85 × 0.85 × 38.9 or about 28 per cent.

M. Deprez, however, did not rest contented with these experiments, but followed them up

FIG. 15.—DEPREZ INSTALLATION AT THE MUNICH EXPOSITION, 1882.

by others in 1883 from the depot of the Chemin de Fer du Nord, Paris, to La Chapelle, a distance of 8,500 metres—about 5¼ miles; and another from Vizille to Grenoble, a distance of 14 kilometres (7¾ miles). The generator used is shown in Fig 46. The most recent, and perhaps the most important, of M. Deprez's experiments in long distance transmission was undertaken in October, 1885, and the object aimed at was to demonstrate the practical application and distribution of power transmitted over a long distance. For this purpose the apparatus was

metres from each other. Each possessed, like the generator described below, two rings; they were each 0.58 metre in exterior diameter and had an electric resistance of 18 ohms.

Our illustration, Fig. 47, shows one of the generators.

In the generating machine the field is produced by 8 electro-magnets of horseshoe form, and the pole pieces embrace the armatures over very nearly their entire circumference. The field is excited by a separate dynamo, and the current is passed through the different electro-

FIG. 46.—DEPREZ GENERATOR AT THE CHEMIN DE FER DU NORD, PARIS, 1883.

intended to operate electric light machines, to drive pumps and to run machine tools at the company's workshops.

The distance from Paris to Creil, between which two points the line extended, is 56 kilometres (34 miles), making a total length of conductor of 68 miles. The line consisted of a lead-incased insulated copper wire 5 mm. in diameter, and its resistance was 100 ohms.

The generating machine was situated at Creil. It had two rings revolving in two distinct magnetic fields, each composed of eight electro-magnets. Each armature had a resistance of 16.5 ohms.

The current produced by this machine was utilized at La Chapelle, near Paris, by two receiving machines, situated at some hundreds of

magnets so that a north pole on one side of the armature is opposite a south pole on the other.

The total weight of each of the electro-magnets is 485 kilogrammes. They are wound with copper wire 2½ millimetres in diameter. The wire is covered with two layers of silk, one of cotton, and finally with a layer of shellac. The total length of wire wound on the magnets is 56,496 metres. The winding is done in sections, each having the form of a flat ring. Each section is composed of 11 layers of 25 convolutions each. The core of each electro-magnet carries 12 of these sections, and the ends of the wires are led to the terminal boards, so that they can be coupled up in any manner desired. The resistance of each section is 1.06 ohm, and the

6

total resistance of the magnets grouped in series would be 203.52 ohms.

The radius of the pole pieces is 710 mm, and their thickness 120 mm. All the pole pieces on the same side of a horizontal plane passing through the shaft are in magnetic communication with one another.

wire is 2½ millimetres in diameter and insulated in the same way as the magnet wire. The following are the principal dimensions and weights:

No. of sections per segment, ... 21
Total No. of sections, ... 231
Length of wire per section, ... 52 metres.

Fig. 47.—DEPREZ GENERATOR AT CREIL, 1885.

The armature is wound in sections after the Pacinotti type, which presents the characteristic that it consists of a series of sections which can be removed and replaced in case of accident. The armature frame consists of a hub with a spider at each end. The sections are separately wound on soft iron cores and are then bolted to and between two opposite spokes on the shaft. Each segment of the armature winding is divided into 21 sections, and the

Total length of wire wound on the ring, . 12,012 metres.
Diameter, external, ... 1,386 mm.
  "   internal, ... 1,070 "
Thickness of wire coil above core, ... 33 "
  "   "   "   " below " . . . . . 45 "
Length parallel to axis, ... 606 "
Weight of wire, including insul., ... 552.5 kg.

In a preliminary trial this armature generated an electromotive force of 16 volts per revolution per minute. Each armature has its own commutator and brushes, the latter being attached to a holder movable by a worm wheel which gears with it.

In a note presented to the French Academy of Sciences, M. Deprez gave the results of experiments undertaken with these machines, and they are quoted below:

FIRST EXPERIMENT.

	Generator.	Receiver.
Speed in rev. per minute,	190	248
Electromotive force, direct or inverse,	5469 volts	4242 volts.
Intensity of current,	7.21 amp.	7.21 amp.
Work in field magnets (in horse power)	9 20	3.75
Electrical work (in horse power) . .	53.59	41.44
Mechanical work measured with the dynamometer or the brake (horse power),	02.10	35.10

EFFICIENCY.

Electrical,	77	per cent.
Commercial or mechanical,	47.7	"

SECOND EXPERIMENT.

	Generator.	Receiver.
Speed per minute,	170	277
Electromotive force,	5717 volts.	4441 volts.
Intensity of current,	7.20 amp.	7.20 amp.
Work in field magnets,	10.30 H. P.	3.80 H. P.
Electrical work,	55 90 "	43.4 "
Mechanical work (measured with the dynamometer or the brake),	61 "	40 "

EFFICIENCY.

Electrical,	78	per cent.
Commercial or mechanical,	53.4	"

These results which showed that 40 H. P. had been transmitted with a commercial efficiency of about 50 per cent. have been variously criticised. In the first place the generator and motor were placed side by side, the line being in a loop around them. Evidently if leakage occurred on the line it would be the same in effect as if the line were shortened. Again, the power required to magnetize the field magnets, which were independently excited, is not taken into account, so that the mechanical efficiency cannot be taken as the true one.

It will be noted that M. Deprez uses an electromotive force as high as 6,000 volts, which he reached on another occasion, and which of necessity requires an extraordinary degree of insulation, both on the line and in the machines. But in spite of the precautions he had taken he met with a mishap which destroyed one of his machines, and which was caused, no doubt, by the giving way of the insulation.

As it is only by the employment of very high electromotive force that we can approach to a realization of M. Deprez's theory of electric transmission, the question naturally suggests itself: What will be the ultimate result of M. Deprez's experiments, taking into account existing conditions?

In considering this question, Mr. W. J. Johnston in a paper read before the National Electric Light Convention at Baltimore, in February, 1886, reviews the situation in the following forcible remarks. He says:

"That power can be transmitted we all know. Given, then, that M. Deprez succeeds in his attempt, will that alter the present condition of affairs, as regards the *economical* side of the problem? The cost of his installation, and the interest thereon, will far exceed the similar items, including maintenance of a steam plant of equal power, at the place where it is wanted: Yet, paradoxical as it may seem, the great problem to be solved is not the transmission of 100 horse power, but of thousands and tens of thousands. Now, this can only be done in one of two ways; by increasing either the electromotive force or the current. If the latter plan is pursued, an increase in the size of conductor must necessarily follow with its attending cost, and this is feasible, but not at present economical. On the other hand, can the electromotive force be increased much beyond the limit which M. Deprez is now using? Dynamo builders know how perfect the insulation of the armature must be, and how little it requires to burn one out under but slightly abnormal conditions. Those especially who have experimented with the 40, 50 and 60 arc light machines using two or three thousand volts, have possibly, on more than one occasion, witnessed an effect in the armature as if the latter had been struck by lightning. This effect is one entirely different from what would be produced in a machine in which the armature has been actually burned out by a heating of the wires from too great a current. The break resembles that made by a disruptive discharge, an actual spark; and M. Deprez has already experienced one of these mishaps, in spite of the fact that he uses two layers of silk and one of cotton for the insulation. The fact is, a dynamo of large power which is subjected of necessity to rough influences cannot be made to generate currents of very high electromotive force for a continued service, on account of the impossibility of securing sufficient insulation. The problem bears considerable analogy to that of the steam engine. The use of high pressure steam of, say, 500 or 1,000 pounds to the square inch, would effect great economy, but it is materially impracticable, as the cost of building engines and boilers to withstand these pressures would be out of proportion to the benefits derived, and no working joint could withstand the pressure.

"For the reasons given above it would appear that as regards long-distance *large power* transmissions, substantial improvements are required before it can become a commercial success."

We have dwelt at length upon the long-distance transmission experiments of M. Deprez,—who indeed stands honorably first and alone in this field thus far,—for the reason that his work presents good ground for study and development. M. Deprez's theory while correct in principle, cannot, unfortunately, be realized in practice to a *commercial* extent, under the conditions prevailing at the present time.

Leaving this part of the subject which is still in the tentative state, and directing our attention to actual successful commercial practice, we find that for moderate distances large powers can be transmitted with ease and economy, as evidenced by the numerous applications detailed in the succeeding chapters.

The time will soon arrive, in fact it is already upon us, when electricity will be distributed for power and light as generally as gas is at present, and it becomes necessary to consider the most economical method of distribution.

The problem of the most economical section of conductor to be employed in a power distribution by electricity is included in this aspect of the question, and is perhaps one of the most important points. In discussing it, Mr. Thomas W. Rae, C. E., assumes for the sake of an example that the amount of power to be circulated in the form of current is 500 horse-power, the length of the circuit 4,000 yards, or 2.27 miles, and the cross-section of the copper conductor 3.25 square inches. The potential of the current is restricted to 120 volts.

The current equivalent of 500 horse power being 373,000 volt-ampères, it follows that the theoretical value of the current flowing in the circuit would be 3,108 ampères.

This, however, says Mr. Rae, is subject to correction for the loss involved in the conversion of mechanical work into current; which is due to the frictional and electrical resistances of the generating dynamo. It will be prudent and more in accord with other conditions of the problem—to be stated later—to put this loss at 15 per cent., and consequently there may be considered to be a current of 2,642 ampères flowing in the conductor.

It is evident that the flow of any appreciable current in any practicable conductor must evolve heat. If a uniform temperature of conductor is to be maintained, this development of heat must be got rid of by radiation or conduction; or it becomes cumulative and detrimental. by creating a wasteful resistance in the circuit and, in the case of insulated conductors, sometimes destroying the insulating medium.

The latter class of conductor would seem to offer especial difficulties, and the problem is as yet too new to have invited much investigation or experiment. While it is undoubtedly true that the best dielectrics are, probably without exception, the worst conductors of heat and the rates of their efficiency, in this sense, are practically unknown, it has also been experimentally demonstrated that insulated conductors have even less tendency to augment temperature under the passage of a current than bare wires. This is a deduction from laboratory tests, and must be accepted only within proper limits. The seeming paradox vanishes when one reflects that the worst possible conductor of heat is dry motionless air, and that the larger periphery of the insulated wire radiates the greater quantity of heat in the same time. The result would be reversed if the two types of conductor were exposed to draughts of wind; but the instance is cited to show that the general formulæ are applicable to both classes. Every case, however, may be said to be a special case, and in view of the numberless and unforeseeable influences affecting the temperature of a subterranean conductor two and a quarter miles long, it seems almost finical to be calculating the effect of a few degrees due to current resistance. Nevertheless the investigation is of importance.

It is presumable that a conductor buried in homogeneous earth and well below the frost line would retain about a uniform temperature all the year round, but that temperature would depend upon the nature of the soil.

It would naturally be one thing for clay, another for sand, and another for rock; and none of these could be known except by experiment. In all probability, a conductor of any considerable length would pass through all varieties of soil, across places alternately dry and wet, possibly near steam pipes, and any attempt to assign quantitative temperature to them would be farcical. In such circumstances, the only recourse is to general formulæ, as furnishing — all things considered — as fair an average of the conflicting influences as possible, and one of Clark and Sabine will do as well as any. It is

$$\theta = 0.2405 \, R \, C^2 \, t,$$

in which

$\theta =$ units of heat.

$R =$ resistance in ohms.

$C =$ current in ampères.

$t =$ time of flow in seconds.

Since the current in this case is practically continuous, t will disappear, and the factor R must be deduced, Clark and Sabine again furnishing the means with their formula

$$R = \frac{1002.4}{w},$$

w being the weight in pounds of a statute mile of the conductor whose resistance is sought.

In the case in question the weight of a statute mile is 65,261 lbs., and its resistance at 60° Fahr. consequently .01536 ohm; making the resistance of the entire 4,000 yards .035 ohm.

Carrying out the operations indicated by the formula, it will be found that

$$\theta = 58,756;$$

that is to say, the given current will develop in the given conductor so many units of heat.

If it were conceivable that the substance of the conductor was water and weighed just 58,756 pounds, this would mean that its temperature would be raised one degree Fahrenheit. But its material is copper, whose specific heat is .092 or—familiarly speaking—which requires but .092 of the quantity of heat that water does to affect its temperature equally; and its weight, as has been seen, is 148,320 pounds.

Adapting the result to these conditions, it will appear that the conductor under consideration will have its temperature raised by the current circulating in it, only 4.17° Fahrenheit, above what it would be if out of circuit.

This increment is so trivial with regard to any harmful influence it might exert upon the insulating medium used with the conductor, that search must be made in other directions for the reason which prescribes its seemingly excessive size.

Good gutta percha will endure a temperature of 120° Fahr. before failing, and india rubber 300° Fahr. So the cause is probably the reduction of current, and consequently of merchantable horse power, resulting from an increase of resistance by augmented temperature.

The resistance of copper increases $\frac{2}{100}$ of one per cent. for each additional degree Fahrenheit of temperature, and in the case of the predicated current and conductor, the resistance of the latter will be enhanced but $\frac{8}{100}$ of one per cent. For the purposes of discussion, the loss of current due to this augmented resistance will be ignored for the present. To estimate the effect of such increase of resistance from a financial standpoint, the subjoined method is convenient.

The 2,642 ampères of current flowing in the conductor are subject to a farther diminution, before they appear in merchantable shape, which occurs in their transformation by the converting dynamos into horse power.

It may be safely taken at 17 per cent.; which amounts to the admission that, of the 500 horse power applied to the generating dynamos, but 70 per cent. may be counted upon as returnable, in the same form, from the converting dynamos. One of the postulates of the problem is that at least this proportion of the applied mechanical power should be recovered after having undergone all its transformations, and the loss has been equally divided—which is probably as fair an allotment as possible—between the two conversions.

There should be, then, 350 horse power available for the production of revenue; but, owing to an idiosyncrasy of electric power, there is very much more. An instant's reflection will satisfy one that where an amount of power is distributed among a number of consumers for intermittent use the chance of every one's desiring to avail of his power at the same instant is infinitesimal, and experience, as far as it goes, confirms this. The character of the work done may also originate compensating influences to the same end, as when the power-circuit includes elevators which not only consume no power in their descent, but even reinforce the main current with the counter-currents created by their own dynamos revolving under stress of their downward gravitation. It would be difficult to make too much of this characteristic, to which is largely due the wonderful economy inherent in this system of power distribution, and which is so prominent in the case of electric railways as to elicit the statement from the late Sir William Siemens that two trains on the same pair of rails, one ascending and the other descending a grade, influenced each other through the common current as absolutely as if connected by an actual rope.

It is this instantaneous adjustment throughout the entire circuit of the supply of, to the demand for, power that precludes waste or superfluity of it.

It is, therefore, considered perfectly prudent with an ordinary power plant to contract to deliver about double the total capacity of the generator.

In the problem under discussion, the quantity of marketable power was limited to 500 horse power, which represents an annual rental of $60,000.

It thus appears that every one of the 2,642 ampères of current flowing in the conductor has a market value of $22.71 per annum.

On the other hand, the specified conductor—at an assumed price of copper, say 15 cents per pound—would cost $22,248; the annual interest on which, at 6 per cent., would create an annual debit of $1,335.

Ohm's fundamental law of currents furnishes a useful point of reference at this juncture, viz.:

$$C = \frac{E}{R}$$

in which C = current in ampères,
R = resistance in ohms,
E = electromotive force, or potential, in volts.

As the latter factor is fixed at 120 volts, unity may be substituted for it in the formula, viz.:

$$C = \frac{1}{R}$$

which then signifies that the current varies as the reciprocal of, or inversely as, the resistance. Colloquially, it reads: having the same electromotive force, to double the current, halve the resistance, and vice versa.

In conductors of similar material and equal length, the relative resistances would be inversely proportional to their cross-sections—or to their weights—and the final deduction is that currents of uniform potential moving in conductors of the same material and of equal length vary as the weight of the conductors.

A convenient unit of comparison is the annual market value of the ampère which, as has been shown, is $22.71. As the number of salable ampères in the case in point is a function of the weight of the conductor, the annual interest upon which per pound is $.009, viz.:

$$\frac{1,335}{148,320} = .009,$$

it follows that it would require the annual interest upon 2,523 pounds of copper, viz.:

$$\frac{22.71}{.009} = 2,523,$$

to equal the annual value of one ampère.

Supposing the conductor to be reduced in weight by this amount and applying the rule deduced for this especial case—of the current varying as the weight—it appears that such a reduction would diminish the current some 45 ampères, viz.:

$$\frac{148,320}{2,523}$$

148,320 : 145,797 :: 2,642 : 2,597
2,642—2,597 = 45,

whose annual value is $1,022.

It thus becomes evident that the diminution of weight would entail vastly greater loss of revenue than the annual saving achieved thereby. Reducing the two opposing quantities to a common unit will give useful constants for the case under discussion, viz.:

Annual interest at 6 per cent. on 1 lb. copper = $.009.

Annual revenue from 1 lb. copper = $0.40.

Weight of copper per ampère = 56 lbs.

From which it appears that until the price of copper rises forty-five times above its present figure, or the value of the ampère falls the same number of times below that assigned to it in the comparison, or a change occurs in both, producing a similar mutual relation, any reduction of weight in the conductor—all other factors remaining constant—would be a source of loss rather than of profit. Since it seems incontestably proven that for the stated case and specified conditions any diminution of the conductor would be prejudicial, it becomes of interest to know if the weight might be profitably increased.

It will be remembered to have been shown above that the given current raised the temperature of the conductor 4.17° Fahr., which increased its resistance .86 of one per cent.

Since current varies inversely as resistance, viz.,

$$100.86 : 100 : : 2642 : 2619$$
$$2642 — 2619 — 23,$$

it appears that this trifling increment reduces the flow by 23 ampères whose annual value is 8522.33. In the given conductor and under the prescribed conditions, each ampère requires 56 pounds of copper, and the addition of 1,288 pounds (56 × 23 — 1,288), the interest on whose cost is but $11.59, would make good this very considerable annual loss. Increasing the conductor by this amount of copper would enlarge its cross-section from 3.25 to 3.30 square inches —a barely appreciable area.

It is evident that the method employed is only approximative, and may be continued to any desired degree of precision. There is no pretence of close accuracy, and it is even less than it might easily be, on account of ignoring fractional quantities and the ordinary small errors in the deduced factors made use of. The idea has been rather to suggest a method of dealing with such questions than to furnish absolute results.

Sir William Thomson has given a formula for computing the most economical section of conductor, which may in some cases be used with advantage, although it is not adapted to all cases; but here again it must be left to the engineer to decide when to use or how to modify the formula.

The resistance of any circuit may be expressed by the formula,

$$R — \frac{l}{s} r,$$

l representing the length of the conductor, s its cross-section, and r being the specific resistance of the metal used—that is, the resistance of a bar 1 metre long, and having a cross-section 1 square millimetre.

The loss of energy through heat, expressed in metre-kilogramme-second units is

$$H J — \frac{C^2 R}{9.81} — \frac{C^2 l r}{s \, 9.81},$$

or for unit length (1 metre),

$$H J — \frac{C^2 r}{s \, 9.81}.$$

Now there are 31.5 × 10⁶ seconds a year, but the current is only used part of this time—that is, for 31.5 × 10⁶ × p seconds only (where p represents a fraction greater than zero and less than 1).

The amount of energy, therefore, lost in one year is:

$$A — \frac{31.5 \times 10^6 \times p \, C^2 r}{9.81 \, s}.$$

Let P stand for the cost of one horse power per year; then the cost of one unit of work (M. K. S. system) will be

$$\frac{P}{31.5 \times 10^6 \times 75},$$

and the cost of A units—that is, of the amount of energy lost in a conductor 1 m. long and with s square mm. cross-section—will be

$$L — \frac{P \, p \, C^2 r}{9.81 \times 75 \times s}.$$

Now, if interest on investment capital be taken at c per cent. a year, and the cost of one cubic metre of the material be v, then the cost of a conductor one metre in length and s square mm. cross-section will be

$$L^1 — s \, v \, 10^{-6} \, c.$$

The total is:

$$L + L^1 — \frac{P \, p \, C^2 r}{9.81 \times 75 \times s} + v \, 10^{-6} \, c \, s.$$

This will be a minimum if L is equal to L^1—that is,

$$\frac{P \, p \, C^2 r}{9.81 \times 75 \times s} — v \, 10^{-6} \, c \, s;$$

from which equation we get

$$s — \sqrt{\frac{P \, p \, C^2 r}{9.81 \times 75 \times 10^{-6} \times v \, c}} — C \sqrt{\frac{P \, p \, r \, 10^6}{9.81 \times 75 \times v \, c}}$$

or if $c — 5$,

$$s — 162.56 \, C \sqrt{\frac{P \, r \, p}{v}}.$$

CHAPTER V.

THE MODERN ELECTRIC RAILWAY AND TRAMWAY IN EUROPE.

SOME experiments were tried in 1867, at Berlin, in electric railways, by Dr. Werner Siemens, but the work was abandoned because the armature of the Siemens machine then used became heated too quickly and too greatly to be of practical service. Under conditions of more promise, the experiments were resumed by Siemens & Halske in 1879, and carried to a successful issue.

The first step which Messrs. Siemens took towards a practical demonstration consisted in the building of a short line of about 500 metres length at the Berlin Exhibition of 1879. In this they employed their well-known type of machines as generators and motors, of which the illustration shows one connected to a Dolgorouki rotary engine, Fig. 48, and a central rail led the current to the machine, the outer rails acting as a return circuit. Prompted by the success with this venture, which was the first of its kind, similar attempts were made in Brussels, Düsseldorf, and Frankfort, for exhibition purposes, with a like result. The first permanent undertaking executed on the Siemens system, however, did not take place until two years later, when, on the 12th of May, 1881, the line between Lichterfelde and the Central Cadetten Anstalt, near Berlin, was opened to the public. This installation differed somewhat in detail from the first attempts in the manner in which the current was led; for whereas in the latter a third central rail was used, the former employed only the two existing rails, one as a lead and the other as a return circuit.

Since this road was put in operation Messrs. Siemens & Halske have built numerous others, and we need only to mention those of the Paris Exhibition, at Vienna, at the Zankeroda mines in Saxony, at Offenbach, near Frankfort-on-the-Main, and the use of the system on the Portrush Railway in Ireland, in order to show the enterprise of a firm which American electricians might well take as an example.

The various methods which the Siemens have employed for conducting the current to and from the motor deserve some attention, as they have by no means restricted themselves to the rails as conductors, but have devised various methods for the purpose. As stated above, their first road employed a central rail, while the second used the main rails only. In order, however, to avoid the danger of giving electric shocks to persons and animals coming in contact with these unprotected rails, and also to avoid the loss due to leakage between the rails in wet weather, recourse has been taken to overhead conductors which lead the current without the objections just named. The first and most natural way out of this difficulty was to string two wires overhead upon which small trolleys travelled. These latter were connected to the locomotive by wires which pulled them along, and thus a constant circuit between motor and generator was maintained. This method, though simple, was found not to work well in practice on account of the vibration and the varying sag in the wire, and another was consequently devised, which was first tried at the Paris Exhibition of 1881, and gave satisfactory results. On this occasion the overhead conductors consisted of brass tubes, slit longitudinally and laid along small stringers of wood, so that the slit was turned downward. Within the tube there was placed a short metal cylinder, at the ends of which there projected two lugs, which passed through the slit in the tube and were connected to a small framework carrying a wheel. The latter was pressed against the lower side of the tube by springs, and thus a good contact was obtained when the wire from the locomotive was attached to the device.

At the Zankeroda mines still another method is employed, well suited to the locality. This consists in suspending two ⊥ beams with their flanges facing downwards. The lower flanges thus present a good means for the conveyance of a trolley, and the latter being provided with a brush, takes off the current, with very little loss of power. It will be seen that for mines this disposition of the conductors is a very happy one. The rough usage to which the rails are subjected, together with the frequent presence of water, makes the employment of the former as conductors impracticable; but the ⊥ rails suspended close to the roof, and attached to hard rubber insulators, leave little to be desired in the matter of efficiency. The motor at

application of electricity as the chief motive power for propelling the tram-car; and 5th. The use of water-power as the actual source from which the motive power is derived. The line is a continuous series of long inclines. Gradients one in forty-five and one in forty are frequent for upward of a mile in length, while steeper gradients of one in thirty exist for shorter distances, the worst gradient being one in twenty-five, the total rise from the depot at Portrush to the summit being 203 feet. The system finally adopted of utilizing electricity

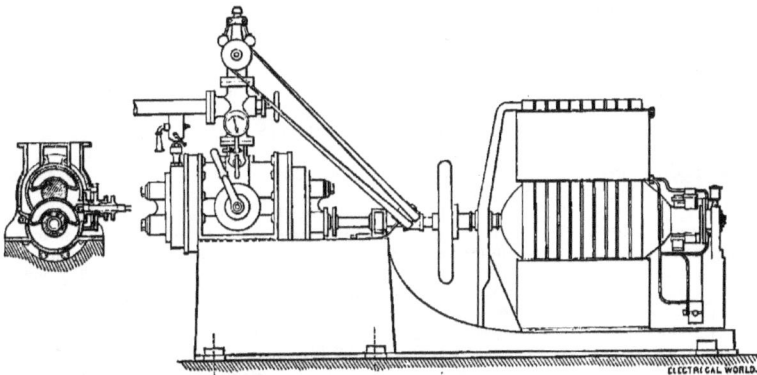

Fig. 48.—Siemens' Dynamo with Dolgohouki Engine.

these mines weighs about 1½ ton and is capable of hauling a load of eight tons at the rate of seven or eight miles an hour.

The main features of the Portrush road were described by Dr. Anthony Traill, LL. D., chairman of the Portrush Electric Railway, at the Montreal meeting of the British Association for the Advancement of Science, in 1884. This line, he said, was specially constructed with a view to the application of electricity as a motive power. The chief distinctive features by which it differs from tramways, as usually constructed, are: 1st. The gauge, which is three feet; 2d. The position of the tramway in respect to the side of the road, viz., it being placed alongside of the road, and not in a central position, and being raised slightly above the surface of the road; 3d. The form of the rail, a flange being substituted for a grooved rail; 4th. The motive power, the

as the motive power differs from the system used on the Lichterfelde, the Charlottenburg and the Paris electric tramways, where overhead electric conductors or storage batteries were used. The track being laid along the side of the road, a third rail or rigid electric conductor is placed along the toe of the fence or ball, consisting of from twenty to thirty foot lengths of T irons, weighing nineteen pounds to the yard, supported on short wooden posts with insulating caps of "insulite," the top surface of the conducting rail being three inches wide and eighteen inches above the level of the tramway rails. The ordinary track rails constitute the "return," completing the circuit. The electricity is now generated by water power on the River Bush, situated at a distance of 1,000 yards from the nearest point of the tramway, and five and a half miles from Portrush, Fig. 49. A fall of twenty-six feet head of water is used to drive two of Alcott's turbines, each capable of working up to fifty-two horse power. These drive on a single shaft, which communicates by belting with a generating Siemens dynamo, giving a maximum current of 100 ampères with 250 volts E. M. F. Each electric car is fitted with a starting handle at each end, and with powerful brakes, and is capable of drawing a second car behind it, with a total complement of 44 passengers comfortably seated. The daily running of the electric cars commenced on November 5, 1883, from which date till July, 1884, upward of 13,000 electrical train miles had been run. The working expenses of the electrical train mile are five cents a mile as compared with eleven and twelve cents per steam engine train mile, and the average cost of twenty cents to twenty-four cents per mile, when horse power is used. The line continues down to the present time in most successful operation. An extension of six miles is in contemplation, and a 12 per cent. dividend is paid. Speaking before the Inventors' Institute in 1885, Mr. Traill, the engineer of the road, stated that 30,000 train miles had then been run, and that 100,000 passengers had been carried. He also said that after repeated trials the management had found an efficient method of making contact between train and conductor. This consists in the use of a steel spring in the shape of a carriage spring; two of these, concave, are fastened at the top and rub along the bottom. The cost and the wear are nominal. The total resistance of the line from the generator and back is 1.9 ohms. Where the rails are crossed by roads, an insulated cable is laid under ground and it connects the two ends of the severed rails, so that the latter, though elevated along the line, do not prevent the crossing of vehicles. As there are two sets of brushes attached to the motor, one in front and the other in the rear, it follows that at short crossings the front brush makes contact before the rear one has left the rail, and thus an uninterrupted current is maintained. Where the crossing is greater than the length of the car, the momentum of the latter carries it over to the opposite end, and in this manner makes connection again.

The Siemens line at Lichterfelde, a suburb at Berlin, has been in operation since May, 1881. The rails rest on insulated sleepers. One rail is positive and the other negative, or return. The gauge is three feet three inches. Each car is driven by its own motor, and has a carrying capacity of 26 passengers. The movement of the motor armature is transmitted to the car wheels by means of a belt working on cylinders outside the wheels. The cars are provided with brakes, which may be put on at either end, so that they will run in either direction without being turned around on the track. Yet another Siemens road was put in operation between Mödling and Brühl, near Vienna, in 1884, for a distance of over two miles. This line has since been in course of extension to Hinterbrühl, making it about a third longer.

The cities of Frankfort and Offenbach are connected by a Siemens electric railway, 6,665 metres (about 4 1-8 miles) long, of 39 inches gauge. It leads from the old "Römerbrücke" Frankfort, through Sachsenhausen, Oberrad, and through the entire town of Offenbach. The trains run over the entire route in about 25 minutes. Two steam engines, of 125 horse power each, drive four dynamo-electric machines and the current is conducted through suitable cables and conductors over the entire line. A switch is provided, regulating, governing, and directing the currents, as may be necessary. The conductors consist of tubes slotted along their entire length at the bottom, and insulated on poles in about the same manner as telegraph wires are arranged. In these tubes a small cylinder slides or runs, from which a conductor extends down to the car and to the dynamo in the same in the usual manner.

We have entered into this detailed description of the methods employed in the Siemens system, because the points involved are of great importance, and may often determine the practical success of similar undertakings. As regards a choice of conductors for electric railways, it is obvious that no definite rule can be laid down, as the method employed must be governed entirely by the exigencies of the case. It is thus seen that at Lichterfelde, which is but sparsely populated and has but little traffic, the rails alone have been used as conductors. In Paris the case was entirely changed, for there the rails lay flush with the street, the city authorities not permitting a raised rail to interfere with the continuity of the pavement. This prevented good insulation, and, in ad-

Fig. 49.—THE WATER POWER OF THE PORTRUSH RAILROAD.

dition, the sunken rail permitted the accumulation of dirt and other matter which would have prevented the wheels from making good contact. All this being foreseen, it became necessary to provide overhead conductors, taking the form described above. At the Zanker-

FIG. 50.—TUNNEL FOR VIENNA ELECTRIC RAILWAY.—
LONGITUDINAL SECTION.

oda mines the form adopted seems eminently practical, and will no doubt be copied in future installations. The exemplifications might be continued, but those enumerated will suffice to show what has been done in that direction, and to point the way to a selection of the best method to be applied to a given case.

FIG. 51.—TUNNEL FOR VIENNA ELECTRIC RAILWAY.—
TRANSVERSE SECTION.

We now come to the consideration of the proposed Vienna system of electric railways which has been elaborated by Siemens & Halske, and which, if carried out, will undoubtedly place that city in the front rank as regards transportation facilities within city limits.

The projectors start out with the proposition that no such railway can be run on the surface of the street, for the reason that contact with the rails would be unavoidable, especially at car crossings, and principally on account of the limited speed which surface railways are obliged to run at. Where quick transit is desirable, therefore, the solution must be sought either in an underground or an elevated system, or in

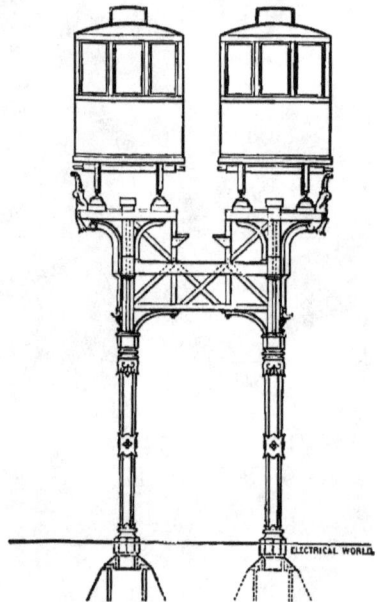

FIG. 52.—ELEVATED STRUCTURE FOR VIENNA RAILWAY.

a combination of both, depending upon the nature of the surface. The last named plan is the one proposed for Vienna, where the rise and fall of the ground would present too heavy grades for either one of the first two systems mentioned. It is therefore proposed to build a road in which tunnels shall alternate with elevated structures, and the manner of building both of these is shown, as they give a good idea of the requirements of the case; Figs. 50, 51, and 52.

As regards the construction of the tunnel, of which we give two views, it will be seen to have a flat roof instead of an arched one. This is necessary, because the latter construction would require a greater height of tunnel, and in addition an increased width. One object in making the tunnel as low as possible is to minimize the grades in passing from tunnel to viaduct, and by making the tunnel as narrow as possible, plenty of room, even in the narrowest

Respecting the viaduct or elevated portion of the road, an essential condition requires it to be as simple as possible. It must be open so as not to obstruct the light, and this requires the parts to be small and simple. Hence the illustration shows both rails supported by only one main girder bracing between them. It is asserted also that this method of construction will diminish the noise of passing trains—a boon that only the dwellers along the lines of

FIG. 53.—THE BRIGHTON, ENGLAND, ELECTRIC RAILWAY.

streets, remains on both sides, for the placing of sewer and other pipes, without interfering with the foundations of houses. Ventilation of the tunnel will be effected by means of perforated plates inserted in the roof and arranged so as to catch any matter that might fall through the holes. They would so act that when a train approached one of them it would force out the foul air through the openings, and as it receded it would draw in a supply of fresh air; plates thus placed at proper intervals would maintain good ventilation. It goes without saying that the cars are to be lighted by electricity.

the rattling and vibrating New York elevated roads can appreciate in its full extent. The course that these lines will take in Vienna demands attention. Although it is not proposed to build the entire road at once, the plan has been so worked out that, when complete, the system will present a network giving access to all parts of the city, from any starting point, in the shortest time. We can best explain the course of the road by asking our readers to imagine a circle drawn: around this inner circle there are drawn others in such a manner that each one shall touch its neighbor and also form part of the circumference of the

FIG. 51.—THE PARIS ELECTRIC STREET RAILWAY OF 1881.

FIG. 55.—THE PARIS ELECTRIC STREET RAILWAY OF 1881.

inner circle. It will thus be seen that any part of the city can be reached, no matter in what direction, in the shortest time consistent with such a comprehensive plan. It will of course be understood that the lines proposed do not take the shape of true circles.

An English railway that has been a notable success, though on a small scale, is that operated during the past two years at Brighton, the well-known watering place, by Mr. Magnus Volk. The line, Fig. 53, is rather under a mile in length, and includes some heavy gradients on sharp curves, the gauge being 2.9. The speed is limited to eight miles an hour, but a speed of over twenty-five miles has been obtained. The current is transmitted along the rails, which are fastened to wooden sleepers resting on the shingle, no special insulation being used. Each car seats thirty passengers, and the motor can draw another if necessary.

The plant comprises two cars fitted with motors; one eight horse power gas engine, one twelve horse power gas engine, one Siemens D_2 series dynamo, one Siemens D_2 compound dynamo.

The working expenses for one year are given as follows (average for two years): Electrical machinery — new commutators and brushes, $48.12; gas engines—refacing slides, etc., $31.17; oil and waste (that used for axles included), $18.30; gas (including that used to light premises, price 78 cents per 1,000 cubic feet), $547.93; attendant, 52 weeks, at $4.32, $224.64; total, $900.16. The gross earnings per car mile are 38½ cents; the gross expenses (all renewals being paid out of revenue), 22¼ cents; car mileage per annum, 23,475; cost of haulage per car mile, 3.84 cents, barely 4 cents; this includes the engine attendant.

Only one car is running, except on bank holidays, etc., on which occasion nearly all the power of the 12 horse engine is used. The repairs to the electrical machinery only amount to 5 per cent. per annum, and to the gas engine only about 2½ per cent. The only repairs to the electrical machinery of the two cars—the work having been nearly equally divided, each car having run about 25,000 miles—have been one new commutator and one spindle bush relined with soft metal; that, therefore, represents the wear and tear for nearly 50,000 miles running. The extra cost of running two cars is very slight, only about two-thirds more gas being

required, and the other expenses being scarcely affected.

Figs. 54 and 55 illustrate the electric car used during the Paris Electrical Exposition of 1881, and referred to above. This car carried 84,-000 persons. The motor was placed under the car body. The second view shows the car turning the sharp curve from the Champs Élysées to the Palais de l'Industrie, where the exposition was held. The first view shows the car from the other side, so as to make plain the method of taking the current from the conductors placed on poles and parallel with the line of track.

The Siemens road that was in operation at the Vienna Electrical Exposition in 1883, is illustrated in Fig. 56. The track was 1528 metres in length. A Siemens motor was placed under the floor of each of the two end cars, and the current was furnished by two Siemens dynamos coupled, so that the current from each armature excited the other's field magnets. One rail was connected with the positive binding post common to both machines, and the other rail with the negative post alike belonging to each. An electromotive force of 150 volts was maintained.

The road most worthy of notice in England after those at Portrush and Brighton is that of Mr. Holroyd Smith, at Blackpool. On the Blackpool tramway every means has been taken to reduce the objections of electric lines in cities to a minimum, as will be seen by our illustration, Fig. 59, which represents a part section of the road-way in perspective. The entire electrical part of the road is below the surface of the street, and the rails are not used as a conductor. The latter consists of two copper tubes, C, of elliptical shape, and having a wide slot for facility of attachment to iron studs, S, which are supported in porcelain insulators. I. The latter are themselves attached to blocks of creosoted wood in the sides of the channel. The tubes are fixed to the studs by the simple device of a wooden pin wedge, W, and they are coupled to each other by two metallic wedges, as shown in our illustration, Fig. 58.

The car which is employed is shown in plan and in longitudinal section, in Figs. 57 and 60, which are so clear that no extended description appears to be necessary.

At each end of the car there is a switch-box with resistance coils placed under the plat-

FIG. 56.—THE ELECTRIC RAILWAY AT THE VIENNA ELECTRICAL EXPOSITION OF 1883.

forms, by which means the strength of the current and speed of the car can be regulated. To reverse the direction in which the car is travelling, the direction of the current through the armature is reversed, the field magnets, which are shunt-wound, remaining always magnetized in the same sense. With this arrangement there is no need to alter the position of the brushes, which in this case consist of two parallel sets of plates placed tangentially to the commutator, and pressed on it by spiral springs. There is only one handle to the two switch-boxes, and that being in possession of the driver, the possibility of accidents caused by interference of others with the electrical connections is precluded. The current is generated by four-pole Elwell-Parker dynamos, Fig. 61, and the motors are also manufactured by that firm. The line is in continuous and highly successful operation. It is about two miles in length.

The dynamos are three in number, two of extra large size for generating the electricity for driving the cars, and the third a much smaller one, for exciting the "generators." These machines were manufactured specially for this work by Elwell-Parker, Limited, of Wolverhampton. The "exciter," Fig. 62, is of their usual type. The length, including pulley, is 3 ft. 3 in.; width, 2 ft.; and height, 2 ft.; its total weight is about 10 cwt. The diameter of the armature is 10 in., with a length of 13 in. The resistance of the field magnets is 60 ohms, and the electromotive force is 150 volts, with a current of 16 ampères. The brushes and commutator are of the usual form. The generators, Fig. 61, are among the largest of this form of dynamo yet constructed, and, as may be seen, differ materially from the Elwell-Parker type; each dynamo consists of two field magnets, with a commutator carrying four sets of brushes in pairs. These brushes are fitted with springs, which are easily adjustable, so that they can be kept just clear of the commutator or instantly dropped into contact. These machines are 7 ft. 3 in. long over all, 5 ft. 8 in. wide, and 2 ft. high, weighing altogether about four tons; they are provided with slides and adjusting screws, are run at a speed of about 650 revolutions, and produce an electromotive force of from 200 to 300 volts (according to speed), and a current of about 180 ampères; the resistance of the field magnets is about 30 ohms, and

that of the armature .004 ohm; the armature is 3 ft. long by 16 in. diameter, and from the illustration it can be seen that it is well ventilated. The machine is massive in construction, and well finished, and it is said to run with the greatest freedom from heating in all parts, a very desirable feature. The dynamo or motor in the cars is run at a speed of 800 revolutions per minute, with an electromotive force of 200 to 250 volts, and a current of about 20 ampères; the resistance of the field circuit is 14 ohms, and that of the armature .074 ohm. There is one point of interest to which it is necessary to call attention, and that is the position of the brushes with regard to the commutator. The cars, which are 13 ft. 6 in. long by 6 ft. 6 in. wide, are arranged so that their position remains unchanged on the line, and consequently it becomes necessary to reverse the "motor" to drive the car either forward or backward. If brushes were fixed in the ordinary way they would be right for proceeding in the one direction, but reverse motion would at once double back the copper wires of the brush. The brushes are therefore fixed as near as possible opposite the centre of the commutator, and are therefore in the same plane and exactly at right angles to a line drawn perpendicularly through the commutator. Their ends press directly against the bars of the commutator, and they are fixed relatively to each other, and so balanced that any alteration in the position of the one produces a corresponding alteration in the other, so that the brushes are kept exactly opposite each other. The switch arrangement of the cars consists of two portions, one a switch for reversing the direction of the current, and the other for starting the car. The former controls the direction of the motion, the latter its starting and stopping, and also its rate of progress.

These switches are contained in wooden pedestals placed under the steps leading to the roof, one at each end. The starting switch consists of a projecting handle, which is moved forwards or backwards; this handle is screwed on to the end of the inside switch, which, moving over certain brass contact pieces, brings into the circuit between the charged conductor and the motor certain resistances which have the effect of diminishing the speed of the armature of the motor. It will be seen, therefore, that a slight alteration of the handle, altering

8

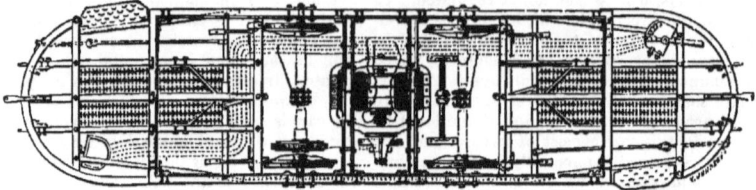

FIG. 57.—PLAN OF CAR.

FIG. 58.—DETAILS OF CONDUCTOR IN CONDUIT.

FIG. 59.—SECTION OF ROADWAY.

FIG. 60.—BLACKPOOL, ENGLAND, ELECTRIC STREET CAR.

the resistance in circuit, produces the required effect in increasing or decreasing the speed of the motor. These resistances are made of carbon rods (similar to those used in an electric arc lamp) firmly attached to brass connecting pieces.

The Besspool electric tramway was inspected and passed without alteration by the English government authorities in 1885, and has lately been accepted from the contractors as satisfactory. It has been constructed to form a link between the mills and granite quarries of the Bessbrook Spinning Company and the railway at Newry, the distance between the two places

The flangeless wheels run upon those outside rails. The maximum gross load of a train is twenty-six tons, consisting of six wagons, which carry about two tons each, and the electric locomotive, weighing eight tons, which also forms the passenger carriage, and is capable of accommodating thirty-four passengers. This load can be drawn up inclines averaging one in eighty-five at a speed of seven miles an hour, and up the stiffest incline of one in fifty at a speed of six miles an hour. The train can be started at any point of the line without difficulty. The motive power is electricity furnished by dynamos situated about two miles

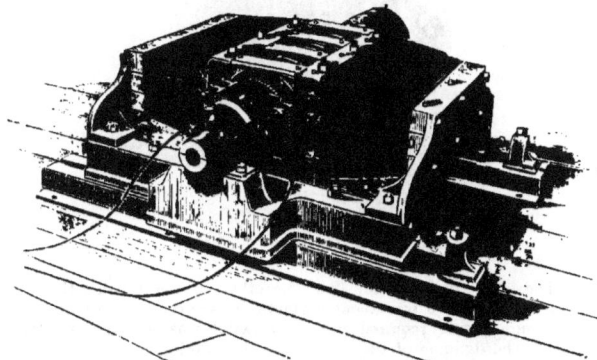

FIG. 61.—GENERATOR AT BLACKPOOL.

being three miles, and the annual traffic, which has hitherto been carried in carts, being about 28,000 tons. The tramway differs from others in that the vehicles are equally well adapted to run on the rails and the ordinary roads, this facility being required by the difficulty which was found in connecting the line to the railway at one end, and to every department of the works at the other. They are carried on four wheels 2½ inches wide and without flanges; the first pair are on a bogie, which can be fixed to form a rigid wheel base, or have shafts fitted to it, and allowed to swivel after the manner of the leading axle of a coach. These wagons carry two tons each, and can be drawn by a horse up moderate grades. On the outside of the ordinary tramway rail, second rails have been laid, to which the ordinary rails act as guards.

from Newry, at Millvale, and driven by a turbine, constructed capable of developing sixty-five horse power. The conductor consists of an inverted steel channel carried on insulators, and fixed midway between the ordinary rails. Both the generators and motors are of the Edison-Hopkinson type, constructed by Messrs. Mather & Platt, and are capable of developing twenty-five horse power. The locomotive is geared to run at a maximum speed of fifteen miles per hour, and this speed is easily attained when there are no trucks attached. The cars are 35 feet long over all, and are carried on bogies at each end, so that they pass readily around curves of 55 feet radius.

M. Lartigue, the well-known French engineer, has applied electricity to the traction of the panniers or cars of his single-rail tramway.

This tramway is employed in Algeria for transporting esparto grass from the interior by the traction of camels. It was an easy step from animal to electric traction, and M. Lartigue has successfully taken it. At a recent Agricultural Exhibition in the Palais de l'Industrie of Paris, a line was shown on which five iron panniers, or double cars in the form of seats, were drawn by an electric locomotive at the rate of seven was carried by a platform car or pannier, and geared with a grooved driving-wheel thirty centimetres in diameter, which ran upon the rail. A rheostat to graduate the speed, switches to stop, start, and reverse the motor, and a seat for the conductor, were also carried by the locomotive, and ran on small grooved wheels. The current was brought to the dynamo by two insulated conductors, one connected to the rail,

FIG. 62.—EXCITER, BLACKPOOL ROAD.

miles an hour. The total weight of the five cars and the electric locomotive was about a ton, and the maximum power required was three horse power. The dynamo of the locomotive was a Siemens D_6, and the generator, which stood about 100 yards from the line, was a Siemens D_4 dynamo capable of developing from five to six electric horse power. It was driven by a Herman-Lachapelle steam engine. The total length of the line was 123 metres. It was built of forty-one rails, each three metres long, and comprised curves of seven and one-half metres radius. The locomotive dynamo the other to the dynamo through small contact rollers in connection with the commutator. One switch was employed to start or stop the train by making or breaking the circuit; the other to reverse its motion by reversing the current. The rheostat, by interpolating resistance into the circuit, allowed the strength of the current to be varied and the speed of the train to be increased or diminished as the case may be. The work was carried out by Messrs. Siemens, and under the direction of M. G. Boistel. The economy of the working is of course largely dependent on local circumstances.

CHAPTER VI.

THE MODERN ELECTRIC RAILWAY AND STREET CAR LINE IN AMERICA.

THE narrative of invention and experiment given in Chap. III. has already detailed the numerous efforts made in America to work out the solution of the many difficult problems encountered in electric railroading. It has also shown beyond question that down to the time of the discovery of the reversibility of the dynamo, such work, no matter how ingeniously and persistently carried out, was doomed to failure. But with the advancing efficiency of the dynamo as a generator or as a consumer of current, and with the success of the Paris Electrical Exposition in 1881, came a revival of interest in the subject, and such a display of energy and ability in this field in America as to have brought the idea to triumphant realization within the brief period of five years.

To Mr. Stephen D. Field, a member of the distinguished Field family, the United States Patent Office has awarded priority of invention in electric railways in America. The papers of Mr. Field were filed in Washington on March 10, 1880 (a caveat was filed May 21, 1879); those of Dr. Werner Siemens on May 12, 1880, and those of Mr. Thomas Alva Edison, on June 5, 1880. These inventors were placed in interference, and it was not until last year that a declaration was made in favor of Mr. Field, on the combination of an electric motor operated by means of a current from a stationary source of electricity conducted through the rails. In view of what had been accomplished before Mr. Field secured his patent, it is altogether unlikely that he will be left in quiet enjoyment of the rights thus conferred. In fact, it is one of the peculiarities of electrical invention to develop colossal litigations, and we do not believe that electric railways will be any exception to the rule. Probably the electric motor companies now springing into vigorous existence will depend for future life not so much upon such fundamental patents as upon the

control of important details of construction and application.

A very interesting and authoritative account of the work of Mr. Field in this line appeared in the New York *Mail and Express* of August 2, 1884, and is here quoted in part:

"In his boyhood he showed a taste for mechanics and for electrical experimenting. He became a telegraph operator before he was sixteen, up among the hills of Berkshire county, where, after two removals across a continent, he at last proved the electric railway a success. In his 17th year the boy went with his family to San Francisco. He was first employed in California as an operator for the California State Telegraph Company. In 1865 he assisted in the construction of the Russo-American Telegraph Line, under the direction of Mr. Frank L. Pope, whom he had known as a boy in Massachusetts, and who afterward, as a patent solicitor, entered his application for a patent on the electric railway. It was in this year that he first learned of the successful solution of the problem of producing electric currents by mechanical means in the magneto-electric, and later in the dynamo-electric machine. During the year 1868 he constructed two electro-motors. The first—a rough model only—was made from an old magnet, some clock wheels, and stray pieces of iron he had picked up in the office. This model worked, and its success encouraged him to have a larger one constructed under his direction. Experiments with the first model proved to him that a galvanic battery would be too cumbersome and costly a means of producing a current ever to become practically useful, and he then endeavored by correspondence to find out the possibility of procuring large power machines. The object in constructing these first motors was to run street cars in San Francisco. Mr. Field's efforts to obtain a dynamo-electric machine were unsuccessful at the time. In 1871 he associated himself with Mr. Geo. S. Ladd and others in the organization of the Electrical Construction and Maintenance Company, of which he acted for nearly seven years as secretary and electrician. In 1877 he went to Europe, and there, at the workshop of M. Breguet in Paris, he saw some Gramme machines, which were exactly suited to the purpose of furnishing a current for his projected motor. On his return from Europe, two of these machines were ordered by Mr. Ladd, but after considerable delay the Californians were informed that Breguet would not send the machines, being afraid of invalidating some of his patent rights in the United States. Upon receipt of this information, a machine was immediately ordered from Siemens Bros., of London. At last their hopes seemed about to be realized and the long cherished, project of applying electricity to locomotion was in a fair way to be practically tested, but another disappointment awaited them. The hoped-for dynamo was lost at sea on the way to San Francisco. Nothing daunted by this misfortune, they promptly ordered another machine, and at last, in the fall of 1878, it arrived. At the same time two Gramme machines were placed at their disposal by the Hon. Milton S. Latham. With these, experiments were made in a loft on Market street, and an elevator loaded with 1,500 pounds of coal was made to ascend and descend by their agency. The possibility of moving a load and of controlling and reversing the motion by means of an electric current supplied by a dynamo machine was now proved beyond a doubt. In February, 1879, Mr. Field elaborated his plans for an electrical railway and made drawings of a motor, which is substantially the same as that afterward put in operation by him at Stockbridge. In May of that year a caveat was filed by him in the United States Patent Office, which covered a claim for an electric tramway motor, the current to be supplied by a stationary source of power and connected with the rails. By means of a secondary machine on the motor the current would be converted into power to be supplied to the axles of the car by suitable gearing, and so the car would be propelled. The claim also covered a method of reversing the motion of the car by reversing the direction of the electric current.

"This was the very first official record of a plan for a dynamo-electric railway. In July, 1879, at the solicitation of friends, Mr. Field came to New York, and placed the matter of filing an application for a patent in the hands of his old friend, Frank L. Pope. This application was filed March 10, 1880.

"Up to this time the electric railway was all on paper. It was considered advisable by Mr. Pope that a working model should be constructed and operated, so in May of that year Mr. Field began at Stockbridge the building of a railway and an electric motor. The machines he first experimented with proved worthless, and he was obliged to stop work until he could procure funds to buy others. This he finally succeeded in doing, and in 1881 the road was in successful operation. Meantime the Siemens Brothers had been experimenting on electric railways in Europe, and had constructed several that worked well on a small scale. Mr. Edison had also been at work on the same problem, and built his well known little railway at Menlo Park. The inventions of the three investigators clashed, and on applying for patents they came in collision. It was soon found that the application of Siemens was subsequent to those of both Edison and Field, and it was thrown out. The trial between the two remaining competitors was long and tedious. The testimony was taken two years ago, but it was only last week that the final decision, awarding priority of claim to Field, was announced. The caveat filed in 1879, with its rude accompanying sketch, was shown to be, as has already been stated, the very first official record of a plan for operating an electric railway. Pending the result of the contest in the Patent Office, a consolidation of the Field and Edison interests was effected in 1883, and the Electric Railway Company of the United States came into existence."

The company here named was organized about the beginning of May, 1883. The project of exhibiting at the Chicago Railway Exposition of that year had been entertained, but it was not until two weeks before the opening of the exposition that it was definitely decided that the electric railway should form one of its features. Everything remained to be prepared. The locomotive was scarcely begun, and the track was not laid. But the work was put under way and pushed with vigor. It is needless to say that the electric railway, under the circumstances, did not fairly represent the inventions of Messrs. Field and Edison, because the short time allowed to complete preparations left no other alternative than to make use of such electrical apparatus and material as could be readily and conveniently procured in the market without stopping to inquire too closely into its fitness for the purpose, and then

force of about seventy-five volts, with a current of about 150 ampères, through its normal circuit resistance of .5 ohm. Its weight was 2,700 pounds. The locomotive itself is shown in Fig. 63. It was named "The Judge," after Chief Justice Field, the uncle of its designer. The track ran around the gallery of the main exhibition building, curving sharply at either end on a radius of fifty-six feet. Its total length was 1,553 feet, or nearly one-third of a mile.

FIG. 63.—THE ELECTRIC LOCOMOTIVE, "THE JUDGE."

to design everything else to suit its electrical and mechanical peculiarities. Under the circumstances, it was impossible to hope for great efficiency or economy of results. It was, indeed, a matter of surprise that an electric railway was produced at all from the resources available at the time; and the execution of the novel task reflected great credit on Mr. Frank B. Rae and his assistant, Mr. Clarence L. Healy, who together attended personally to the many details. The same type of Weston shunt-wound machine was obtained from the United States Electric Lighting Company for generator and motor. At its normal speed of 1,100 revolutions, the machine had an electromotive

The track, Fig. 64, was of three-foot gauge, and had a central rail for conveying the current, the two outer rails serving as the return.

In order to secure a low resistance and proper connections between all the rails, a precaution made necessary by the low electromotive force of the generator, wires were laid under each rail. The inside rail was wired with No. 6 (B. & S.) bare copper wire and the outside rail with No. 8 iron wire. The central rail was also wired with No. 8 copper wire. A good contact was made with each rail by proper fastenings at the joints and also by laying the wire under the rails in the supporting plates, so that the weight of the rail rested upon it. These pre-

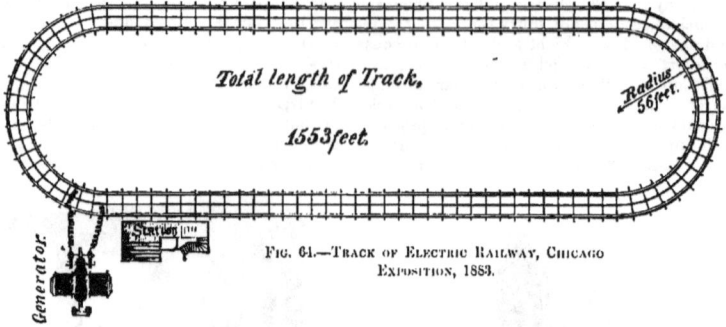

FIG. 64.—TRACK OF ELECTRIC RAILWAY, CHICAGO
EXPOSITION, 1883.

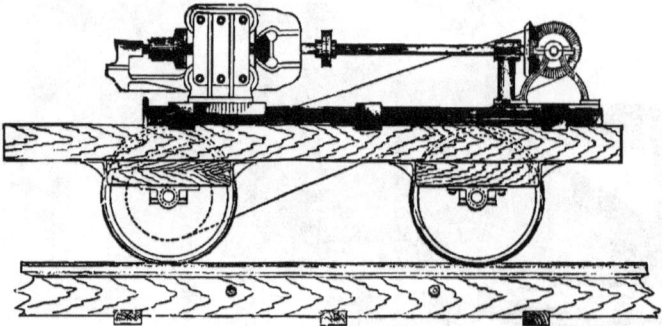

FIG. 65.—"THE JUDGE"—SIDE ELEVATION.

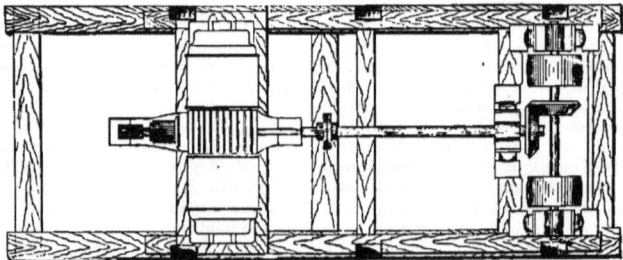

FIG. 66.—PLAN VIEW OF "THE JUDGE."

cautions practically reduced the resistance of the line to a value so small as to be inconsequential. It need scarcely be said that if the electromotive force of the current used had been higher these precautions would not have been required, and a simple connection of the rails together without wires would have sufficed. With an electromotive force of 300 volts, for instance, to produce the same current, the addition of .25 ohm to the circuit resistance by the rails would not have produced a very marked lowering effect on the current delivered by the generator, but in this case it meant a reduction of one-third.

The internal construction of the locomotive will be readily understood from the accompanying Figs. 65, 66, 67, and 68. Fig. 65 shows the

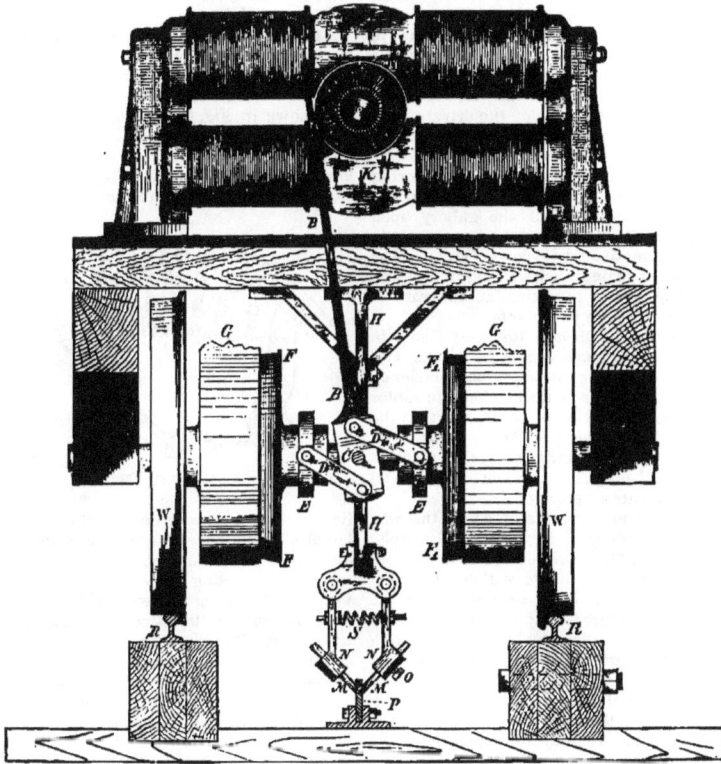

Fig. 67.—Rear Elevation of "The Judge."

locomotive in side elevation with its cab removed. Fig. 66 is a plan view of the same. These figures show the manner of transmitting the power from the armature of the motor to the driving wheels. The motor was placed crosswise upon the frame. Its armature shaft was coupled to an extension shaft which was pro-

9

longed forward and transmitted motion by means of bevel gearing to a counter-shaft carrying two pulleys. From these pulleys the power was transmitted by means of belts to the loose pulleys on the axle of the drivers. It will be noticed that this arrangement threw the greatest weight directly over the driving wheels. The gearing up of the armature extension-shaft to the counter-shaft was made so as to reduce the speed three times. The pulleys on the counter-shaft were twelve inches in diameter, the driven or loose pulleys on the axle of the drivers were twenty-six inches in diameter, and the car wheels, or drivers, were thirty inches in diameter. The maximum speed which this gearing would produce was about twelve miles per hour, but the weakness of the gallery upon which the track was laid made it necessary to run the locomotive at a lower speed. The average speed maintained was eight miles an hour, the armature revolving at the rate of about 750 revolutions per minute.

It was found by Messrs. Rae and Healy in their preliminary experiments with the two machines that the condition of best efficiency of the generator was realized when the motor had attained its maximum speed, and that the power developed was greater at that time. From this they inferred that the proper moment to put the load on the motor was when it had reached its greatest velocity.

The mechanism by means of which this was accomplished was quite ingenious and simple, as seen in Fig. 67, which represents a rear end elevation of the locomotive without the cab. The loose pulleys G G ran on the axle of the drivers W W, as previously stated, motion being transmitted to them by belts from the pulleys of the counter-shaft, as already shown. F F F_1 F_1 were cone friction pulleys fitting into the interior of the rim of the loose pulleys. These friction pulleys revolved with the shaft, being connected thereto by means of keys and keyways, which were loose, however, so that the friction pulleys could be free to slip lengthwise on the axle as they revolved. The hub of each friction wheel F_1 F_1 carried a collar E E which was connected by arms D D to a lever B fulcrumed at C on a projection from another collar fitted around the shaft. The operation of this form of friction clutch will be readily understood. In the position shown, the friction cones F F F_1 F_1 are removed from the

pulleys, which are free to move loosely upon the axle. But upon moving the lever B to the right the friction cones are both moved outward from the centre, and caused to engage the inner surface of the pulleys, and thus the motion of the loose pulleys is communicated to the drivers W W.

As already stated, the central rail of the track formed one conductor and the two outside rails R R the other. The device for "picking up" the current was quite ingenious, and is also shown in Fig. 67. It consisted of a kind of inverted vise firmly bolted to an arm H H projecting downward from the frame of the locomotive. The jaws N N of the vise were each perforated with three holes directed ob-

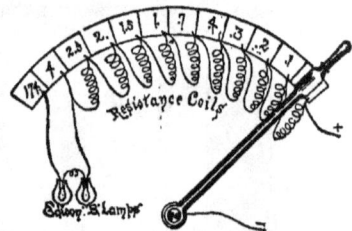

FIG. 68.—SPEED REGULATOR.

liquely downward and inward, through which bundles of phosphor-bronze wire passed, being securely fastened and held by a screw O. A spring S extending between the arms of the vise served to bring the two brushes M M into close and sure electrical contact with the central rail P. The wire being stiff and firm, the contact was equally certain whether the locomotive was moving forward or backward.

An important feature of this electric locomotive was the means by which the speed could be regulated at pleasure by a "throttle-valve" arrangement resembling that found in every American locomotive, and which answers the same function; namely, to control the amount of force put in action. The nature of the device was as simple as it was efficient, and consisted simply in a lever, Fig. 68, by the motion of which the resistance of a suitable rheostat could be thrown in and out of the main circuit with the evident result of controlling the amount of current flowing therein. The lever was placed horizontally and could move over contact segments disposed in a circle. These segments were insulated from each other, but they were connected by coils of iron wires so as to make a certain resistance, and possessing sufficient area of section to remain cool. Thus, when the lever was in the position shown in the figures, there was no resistance included in the circuit. When the lever was moved forward to the next segment, the resistance added to the circuit was .1 ohm. On moving to the next the resistance added was .2 ohm, and so on, the amount of resistance included in circuit when the lever touched each segment being in Fig. 68 indicated by the figures thereon. The rheostat comprised two Edison *B* lamps of 85 ohms each, so that the total resistance when the lever touched the last segment was 174 ohms. It will be readily understood that by means of this device the amount of current in the main circuit could be easily varied.

The high resistance necessary at starting to cause the magnetic field of the generator to magnetize itself was readily afforded by this rheostat on closing the circuit. Another interesting feature of the Chicago electric locomotive was the device by means of which the current was reversed at the motor when it was desired to make the locomotive move backward (Fig. 69). The lever *J* caused the wheel *II II* to turn on the armature shaft *G*. This wheel geared with two wheels *E F*, to which were fastened arms *C C*, *D D*. These arms carried brush-holders and brushes A_1 A_1 B_1 B_2. The function of the device was simply to change the relative direction of the current through the armature of the motor. In the position shown in the figure the positive brush B_1 touched the commutator at the left-hand side, while the negative brush A_2 touched it at the right-hand side. On moving the lever *J* as far as *K*, the brushes A_2 B_2 broke contact. In this condition the motor was out of the circuit and received no current. This is just the same as when the reversing lever of a steam locomotive is moved half way, thus cutting off the steam entirely by preventing the motion of the slide-valve. On moving the lever still further, the brushes A_1 B_1 came in contact with the commutator of the motor, so that the positive brush B_1 touched the right-hand side of the commutator and the negative brush A_1 the left-hand side, instead of the opposite. The re-

sult was necessarily a reversal of the direction of rotation of the armature of the motor.

The locomotive was also provided with an electric bell. This bell had a resistance of 350 ohms, and was rung by a switch which

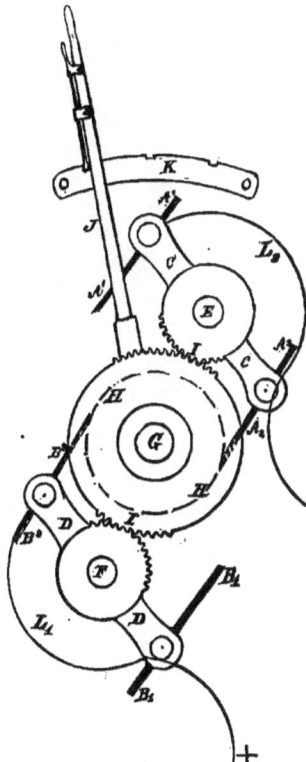

Fig. 69.—Reversing Mechanism.

placed it in parallel circuit with the motor. Its high resistance prevented the diversion of the current from the motor in any appreciable quantity.

The locomotive was twelve feet long and five feet wide. Its total weight was about three tons. It was intended to be run with two

FIG. 70.—EARLY EDISON ELECTRIC LOCOMOTIVE.

passenger cars, but it was found upon inspection that the gallery was too weak to stand the strain. Even after the gallery had been strengthened, it was not deemed expedient to exceed a speed of nine miles an hour.

The Chicago electric railway was the first constructed in this country for business purposes, and, considering the short lease of active life which was left to it after it was finally completed and put in operation, its success was most surprising. Owing to delay in receipt of the report of the engineer intrusted with the task of strengthening the gallery, the road was not permitted to operate for business until June 9, 1883, but experimental trips of the electrical locomotive were made daily from June 2.

Upon June 5 "The Judge" and its attached car loaded with sixteen passengers was started around the track. The railway was opened for business on June 9 and closed June 23, having run in all 118¾ hours and 446.24 miles. It carried 26,805 passengers. It was afterwards sent to the Louisville Exposition during the same year, and there carried a large number of passengers.

Mr. Thos. A. Edison's work in electric railroading dates back to the spring of 1880, although his ideas on the subject had been made known at least a year earlier. In 1880, he built a track at Menlo Park, N. J., near his laboratory. This line was less than half a mile in length, and no special pains were taken in the preparation of the road-bed. The early locomotive employed is shown somewhat roughly in Figs. 70 and 71. The generator and motor were of a type then made by Mr. Edison, but not now in use. The current was led to the track by two copper wires, one to each of the rails, which were thus positive and negative, and were insulated from each other. The armature of the motor was

FIG. 71.—EARLY EDISON ELECTRIC LOCOMOTIVE.

connected in the usual manner with the driving wheels, and made four revolutions to their one. Fig. 72 shows the Edison locomotive as afterwards improved.

Towards the close of 1883, the experiments of Mr. Leo Daft, the electrician of the Daft Electric Light Company, began to attract attention; and the ability and perseverance since exhibited by

curve having a radius of about 290 feet; but the tests as a whole gave great encouragement to Mr. Daft's friends.

We show in Figs. 74 and 75 a plan and elevation of the locomotive "Ampère," used on the Saratoga road, and still in operation elsewhere. It is about ten feet in length, and rests upon four wheels. The motor is situated at the rear

FIG. 72.—IMPROVED EDISON ELECTRIC LOCOMOTIVE.

Mr. Daft in the various application of electric motors have given him great and deserved prominence to-day. At first the trials of the Daft motor were made on the grounds of the company's works at Greenville, N. J., and there proving successful they were renewed on the Saratoga and Mount McGregor Railroad in November of 1883. The line is about twelve miles long, and abounds in sharp curves and steep grades. During its first trip there, the motor "Ampère," Fig. 73, being too light for the pull given it, jumped the track on a

end of the platform, and is incased in a box to prevent injury from dust.

The armature of the motor has a special construction in order to deliver the large current required. Mr. Daft achieves this by grouping equidistant along the periphery of his annular revolving armature as many superposed bobbins as the case in point demands, and fagoting their terminals before leading them to the segments of the commutator to which they respectively belong. Each one of a group of bobbins, on the armature, crosses the magnetic field at approximately the same instant, and the current developed in them collectively is of a potential due to their few convolutions and of a quantity proportional to the enlarged channels of escape.

At each end of the armature shaft there is keyed a pulley, from which belts run to the large central pulleys mounted on the counter-shaft, which is situated midway between the driving wheels. Another set of belts connects the latter with the counter-shaft, and the reduction of speed from the armature pulleys to the drivers is in the ratio of eight to one. The two small wheels shown as resting on the rail are phosphor-bronze contact-wheels; they bear upon the central rail, and through them the current is taken up. They are fixed upon springs so arranged that they can mount any small obstacle on the rail; but, being two in number, one of them is always in contact, thus insuring an uninterrupted current.

The driver of the locomotive is provided with a seat, and situated directly in front of him are three switch-boxes. The one toward the right is the main switch by which the current can be turned on or off. The regulation of speed in the "Ampère" is effected entirely by means of the "multi-series switch," constituting the middle box, which is so arranged as to effect an almost endless variety of changes in the intensity of the field of force, without the use of idle resistance. In order to accomplish this, some iron wire is employed in the outer coils of the field magnets so as to obtain the required resistance without too greatly exceeding the magnetic limit.

The brakes of the "Ampère" are controlled by the third switch shown, and a mere turn of the hand causes them to act with any required degree of intensity. The brakes themselves are of the so-called "pendulum" type. As will be seen, they are suspended from the frame of the car and swing freely on a bolt passed through an eye at their upper ends. The suspended frame carries an electro-magnet wound with stout wire, and when the current is switched into the latter the magnet immediately follows the attraction of the wheel, pressing against it and exerting a powerful grip. Situated at the side of the driver there will also be seen a lever, by means of which the locomotive can be sent forward or reversed at will. This is accomplished through the medium of

Fig. 73.—Daft Electric Locomotive "Ampère."

four brushes arranged about the commutator and placed equidistantly apart. The lever has two rods at its lower extremity, which connect with two brushes each, and by this means either

starting point. The track was laid with thirty-five pound rails, in addition to which a central track of similar rails was laid upon blocks of hard wood, saturated with resin, which were

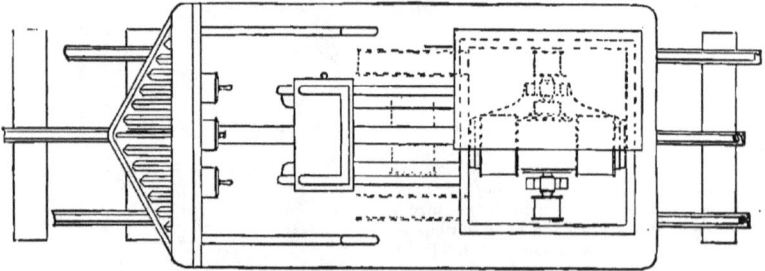

FIG. 74.—PLAN OF "AMPERE."

pair can be put in contact with the commutator with the result above mentioned.

From the above description it will be seen that the locomotive presents all the features of a well executed system, and some details of the test at Saratoga, mentioned above, will there-

spiked down to the ties at intervals of six or eight feet; upon this ran the phosphor-bronze contact-wheels.

The generators consisted of two of Mr. Daft's old type No. 8 series machines, which were operated about 100 yards from the track by a

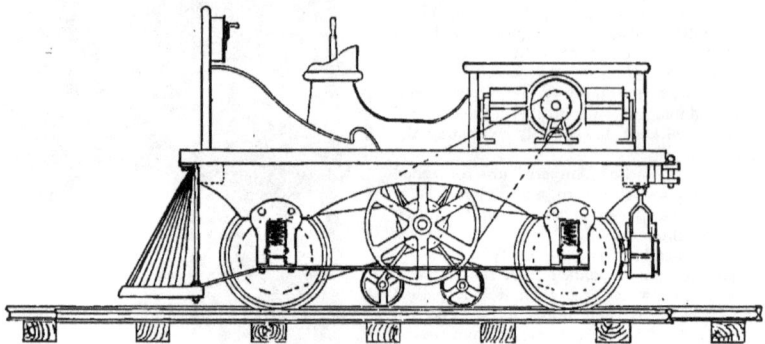

FIG. 75.—ELEVATION OF "AMPERE."

fore prove of interest. On that occasion the track upon which the "Ampère" ran was about one and one-quarter miles long, being part of the main track of the Saratoga and McGregor Railroad, and included a very sharp curve and grade combined, about one mile away from the

twenty-five horse-power Buckeye engine situated in the Saratoga Rubber Works. The Daft machines will be found fully described in Chap. IX.

The calculated resistance of the line was about two ohms. The actual resistance could not bo measured, owing to the continual earth disturbances. By other tests, however, it was evidently very low. The resistance of insulation between central and outside rails, when the ground was wet from a recent rain, reached about 130 ohms, showing a leakage, from that source, of about one and one-half per cent. of current with motor under maximum load. The resistance of motor, as given below, is taken while at rest; it is of course higher with the motor in motion, as will be readily understood.

The following are some of the results of the tests made at that time:

Internal resistance of primary machines in parallel, ... 0.42 ohms.
Resistance of motor, ... from 1.04 to 5 "
     "   " line (calculated), ... 0.2 "
     "   " track-insulation, ... 130 "
     "   " motor arranged for low speed
and great traction at start—at rest, . . . 3.10 "
Resistance of motor arranged for highest duty, 1.15 "
Mean electromotive force over high resistance
shunt at start, ... 100 volts.
Electromotive force, with lowest external resistance employed, ... 130 "
Current, when ascending grade, ... 80 ampères.
Revolutions per minute, ... 1050

The actual performance of the "Ampère" consisted in hauling an ordinary railway car weighing ten tons, containing sixty-eight persons in addition to the motor, which weighed two tons, and had five persons upon it. The speed obtained was eight miles per hour upon a track having a gradient of ninety-three feet to the mile, and included a curve of about 20°. This showed a maximum duty of about twelve horse power, and, although the actual efficiency was not determined, it ought to be mentioned that the twenty-five horse-power engine, which actuated the primary machines, was also doing other duty in the factory.

During 1884 Mr. Daft built and equipped a small line on one of the piers at Coney Island. This did not go into operation until part of the season had passed, but it carried 38,000 passengers. A little later another Daft road at the Mechanics' Institute Fair in Boston carried between 4,000 and 5,000 passengers weekly for more than a month. Its motor "Volta" was then taken to t'ie New Orleans Exposition, where a Daft road was put in operation between the main building and the government building, a

distance of nearly a fifth of a mile. It was there run regularly by a No. 4 Daft generator, driven by a Payne engine, and again carried several thousand passengers.

In the early part of the spring of 1885 the Baltimore Union Passenger Railway Company, hearing of the rapid progress of the Daft Electric Light Company with its system of electric railways, and wishing to increase its carrying capacity, investigated the matter. Satisfied with the completeness of the system, an order was at once given to construct two motors and equip the Hampden branch of the lines named. It was some time, however, before definite plans were settled upon; but about the middle

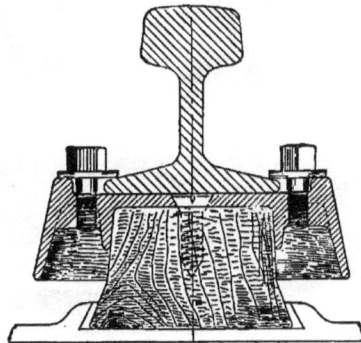

FIG. 76.—METHOD OF RAIL INSULATION.

of April work was begun both at Baltimore and at the Daft works. On June 10 the first motor was shipped. The Baltimore Union Passenger Railway Company, Edgar M. Johnson president, T. C. Robbins general manager, is one of the largest in the city. It operates twenty-five miles of roads, and has within its stables nearly 400 horses. The Hampden branch is just two miles long, runs through the villages of Hampden, Mt. Vernon, and Woodbury, aggregating some 15,000 inhabitants, and is one of the hardest bits of line the company operates. Starting from the main terminus on Huntingdon avenue, there is scarcely 300 feet of level road the entire length. The village of Woodbury, though not two miles distant, is 150 feet higher than Baltimore. Grades and curves

10

constitute the main features; in fact Mr. Daft appears to disdain an ordinary track on the level. The heaviest grade on a tangent is 319 feet, and on a curve 352 feet per mile. The sharpest curve has a radius of but 50 feet, the largest 80 feet.

To equip this road the joints of the outer rails were perfected, and a third rail, an ordinary 25-lb. T rail similar to the outer rails, was laid, with the Daft patent insulator, midway between the outer rails.

The insulator, Fig. 76, consists of an iron shoe of diamond shape, eight inches long, three

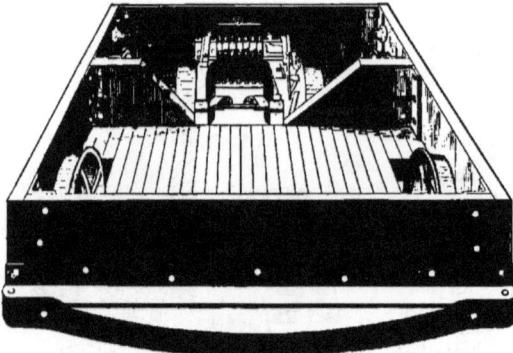

FIG. 77.—DAFT MOTOR, "MORSE."

and one-half inches wide and one-quarter inch thick, with two converging ways upon one of its surfaces.

Wedged between these ways is a round block of wood of truncated cone shape, with a height of two and one-half inches. Upon this block is screwed a round iron cap. This is four and one-half inches in diameter ,and two inches deep. Coming within three-quarters of an inch of the iron shoe, it thoroughly protects the wood block. The rail placed on the cap is held in position by two bolts screwed into the cap. The difficulties of constructing such a work, it being all entirely new, were many, but they were met and successfully overcome.

The centre rail forms the outgoing lead, the two outer rails with the ground being the return. The resistance of such a line averages less than .3 of an ohm, with perfect joints. At

the main terminus a new building, forming one room 20 by 40 feet, was built for the engine and dynamo. The engine is a 16 by 24 inch Atlas engine, made at Indianapolis. The boiler and all fittings are from the same firm.

The dynamo is one of the Daft Company's largest. Its total weight is 4,200 pounds and its maximum capacity is 300 ampères at 125 volts electromotive force. A nine-inch double belt connects direct from the ten-foot fly-wheel on the engine to a fifteen-inch pulley on the dynamo. Switches, regulators, automatic cutouts, and all other safety devices necessary for a complete system were put in, as precautionary measures against every possible form of danger or trouble.

The construction and appearance of the motor for this line is fairly represented by Figs. 77 and 78. Its name, as will be seen, recalls the important connection of Professor Morse with the city of Baltimore.

Over all, the motor measures 12 ft. 6 in. by 6 ft. 6¾ in. The frame is constructed of 2¾ by 19 in. ash, bolted together and braced with four-inch angle iron. The inside dimensions are 9 ft. 1½ in. by 5 ft. 11¾ in. The wheels are standard car wheels, but with specially deep flanges and wide treads; they are thirty inches in diameter and have five foot centres.

The cab is built in the ordinary manner. It is finished inside with ash and black walnut, and is very neat and substantial. The receiving machine is a compound series motor capable of delivering eight horse power. Its total weight is 1,100 pounds, the armature being 190 pounds. The compound nature of the field permits of a wide range of resistances, and hence of magnetic strength of field. As the armature-speed depends, in a certain sense, upon this field, a perfect means of regulation of speed is obtainable.

Motion from armature shaft to car wheels is obtained by internal gears. Upon each end of this shaft a three-inch phosphor-bronze gear is keyed. These engage with large gears, twenty-seven inches in diameter, fastened to the axle of the driving wheels. By this arrangement the energy of the armature is utilized practically almost directly upon the periphery of the driving wheels. The speed of armature to drivers is as nine to one. Therefore as the wheels must make 500 revolutions to the mile, the armature makes 4,581 revolutions.

The ratio of peripheral speeds, however, of armature to drivers is as 3.27 to 1.

As high speed is the normal condition of an armature, no real sacrifice is made to gain leverage. The speed of armature for eight miles an hour, the limit the law allows, is 610. To take up all back-lash of gears, the motor is arranged with pivoting bearings at one end and a regulating screw at the other, both resting upon heavy pieces of rubber. This pivoting arrangement is again advantageous in case of repairs, or inspection, through accident; the large gears being held in place by means of a set screw and long spline. On loosening the set screw, they can be easily removed along on the axle, freeing the small gears. The motor then can easily be raised to a vertical position allowing free inspection. The total weight of the "Morse" is about 4,200 pounds.

The wiring and controlling mechanism is equally as simple and substantial. No. 4 B. & S. underwriter's wire is used throughout. It is run in grooved sheathings, and covered so that no wire is to be seen excepting at the motor. Every precaution has been taken to obviate any danger arising from moisture or short-circuits.

The controlling device consists of four heavy brushes bearing upon a stout frame of soapstone, carrying broad and properly shaped contact pieces. This whole is enclosed in an 8 in. by 16 in. iron box, with an ordinary engineer's handle and guide.

Four movements are made, controlling the combinations of the field magnets, which vary from .39 to 3.75 ohms. The resistance of the armature is .24 of an ohm. By proper connections with the switch, it can be readily seen that the motor can be slowly and easily started, stopped, or run. By turning a small handle placed just to the left of the main switch, either to the right or

left, one of two pairs of brushes is brought to bear on the commutator, thus giving the directive motion to the armature, and obviously to the car.

Another switch just to the right of the main switch is a dead cut-off controlling the main current coming from the contact wheel. This is placed underneath the car, and consists of a heavy fourteen-inch wheel of phosphor-bronze, free to slide four inches to the right or left, and rotating freely upon its shaft. A deep groove is cut into the rim, fitting the centre rail. By a

Fig. 78.—Details of the Motor "Morse."

lever and a heavy spring a constant pressure tends to keep the wheel down on the rail. By this arrangement the wheel can adapt itself to every curve or change of level of the rail. An ordinary hand brake is placed in the car just to the left of the switches. By this handy arrangement, one man, with a little practice, can easily manipulate the switches and brake, and so control the car.

It may be here stated that for much of the information given of the Daft system, we are indebted to Mr. G. W. Mansfield, to whose efficient hands Mr. Daft has generally intrusted the execution of his plans.

This equipment went into service on August 8, 1885, from when until now the road has been dependent upon electricity as its sole motive power.

During the first six months of operation several storms of peculiar severity visited the region, but contrary to the predictions of all the visiting electricians the road did not suffer any more interruption to travel than was experienced on ordinary roads from the same

drivers that the service is better at such times than in fair weather.

The road was at first supplied with two motors only, the "Morse" above described and the "Faraday," but these being found insufficient. both in size and number, for the greatly increased traffic, the more powerful motors "Ohm" and "J. L. Keck" were added early

Fig. 79.—Curve on the Baltimore Electric Street Railway.

cause. Several times parts of the track have been actually submerged by rain, when the spectators have been treated to the extraordinary spectacle of an electric motor hauling a heavily loaded car, with the flanges of the driving wheels deep in the water! At such times, of course, the insulation was somewhat impaired, but never so much as to cause any marked change either in the speed and capacity of the motors or the load on the station dynamo. Indeed it is a cherished illusion of the motor

in 1886, and the trains are now despatched at intervals of twenty minutes, instead of half hourly, as before.

Each motor performs an average daily run of seventy-five miles, which, considering the extraordinary grades and curves of the road, is very heavy duty for a mechanical tractor. The General Manager of the Baltimore Union Passenger Railway Company, Mr. T. C. Robbins, has made many improvements in the road since the introduction of electricity, and has throughout shown a rare intelligence which, together with his enthusiastically progressive temperament, has largely contributed to the successful prosecution of the work. A late survey of the road shows the heaviest grade to be that entering Roland avenue, about one and one-half miles from the station, of 348 feet per mile on a curve of 75 feet radius, but there are several gradients of over 250 feet per mile, and curves ranging

As a test for himself, Mr. Robbins once sent to the city for one of their heaviest cars, ... lbs.   5,100
And carried a load of eighty-one persons over the road (say 81 × 125 pounds), ... lbs. 10,125
The weight of the motor used was . . . . . lbs.   4,500

Total, ... lbs. 19,725

Thus he says that 19,725 pounds were carried over the road by one motor of . . . . lbs.   4,500

FIG. 80.—OVERHEAD CONDUCTOR ON BALTIMORE RAILWAY.

from 40 to 90 feet radius. Fig. 79, an accurate reproduction of a photograph, represents the motor "Morse" on the curve No. 5, which combines a gradient of 275 feet with a curve of 80 feet radius. Mr. A. H. Hayward, the electrician in charge, also superintended the Daft Electric Railroad at New Orleans. It is now proposed to extend the line, with overhead conveyance of the current.

As some figures may be asked for, the following will be found of interest. They were published several months ago:

His engine and boiler cost, approximately, ... $2,400
His two motors cost, approximately, $3,000 each, .   6,000

Total, ... $8,400

There is also the expense of conducting rails and wires, insulation, protection, etc.

His expense of running per day is 1½ tons of soft coal, ... $1.75
Engineer and fireman at power station, ...   4.50

Or, excepting oil, waste, wear and tear, per day, $0.25

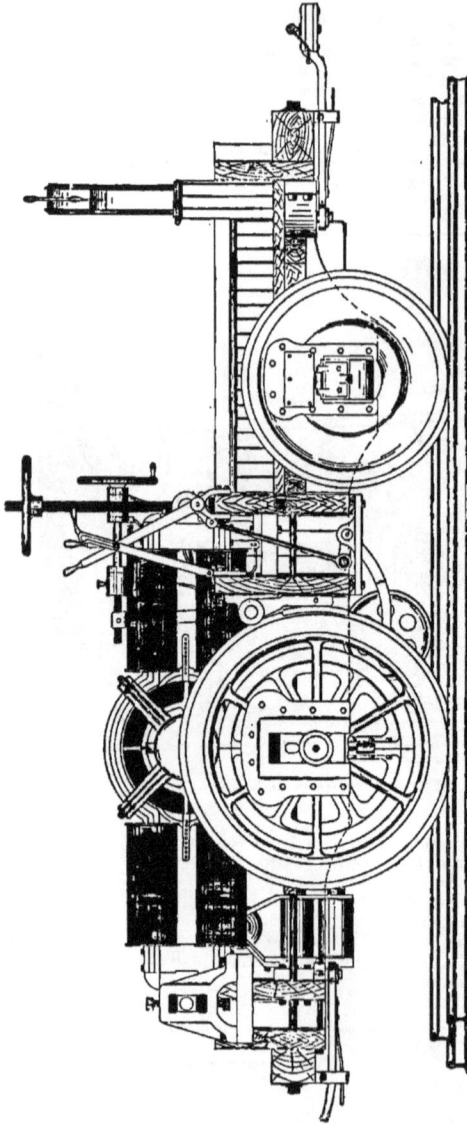

FIG. 81.—SIDE ELEVATION OF DAFT LOCOMOTIVE "BENJAMIN FRANKLIN."

The above represents the cost of his power, and equals the work of thirty horses per day. The average receipts from the cars carried by the two motors are $18 per day, and he has taken (on a Sunday) total receipts of $86 in one day.

As the present work goes to press, extensions and improvements are being made on the line. Fig. 80 represents the new overhead conductor attachment which has recently been added to the Daft motors on the Baltimore & Hampden Electric Railroad. As will be seen by reference to the cut, the contact mast is so attached as to be readily operated from the inside by the engineer; this was deemed necessary as the overhead conductors are at present only used at street crossings, thus leaving the track at such places free from the third rail and the raised guard, which were found to be objectionable. The manner of connecting the motors with the overhead conductor for the California railroads, now being equipped by the Daft Company, differs in some respects from that shown in the engraving, especially in the manner of completing the circuit to the motor; but as this plan has not yet been put in operation no further reference need now be made to it. As an instance of the continued successful working of the Baltimore road, it is worthy of note that on the three recent holidays, the 3d, 4th, and 5th of July, upward of 7,000 passengers were carried, or over 3,400 more than by horses for the same time last year. From figures supplied by Mr. T. C. Robbins, it appears that the line has carried 31,907 more passengers than it did with horses during a like period of nine months, and that the cost per passenger has been only 1.66 cents, as compared with 3.01, or .83 cents per passenger mile as compared with 1.55 cents with horses. The cost of horse power per day was $18; the cost of electric power is $12.

It may be safely asserted that few engineering events within the last few years have attracted more popular and professional attention than the trials of electricity as a motive power on the elevated railways of this city. The daily papers printed columns regarding the event, and a number of illustrated papers produced illustrations of the Daft motor put upon the tracks.

Before entering into a description of the system, as it is operated, it may be well to recall the events which led up to the present state of affairs.

The idea of running the elevated railway trains by electricity was broached several years ago, the many strong points in its favor, over steam, being pointed out. Nothing, however, was done in the matter beyond its mere discussion, until the early part of 1885, when at a meeting of the various electric motor companies, an attempt was made to consolidate their interests, and to test the motors of the various companies represented. A commission was to be appointed (Sir William Thomson being designated as one of the members) to test the motors, and the best system was to be adopted. Several meetings were held, but the scheme finally fell through. This agitation acted as a stimulus, however, for shortly afterward the Daft Company obtained permission to

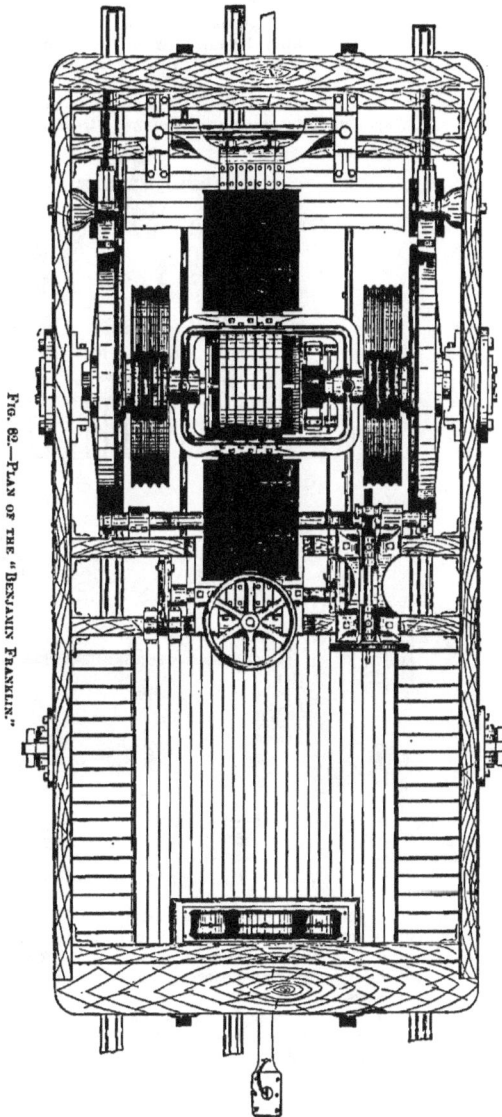

FIG. 82.—PLAN OF THE "BENJAMIN FRANKLIN."

equip a section of the Ninth Avenue Elevated Railway on its system, while the Edison-Field interests were assigned to the Second Avenue road.

From that time until August, the Daft Company was busy equipping a central station,

central rail, through which the current is led to the motor, is elevated above the outer ones, resting upon the insulator shown in Fig. 76.

The central station, in which the generating dynamos are placed, is situated in Fifteenth street, a distance of about 250 feet west of the

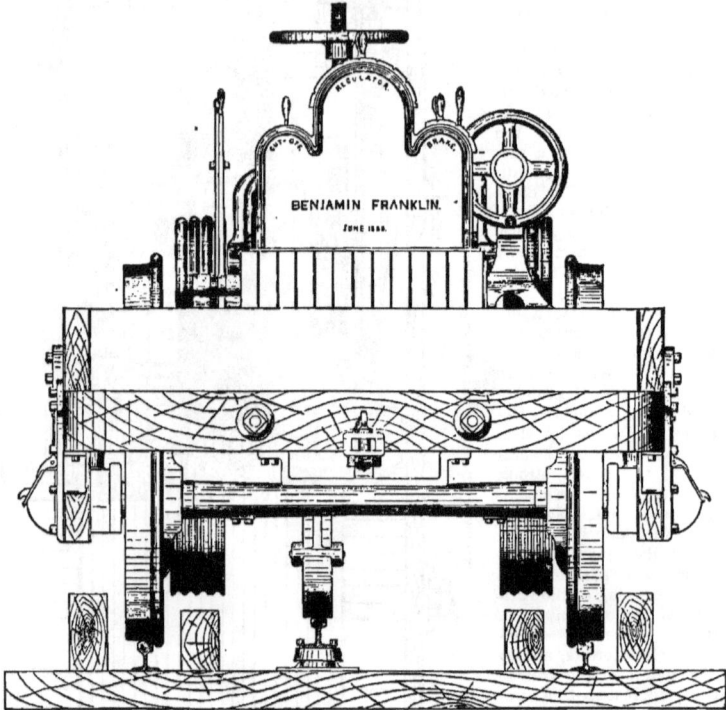

FIG. 83.—REAR ELEVATION OF THE "BENJAMIN FRANKLIN."

building a motor, and laying down the central rail required. It must be understood that the latter operation had to be performed during regular traffic hours, with trains passing every five minutes or less.

The road was and is equipped from the elevated railway station at Fourteenth street, up to Fifty-third street, a distance of two miles, in which a heavy grade is encountered. The

tracks, and connected with the latter by a stout conductor. The station contains a Wright steam engine and three generators. In addition there is a small dynamo which runs the Daft arc lamps, by which the station is lighted at night.

The motor used is named "Benjamin Franklin," with which a speed of twenty miles an hour has been attained.

The illustrations, Figs. 81, 82, and 83, show the arrangements of the motor in detail, as it stood on the track, the cab being removed for the sake of clearness.

In the arrangement of this motor Mr. Daft has entirely avoided the use of belts, power being transmitted by friction from the armature to the drivers, and the amount of it can be regulated at will according to the load. As will be seen from the plan, Fig. 82, the motor-dynamo is supported at the rear on a shaft resting in bearings; its front end is supported by a threaded eye through which passes a long screw, which is turned by a hand wheel, as shown in the side-elevation, Fig. 81. The armature shaft carries a friction wheel nine inches in diameter, which bears upon a larger friction wheel three feet in diameter, keyed to the axle of the main drivers. With this arrangement it is obvious that by turning the large screw the upper friction wheel can be pressed against the lower to any desired degree, thus preventing slip, even with the heaviest loads. By means of the screw also the entire motor-dynamo can be raised clear above the driving wheels, so that the armature can be taken out and inspected with convenience.

The bronze contact-wheel which bears against the central rail is 15 inches in diameter and is raised and locked by the two levers at the side shown in the view, Fig. 81. Another lever on the other side constitutes the "reversing lever," by which the brushes of the dynamo are set so as to give the motor a forward or backward motion. There are two pairs of these brushes and the motion of the lever alternately puts either pair in contact with the commutator. Like all of Mr. Daft's railway motors, this one is provided with his electric brakes. These consist of large electro-magnets which, being energized, are attracted by the wheels and press against them like the ordinary brake. The motor-man occupies the clear space in front of the motor-dynamo, and before him is placed the case containing the regulating, brake, and cut-off switches, as shown in the end view, Fig. 83. The switch at the right controls the brakes and that to the left makes or breaks the current as desired. In the centre is placed the "regulator," by which the speed of the motor can be altered at will. There the terminals of the compound winding of the motor-dynamo are brought, and by moving the lever

to the different notches, the resistance of the field magnets is altered, which changes the speed correspondingly. The driving wheels are 48 inches in diameter; the trailing wheels 36 inches. Their shafts supporting the motor-dynamo rest in specially designed resilient bearings, so as to reduce any shocks to a minimum. The motor is designed for 75 horse power and a normal speed of eighteen miles per hour, with a possible speed of forty miles. The motor complete weighs nine tons, and measures fourteen feet six inches in length, over all.

With the "Benjamin Franklin" several runs were made, to the satisfaction of all apparently, except Mr. Daft himself, who became convinced that the motor was too light for its work, i. e., that its weight was not sufficient to give it a grip upon the track adequate to the load it could pull. This, however, is a defect in the right direction. Mr. Daft, having temporarily withdrawn the locomotive, has now rebuilt it, making it much heavier, and the demonstration on the elevated road will be resumed as this volume goes through the press.

A recent comer in the field, but one whose operations are destined to be of the first importance, is the Bentley-Knight Electric Railway Company, owning and using the patents of Messrs. Edward M. Bentley and Walter H. Knight, whose system was put to initial experiment in August, 1884, on the tracks of the East Cleveland Horse Railway Company, Cleveland, Ohio. The plant consisted as usual of stationary engines and dynamos (Brush). The conductor was placed in a conduit between the rails and running the entire length of the road. The current was taken up by a conductor brush passing through a slot in the conduit and in contact with the conductor there—thus maintaining unbroken connection with the source of power. The road equipped in this way was two miles long, with a branch track, a turnout and two curves of 45 feet radius; and a railroad crossed it at an angle of 45° on the level. Two motors were employed. The line was operated experimentally for a year, and during that period demonstrated the entire feasibility of electric street railways. Our illustrations Figs. 84 and 85 show the cars as they appeared running through Cleveland. In the one instance, the car is ploughing its way through the unusually deep snow of the winter of 1884-5. At no time was the snow deep

11

FIG. 84.—SUMMER VIEW ON BENTLEY-KNIGHT ELECTRIC STREET RAILWAY, CLEVELAND, O.

enough to cause any interruption of traffic so far as electricity was concerned. In the other instance, the line is shown under normal summer conditions.

Since that time, the Bentley-Knight Company, having its offices in New York, has made arrangements with the Rhode Island Locomotive Works of Providence, R. I., to manufacture its apparatus, and is now busily engaged making preparations for the work it proposes to undertake. During the past year extraordinary popular interest has been manifested in everything pertaining to increased rapid transit facilities. The extension of the New York and Brooklyn elevated lines, the proposed introduction of the same system throughout our greater cities, and the preliminary work of the several sub-surface railway companies, have stimulated the discussion on the merits of such systems of motive power as may satisfactorily accomplish the work of the steam locomotive without the numerous disadvantages attendant upon its intramural use.

In this connection, the illustration, Fig. 86, of the electric locomotive designed by the Rhode Island Locomotive Works cannot fail to be of interest. This illustration is taken from the working drawings of the Bentley-Knight Company, accompanying various estimates lately submitted by its engineers, Messrs. Bentley and Knight.

This locomotive is especially designed for light passenger work. It is standard gauge, has a wheel-base of twelve feet, and weighs about 48,000 pounds, all of which weight is equally distributed upon its six-coupled sixty-eight inch driving-wheels. The nominal electrical capacity of its twin motors is 500,000 watts. The motor armatures are thirty-six inches in diameter, and exert their force upon the drivers without the intervention of any of the various forms of gearing which some think have seriously impeded the successful introduction of very large electric railway motors. It is equipped with electric headlights, bells, and automatic tubular electro-magnetic brakes, and is fitted with electric connections for incandescent lamps and brakes throughout the train. It has no reciprocating parts, and is equally adapted for use with overhead, surface, or sub-surface connection with the central power station of the line. The company's motors are built so as to be exceedingly power-

ful, solid, and compact, and are balanced to a nicety; and the care with which their interchangeable parts are manufactured renders impossible any serious interruption of work from ordinary accident.

Some noteworthy plans have been prepared by the New York District Railway with a view to the use of electric locomotives, of the kind just described, on the proposed underground railway for Broadway, New York. With such a road Broadway can be utilized for legitimate passenger traffic. It has long suffered, and until recently was suffering, from the diversion of travel to streets parallel with it. The slow and clumsy stages, rattling and jolting over the rough cobble stones, could never accommodate its frequenters satisfactorily. Horse-car lines with which the thoroughfare is now afflicted can never be anything but an impediment to traffic; while they are much too slow to carry the hundreds of thousands of persons to whom, in modern New York, rapid transit between the Battery and the Boulevards has become an absolute necessity. But while horse-car lines along Broadway are a disgrace to the city, the erection of an elevated railroad would be scarcely short of sacrilege. Broadway, with its natural advantages, ought to be the finest street in the world. But that it cannot be if given up to the horse-car or to the abomination of ugliness called an "elevated road." All the requirements of rapid transit for Broadway and of burying the wires are met in the proposed "scientific street." Some objections may already exist, or will perhaps arise when the work of carrying out the scheme is actively prosecuted. But taken in its entirety the idea is one that recommends itself to us as much for its practicability as for its brilliancy. The underground railway would give additional value to property all over Manhattan Island, would relieve the present avenues of traffic that are now so sadly crowded morning and night, would pay handsomely as an investment, and would preserve Broadway in a renovated condition, picturesque and beautiful, for those grandiose demonstrations of a civil and military character in which, as a people, we take so much pleasure. Operated by electricity, after the manner described, the underground road would be very pleasant for travel. Ventilated perfectly, cool, regular, speedy, without noise or dust or smoke, it would compare most favor-

FIG. 85.—WINTER VIEW ON BENTLEY-KNIGHT ELECTRIC STREET RAILWAY CLEVELAND, O.

ably with any other kind of locomotion on the surface or on elevated tracks. It would be a road worthy of Dr. Richardson's ideal City of Hygeia, as well as of the metropolis of the Western Continent.

The main features of the road, as set forth in the carefully prepared plans of the company are the following:

1. Two express tracks, throughout the line from the Battery to the Harlem river, forming a "through," standard-gauge, rapid-transit road of enormous capacity and capable of great speed, with easy access and egress at a few commanding points.

2. Two "way" tracks, throughout the line from the Battery to the Harlem river, forming a rapid-traffic, standard-gauge line between frequent stations.

3. Continuous galleries on either side of the railways, arranged to house all the present water, gas, pneumatic, steam, and other pipes which occupy the street below, together with all the electric cables and wires now arranged upon poles and house-tops above the streets, all service-pipes being in immediate contact with the vault wall of every house on the line, where they will everywhere and at all times be accessible for alteration, repair, replacement, and inspection.

4. The whole to be built and operated (as to the standard section) between the curb-lines and (except at Canal street) above mean high-water, for the purpose of avoiding the invasion of the valuable vaults of Broadway, and for the further purpose of compensating existing vested corporation rights, without encroaching upon vested private rights, or private property.

The roadway of lower Broadway, between the curbs, furnishes all the accommodation required for every purpose. It is divided into two sections; the one centrally placed affords accommodation for the way and express lines; the section on either side disposes of the existing impedimenta of the street at the point of access to the abutting houses. By this disposition of the street all requirements are fulfilled. (1) A smooth, noiseless, and unobstructed surface is provided for pedestrian and vehicular traffic. (2) Express and way trains for through "rapid transit," and for rapid transit from station to station. (3) Permanent housing for sewer, water, gas, steam, pneumatic, and electric conductors and pipes, with access through-

out for inspection, and in all cases in immediate contact with the premises where the connections are to be made. In neither express nor way stations is private property taken, nor at any point does the structure abut private premises, even during construction.

The method of construction is as follows: Street excavation is effected in sections, and is governed by the extent and character of the traffic, travel being maintained unobstructed by a system of movable bridging. A uniform platform of concrete a, Fig. 88, about two feet in thickness, floored by a half inch of Trinidad asphalt, extending across the street at a maximum base depth of about seventeen feet, forms a foundation for the whole structure. Upon this is erected the external vault wall b, securing to the abutting proprietor the permanent use of the whole vault and area undisturbed throughout the standard section. This vault wall is fitted while under construction with suitable connections for gas, steam, electricity, sewer, and water at every house. This wall is also the external wall of the pipe galleries c, arranged adjacent to both curbs. The galleries are subdivided longitudinally and continuously by beams riveted to their internal and inserted in their external walls, which support the sewers and other pipes. Access throughout is provided at the termini and stations, and they are calculated for access to, housing, and inspection of, the tubes, pipes, and wires. The electrical conductors d, of the various telegraph, telephone, lighting, burglar-alarm, messenger, and time companies are arranged anti-inductively, upon shelves riveted to the roof and upper gallery beams. There being no permanent floor above the foundation, the pipes in either gallery are accessible from above or below. Street opening for repair, replacement, or connection is thus wholly obviated. The internal wall supporting the galleries is formed by iron columns e, placed four feet apart, and coincident with those forming the outer wall of the "way" railways. These columns are composed of two angle irons riveted, and rest upon a continuous granite foundation f. The galleries contribute largely to the cost of construction, but are indispensable to a safe, convenient, and equitable replacement of present impedimenta enjoying vested rights, and to access thereto at every house on the route. The space remaining between the pipe galleries is dis-

FIG. 87.—STANDARD SECTION OF THE PROPOSED UNDERGROUND ELECTRIC RAILWAY FOR NEW YORK.

Fig. 89.—STATION NEAR FOURTEENTH STREET ON PROPOSED UNDERGROUND ELECTRIC RAILWAY FOR NEW YORK.

HOUSE VAULT
INTACT.

FIG. 89.—METHOD OF MAKING PIPE AND WIRE CONNECTIONS.

posed in four railways, for the accommodation of an up-way and express and a down-way and express train g. These ways are formed by five rows of columns, each composed of four angle irons, arranged longitudinally four feet apart, resting on a continuous granite base, the spaces between the columns at the foundation and the roof being filled by a panel composed of a tough, non-resonant material, "ferilax" h, composed of steel wire, vegetable fibre, and solidified oil compressed into a solid panel by

the permanent street, upon which the pavement will be relaid. This structure as a whole contemplates the minimum of excavation, the maximum of capacity, the greatest number and most equal distribution of points of support, and consequent maximum of strength and stiffness in use.

The railways form open cylinders from station to station, and the trains being of approximate cross-section constitute loose pistons always moving in the same direction; the ob-

FIG. 90.—VAN DEPOELE GENERATOR.

hydraulic power. This panel fulfils a double function: it completes the enclosure for purposes of ventilation, and it prevents resonance, which might be caused by the rapid passage of a train through an enclosure with metallic walls. The roof is supported and the whole structure tied by beams i placed four feet from centres which extend across the entire span, bolted at every eight feet to the columns, the ends being inserted in the vault wall. Upon these beams the steel ten-inch span, buckle-plate roof k is laid and bolted; over this is a two-inch skin of Trinidad asphalt, as a protector from chemical contact and dampness and as a slight cushion. Above this is placed six inches of concrete, which completes

12

vious effect is the establishment of a ventilating current, dependent for its force upon the approximation of cross-sections, the speed of the trains, and the integrity of the tunnels; as the products of artificial combustion are excluded from the tunnels, the requirements of ventilation are reduced to a minimum, and perfectly performed. The traffic-rails, the electrical conductor conduit l, and the guard-plate are bolted to the same steel tie, which arrangement secures perfect alignment, the tie being permanently set in the concrete foundation. A deflecting-plate m attached to the structure at the cornice line and the guard-plate external to the rail render destructive derailment impossible.

Roads and galleries constructed in this way have the incidental advantage of being accessible from one to another at any point and across the whole system, from curb to curb, performing the vital functions of ventilation, and of guaranteeing complete immunity from collision or derailment, without obstructing transverse communication when it is required.

This plan was first brought to the attention of electricians in 1884, when Col. Rowland R. Hazard presented a paper on the subject at a meeting of the American Institute of Electrical

Fig. 91.—Large Van Depoele Motor.

Engineers; and it is now familiar to the public. Fig. 87 shows a standard section of the line on the Broadway division. Fig. 88 is a view of the road at the proposed station near Fourteenth street and Union Square. Fig. 89 gives an illustration of the method of making connections in the pipe and wire galleries for mains and way service of every description.

The Arcade Railway Company, also proposing to construct an underground railway along Broadway, announces, too, its intention to use electricity for locomotive purposes. The work, whenever carried out, and by whomsoever, will be one of great profit, utility, and convenience.

The Van Depoele electric railway system, now in successful operation at so many places in this country, as well as in Canada, is the invention of Mr. Charles J. Van Depoele, the electrician of the Van Depoele Electric Manufacturing Company, of Chicago, Illinois, and it is the result of constant experiment in generators, motors, and the transmission of power, beginning in 1874 and running down to the present time.

The generator, Fig. 90, is a model of simplicity. The motor is changed slightly from the ordinary Van Depoele dynamo to adapt it to the work of transmission of power. These machines are of various sizes and styles, from a motor weighing one pound to the eighty horse-power motor weighing eight thousand pounds. The accompanying cut, Fig. 91, illustrates the large motor for running railway trains.

The first railway operated under the Van Depoele system was laid in Chicago in the winter of 1882-3, and the current was conveyed by a wire. In the fall of the same year a car was run at the Industrial Exposition in Chicago from an overhead wire.

In 1884 a train was run at Toronto, Ontario, by the Van Depoele system, using an underground conduit. This road was operated successfully and carried the passengers from the street car line to the exposition grounds, and was a perfect success. It was operated as long as the exposition lasted. This train averaged 200 passengers per trip; the speed was about thirty miles per hour.

In the fall of 1885, at Toronto, the road connecting the exposition grounds with the street railway, a distance of one mile, was equipped with a Van Depoele motor, Fig. 92. This train consisted of three cars and a motor-car. As there was only one track, it was necessary to run at a high rate of speed. An overhead wire was used as a conductor, it requiring but a few days to put it in operation; an ordinary forty-light dynamo was used, driven by a Doty 10 × 16 engine.

The average speed of the train was about thirty miles per hour. The trains carried from 225 to 250 people, and the average number of passengers per day was over 10,000. The amount of coal consumed was 1,000 pounds per day. This road carried all the passengers that could be gotten on and off the cars.

FIG. 92.—VAN DEPOELE ELECTRIC RAILWAY, TORONTO, CAN.

FIG. 93.—MINNEAPOLIS RAILWAY—VAN DEPOELE SYSTEM.

FIG. 94.—ELECTRIC STREET RAILWAY, MONTGOMERY, ALA.—VAN DEPOELE SYSTEM.

FIG. 95.—ELECTRIC STREET RAILWAY, MONTGOMERY, ALA.—VAN DEPOELE SYSTEM.

For the purpose of conducting experiments, a portion of the South Bend Railway line was equipped in the fall of 1885 and several independent cars were run with small motors, the generator being driven by water-power. It was a distinct success, the cars travelling in different directions from the same conductor. This road has not yet been equipped, however, owing to change in management.

quite a distance from the track, and is driven by an old slide-valve engine, 12 x 18 cylinder, making 125 revolutions per minute. The consumption of coal is about 3,000 pounds for seventeen hours' run. Forty-eight trains are run each way daily, running from 6 A. M. to 11.30 P. M. Trains are composed of from three to four closed railway coaches weighing eleven tons each, or of a larger number of open cars weighing six tons each. As many as eight of these cars have been hauled at one time, and this up a grade of three and one-half per cent., and the cars crowded to their utmost capacity with passengers, giving a total of ninety-one tons. The motor works perfectly.

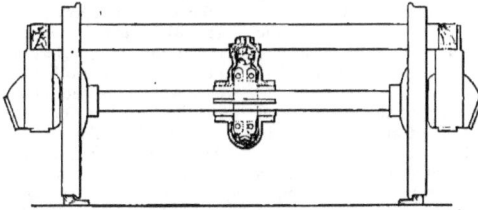

Fig. 96.—Pendleton Method of Attaching Motors to Cars.

directions from the same conductor. This road has not yet been equipped, however, owing to change in management.

At New Orleans, during the late exposition, a train, consisting of three large cars, was run successfully until the end of the exposition.

The Minneapolis, Lyndale & Minnetonka Railway Company, of Minneapolis, have been obliged to discontinue the running of their loco-

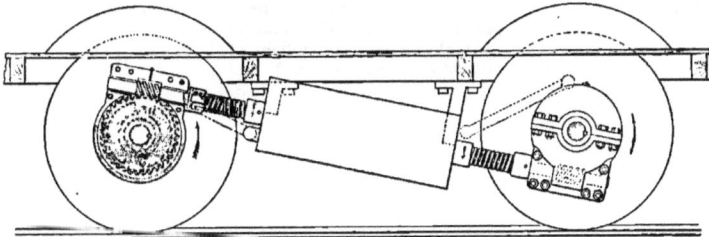

Fig. 97.—Pendleton Method of Attaching Motors to Cars.

motives in the more thickly settled portions of the city of Minneapolis, and an arrangement was made to bring the cars into the city and deliver them back to the steam locomotives. This is being done successfully (Fig. 93). The motor is located upon a cheaply constructed motor-car and takes the current from an overhead copper wire. The generator is placed

At Montgomery, Ala., the Capital City Street Railway have been running two cars for some time (Figs. 94 and 95). The grades are over seven per cent.; the distance is over one and one-half miles. Motors are placed on the platform of each car and do the work well. The speed over the grade is six miles per hour. The cars run sixteen hours per day, and the generator is driven by an old-fashioned slide-valve engine stationed 250 feet from the boiler. The amount of coal consumed per day is 3,000 pounds, including getting up steam from cold water.

At Windsor, Ontario, a train has been running on the track of the Windsor Electric Street Railway Company since June 6, and giving good satisfaction. The distance travelled is about two miles. Roads equipped with the Van Depoele system have recently gone into operation at Detroit, Mich., Appleton, Wis., and Scranton, Pa.

Acting upon the idea that the development of the electric propulsion of street cars might be greatly advanced by a device to allow of a ready and cheap method of attaching the motors to the existing cars, Mr. John M. Pendleton, of New York, has designed an ingenious plan of attachment for this purpose, which we illustrate in the accompanying engravings, Figs. 96 and 97. These show respectively a front and side elevation of a car equipped with the motor, according to Mr. Pendleton's plan. The general arrangement of wheels and axles, it will be observed, is the same as that of the ordinary horse car.

The electric motor is suspended from the floor of the car, and the revolving armature carries a coiled spring extension at each end, terminating in a worm or screw-pinion wheel, held by journals on each side.

The interposition of the spring presents several advantages, for it not only allows for the distortion of the car with varying loads, or from other causes, tending to throw the axis of the motor out of line, but in addition the springs relieve the axles of any sudden strain due to rapid starting or stopping of the motor. The retaining links beside the springs allow of torsion, but limit the extension and contraction of the shaft where heavy strains occur, such as on the ascent of heavy grades.

It will be noted that the two axles are differently geared, one having the worm pinion on the top, and the other at the bottom, of the worm wheels, respectively. By this arrangement the thrust on the motor is equalized and friction on the collars is avoided.

The worms are cut with a coarse pitch so as to allow free movement of the car; but the speed is reduced by the worm wheels attached to the axles, in the ratio of 12 to 1, enabling the motor to operate at the rate of 1,000 revolutions, corresponding to a speed of eight miles per hour for the car.

With the idea of adapting the system to existing rolling stock, the worm wheels are split and securely bolted to the axles and keyed in addition. The hub of the split worm-wheel carries a cover or box, which is made oil-tight and which surrounds the worms. These boxes are filled with oil, which insures a constant and copious lubrication, reducing the friction and wear to a minimum, and preventing the access of dust to the working parts.

(In a later chapter will be found illustrated descriptions of the Sprague, Henry, and other systems.)

CHAPTER VII.

THE USE OF STORAGE BATTERIES WITH ELECTRIC MOTORS FOR STREET RAILWAYS.

In the present chapter we take up a method which, although now looked upon with distrust by many, may yet prove to be one of the most feasible means for the propulsion of railway cars. We refer to the employment of accumulators, the stored energy of which, conveyed to a motor in the form of current, sets it in motion, and with it the car. While this mode of propulsion was until lately in the experimental state, the progress made has been such that a satisfactory solution of the problem appears to have been reached; indeed, the immediate future will see cars propelled by the energy derived from accumulators, with success, judged from the standpoint both of convenience and economy.

If we undertake to examine into the merits and demerits of such a system, it is discovered that the main argument brought forward against the use of accumulators for this purpose consists in a demonstration of the large loss of power which a number of transformations entail. That a number of reducing stages have to be gone through, is obvious, for we have: I. The mechanical energy developed by the engine. II. The conversion of mechanical into electrical energy in the dynamo. III. The conversion of electrical into chemical energy in the accumulator. IV. The reconversion of chemical into electrical energy. V. The final transformation of electrical into mechanical work by the motor. Here, it will be seen, four transformations take place, which must necessarily result in loss, but it is boldly asserted that by good apparatus and economical management these losses are reduced to a point below that experienced with other systems, and with the gain of many offsetting advantages.

There are two principal methods in competition with electricity to supplant the use of horses on tram lines, and they are steam and compressed air. In comparing electricity with steam, we find two ways in which the latter can be applied, viz., by steam locomotives direct, and by an endless cable driven by a steam engine. Using locomotives, there is required a separate engine and boiler for each car or train of cars, and a consumption of fuel between six and seven pounds of coal per horse power per hour, which latter figure may be considerably exceeded where-frequent stoppages occur; and to this must be added the other expenses incidental to engine-running. With cable transmission, there need be only one large engine using two and one-half pounds of coal per horse power per hour; but the cost of construction of a tunnel for the passage of the cable and of intricate machinery for grades and curves is a large item which must be taken into consideration. With compressed air, the use of separate locomotives is necessary, and while the engine may not use more coal than in the preceding case, the large loss of power due to the wasted heat of compression makes it a matter of doubt whether this system can be economically employed for the purpose, often as it has been attempted. Taking up our original system, all that would be required is a good central engine as in the preceding examples, and a dynamo, while each car would be supplied with a small electro-motor and storage batteries fitted into compartments in the car. Objections have also been raised with respect to the power lost in transporting the dead weight of the accumulators and motor, but even this objection appears to have been greatly lessened, so that, as compared with steam and compressed-air locomotives, the former shows up quite favorably.

These are the conditions, roughly sketched, that enter into the problem, the solution of which lies in the choice between a system requiring a large original outlay of capital and one in which the cost of power or running expense is the principal item.

Looked at from the standpoint of *convenience* and *applicability*, the propulsion of tram-cars through the medium of accumulators must be conceded to be second to no other. The batteries occupy no valuable space, being stowed under the seats, while the motor can be placed under the car body as shown in our illustra-

FIG. 98.—LOCOMOTIVE DRIVEN BY ACCUMULATORS, LISIEUX, FRANCE.

tions. Adding to this the absence of smoke, dust, escaping steam and its accompanying noise, it becomes manifest that the points in favor of such a system are of a most decided nature.

While it is hardly probable that the storage battery will supplant the locomotive for heavy and continuous railway traffic, it is evident that such a service is eminently applicable on street car lines within city limits; and from this standpoint we have viewed it here. We now pass on to examine what has been done towards putting the system in a practical shape.

One of the instances in which it has been applied successfully is presented by the arrangement in use at a bleaching establishment at Breuil-en-Auge, near Lisieux, France. Fig. 98 shows a locomotive, the accumulators being carried on a tender which is not shown. The installation is used for the purpose of gathering and folding the sheets of linen which are spread out upon a meadow to bleach. The peculiar nature of the case made the use of accumulators the only method which could be applied. A steam engine was out of the question, since the dust and smoke would injure the linen, while to lead the wires through the tracks laid on a damp meadow might have entailed a large loss of current.

London, Brussels, and Paris have all seen tram-cars run by storage batteries in operation, and Fig. 99 represents a car which for some time ran in Paris. The illustration will give a good idea of the manner of disposal of the accumulators and motor.

Early in 1883 a similar experiment in street car locomotion by storage was made at Kew Bridge, London, on the Acton tramway line. The car used at the Kew Bridge experiment and shown in the accompanying illustration, Fig. 100, was fitted with an accumulator battery consisting of fifty Faure-Sellon-Volckmar cells, each measuring 13 in. by 11 in. by 7 in., and weighing about eighty pounds. The accumulator battery was capable of working a tram-car with its full load for half a day, or in other words seven hours. When charged it contained about 560 ampère-hours, of which 400 were withdrawn with the greatest regard to economy. The accumulators were stored under the seats of the car, and the current was conveyed by insulated wire to a Siemens dynamo machine acting as a motor, and connected with the axle of the wheel. As soon as the communication between the boxes and the machine was effected, the electric current being led into the motor set the armature in revolution, and the power was conveyed to a pulley fastened on the same axle as the armature. The Siemens machine worked most favorably with an electromotive force of 100 volts and a current of sixty ampères, and as 746 watts constitute an electrical horse power, the result was a consumption of eight electrical horse power and a yield on the pulley of five and three-fifths mechanical horse power. The action of the motor could be reversed at will, and the power increased or diminished as required by adding to or taking from the number of cells composing the accumulator by means of a simple switch; while by breaking the circuit the motive power was stopped, and the brake being then applied the car was almost immediately brought to a stand-still.

At the trial trip several noted electricians were present, and the experiment was pronounced by them fairly satisfactory. The car could carry a load of forty-six persons, the total weight being about five tons. The speed attained was six miles per hour, and the car ran smoothly along level road and down hill. The cost of running the car in this manner was estimated at a sum equivalent to $1.50 per day for each car, against $6.25 for horses. The car was lighted by Swan incandescent lamps, and furnished with electric bells, all worked from the same accumulators.

Under the direction of their engineer, Mr. Reckenzaun, the Electrical Power and Storage Company, London, fitted up a car with bat-

tory and electric motor at the close of 1884, and made a tramway four hundred yards long for the car to run upon. The experimental trials with it were carried on for many months, and the results were extremely satisfactory. The whole series of accumulators in the car weighed only one and one-quarter tons, and the motor, gearing, and accessories weighed about half a ton, bringing the total weight of the motive power to one and three-quarter tons. The car,

FIG. 99.—CAR USED WITH ACCUMULATORS, PARIS.

which had been transformed out of an old one, for many years running on the Greenwich and Westminster line, weighed two and one-half tons—the modern cars on the American lines weigh only thirty-two cwt.—and its load of forty-six passengers brought the total up to five and one-half tons. Comparing the weight of this motive power with steam or compressed-air locomotives, which do not weigh less than from eight to ten tons, the comparison speaks well for electricity. The car, moreover, was put on two bogies, each with four wheels, whereby the wheel base was diminished, and the cars could turn corners and encounter curves of very short

13

FIG. 100.—THE KEW BRIDGE EXPERIMENTS.

radius. Another advantage of this arrangement was, that there was no such overhanging, and, consequently, no such rocking in travelling, as there is in the ordinary cars which have their four wheels placed at short distances from the centre.

on ordinary street car lines. The car at Millwall could be run for two hours with one charging of the accumulators, starting, stopping, and reversing every minute. The used accumulators were taken out and the car supplied with fresh charged cells in as short a time as is

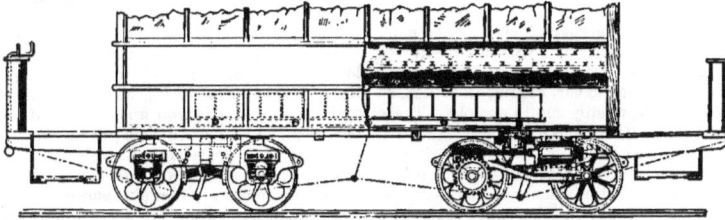

FIG. 101.—RECKENZAUN CAR—ELEVATION.

The next experimental line, at Millwall, was a difficult one. The line made a bend of nearly a right angle, and an actual curve of thirty-three feet radius had to be passed. The inclines varied from a level on the portion from the shed-end to the curve, and rose thence from one in forty to a gradient of one in seventeen at the opposite termination. This steep incline had, consequently, to be faced without a run, a rush being prevented by the sharp curve intervening. The

occupied by the changing of horses. This operation was accomplished with ease by means of a trolley fitted with rollers. The accumulators were placed under the seats completely out of sight ; the motor was placed under the car very neatly, and was only seen when looked for. The interior was furnished with four 20-candle power incandescent lights, and with pushes for electric bells for communication between the passengers and the conductor. The travelling

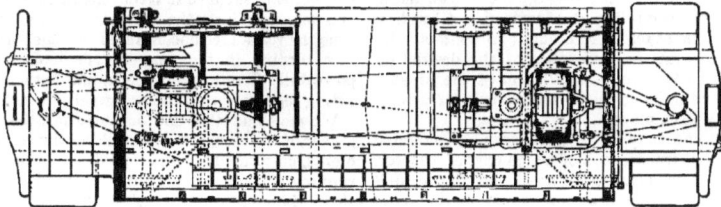

FIG. 102.—RECKENZAUN CAR—PLAN.

new car overcame all these difficulties and made its way with surprising speed and steadiness. It has been considered very adverse to the economical use of stored electricity that so many transformations of energy had to be encountered, but the practical experience of the Millwall experiments was asserted to be that the running expenses, including fifteen per cent. for depreciation of machinery, and fifty per cent. on accumulators, were about half the cost of horses

was perfectly free from vibration or tremor of any kind, and was absolutely faultless in that respect. Every detail, mechanical or electrical, had been well thought of and well worked out.
  Within the last few months, Mr. Reckenzaun has made again a highly successful demonstration—this time at Berlin—with his motor applied to street cars and deriving current from storage batteries. Our illustrations, Figs. 101 and 102, show the car, in part sectional elevation

and in plan, and give the general arrangements which have been worked out with great care and do credit to Mr. Reckenzaun's perseverance and skill. The various arrangements may be classed under the following headings, viz.: 1. The battery. 2. The motors. 3. Reversing and transmitting gear. 4. Speed regulation. 5. The brakes.

1. The battery consists of sixty cells, each weighing forty pounds and with a capacity of 150 ampère-hours. They are placed on a board under the seats of the car, resting on rollers, so that they can be readily run in and out. There are two rows of fifteen cells each under each seat. They are coupled in series, and hence give an electromotive force of from 110 to 120 volts. The storage batteries are changed every two or four hours, according to the length of the trip, and the change can be performed in about three minutes, not occupying more time than a change of horses.

2. The electric motors employed are of the Reckenzaun model. They weigh 420 pounds, and are capable of delivering from four to nine horse power. At 120 volts their efficiency is seventy-five per cent., and at the nominal speed of seven miles per hour they make 1,000 revolutions per minute. But this speed can be raised to ten miles per hour.

3. The reversing arrangement by which the car is run in either direction consists, as in many electric railways, of two pairs of brushes, either one of which is brought in contact with the commutator, according to the desired direction. For this manipulation of the brushes a lever similar to the reversing lever of a locomotive is employed.

The car body, as will be seen, is mounted upon two trucks, each of which carries a motor; and worm gearing is employed to transmit power from the armature shaft to the axles of the wheels. Objections have been raised against this form of gearing, on account of the high friction encountered, but Mr. Reckenzaun's experiments show that only fifteen per cent. is lost in transmission. He has also demonstrated, contrary to the general opinion, that the car runs freely on a down grade, its progress not being impeded by the worm and worm wheel. It was, of course, necessary to select a particular pitch of the screw worm to make this possible, and also to insure excellent lubrication.

4. Changes in speed are effected by different combinations between the whole battery and the two motors. In the electrical car tested at the Antwerp Exhibition, and of which an interesting account will be found a few pages later, the same thing was accomplished by a change of potential, effected by cutting out a corresponding number of batteries. This, of course, prevents the batteries from being discharged uniformly, and is not conducive to their long life. Mr. Reckenzaun's method of employing all the batteries during all speeds evidently overcomes this objection and allows of three combinations, viz.: All cells connected with one motor ; all the cells connected with the two motors joined in series, or all cells connected with both motors joined in parallel circuit. These three methods of coupling suffice to give the car a speed corresponding to the walk, the trot, and the sharp trot of a horse. The switch which accomplishes these changes is very simple, and the running of the circuits is shown in the plan, Fig. 102.

Two forms of brake can be brought into play on the car ; the ordinary mechanical and the electrical brakes. The latter are called into action automatically when the switch cuts off the battery current. The motors are then converted into dynamos which generate a current that is sent into the coils on the brake-shoes, magnetizing them so that they are attracted by, and press against, the wheels. At the same time the resistance encountered by the armature turning in the magnetic field also acts powerfully to retard the speed, and both these acting together bring the car rapidly to a halt.

We may add that Mr. J. Zacharias, the engineer of the company undertaking the experiments, calculated from accepted data that the running of such a tramway by electricity instead of horses would bring about a saving of fifty per cent. in the yearly expenses.

At the Antwerp Exhibition, of 1885, a series of most interesting tests were carried out on tramway motors, as mentioned above, and after four months of trial the first prize was awarded to the electric car driven by accumulators. In a paper read before the Society of Arts, early in the present year, Capt. Douglas Galton, the English juror upon the testing committee, gave a résumé of the experiments, which rank among the most interesting and important made on this class of motors. There were five different motors which entered upon the tests, and they may be divided into two classes as follows: Three were propelled by the direct action of steam, and two were propelled by stored-up force supplied from fixed engines.

Propelled by the Direct Action of Steam.

1. The Krauss locomotive engine separated from the carriage.
2. The Wilkinson locomotive, also separated from carriage.
3. The Rowan engine and carriage combined.

Propelled by Stored-up Force.

4. The Beaumont compressed air engine.
5. The electric carriage.

We give below the principal results in so far as they relate to the electric car, and also the tables of comparison between all the motors.

In the electric tram-car the haulage was effected by means of accumulators. The car was of the ordinary type, with two platforms. It was said to have been running as an ordinary tram-car since 1876. It had been altered in 1884 by raising the body about six inches, so as to lift it clear of the wheels, in order to allow the space under the seats to be available for receiving the accumulators, which consisted of Faure batteries of a modified construction. The accumulators employed were of an improved kind, devised by M. Julien, the under manager of the Compagnie l'Electrique, which undertook the work. The principal modification consists in the substitution, for the lead core of the plates, of one composed of a new unalterable metal. By this change the resistance is considerably diminished, the electromotive force rises to 2.40 volts, the return is greater, the output more constant, and the weight is considerably reduced. The plates being no longer subject to deformation, have the prospect of lasting indefinitely. The accumulators used were constructed in August, 1884.

An experiment was made on October 21, 1884, to ascertain, as a practical question, what was the work absorbed by the Gramme machine in charging the accumulators. The work transmitted from the steam engine was measured every quarter of an hour by a Siemens dynamometer ; at the same time the current and the electromotive force given out by the machine, as well as the number of the revolutions it was

making, were noted. It resulted that for a mean development of four mechanical horse power, the dynamometer gave into the accumulators to be stored up 2.28 electrical horse power, or 57 per cent. The intensity varied between 25.03 and 23.51 ampères during the whole time of charging. Of this amount stored up in the accumulators a further loss took place in working the motor ; so that from thirty to forty per cent. of the work originally given out by the steam engine must be taken as the utmost useful effect on the rail. It was estimated that to draw the carriage on the level .714 horse power was required, or if a second carriage was attached, .848 horse power would draw the two together. This would mean that, say, two horse power on the fixed engine would be employed to create the electricity for producing the energy required to draw the carriage on the level. The electric tram-car was quite equal in speed to those driven by steam or compressed air, and was characterized by the noiselessness and ease with which it was manipulated.

It should be mentioned that the car was lighted at night by two incandescent lamps, which absorbed 1.5 ampère each ; and the brakes also were worked by the accumulators. The weight of the tram-car was 5,654 pounds ; the weight of the accumulators was 2,400 pounds ; the weight of the machinery, including dynamo, 1,232 pounds. The car contained room for fourteen persons inside and twenty outside.

The original programme of the conditions which were laid down in the invitation to competitors, as those upon which the adjudication of merit would be awarded, contained twenty heads, to each of which a certain value was to be attached ; and, in addition to these special heads, there were also to be weighed the following general considerations, viz.: *a.* The defects or inconveniences established in the course of the trials. *b.* The necessity or otherwise of turning the motor, or the carriage with motor, at the termini. *c.* Whether one or two men would be required for the management of the engine.

As regards these preliminary special points, the compressed air motor, as well as the Rowan engine, required to be turned for the return journey, whereas the other motors could run in either direction. In regard to this, the electric car was peculiarly manageable, as it moved in either direction, and the handle by which it was managed was always in front, close to the brake. The carriage was the only one which was entirely free from the necessity of attending to the fire during the progress of the journey, for even the compressed air engine had its small furnace and boiler for heating the air. Each of the motors under trial was managed by one man.

The several conditions of the programme may be conveniently classified in three groups, under the letters A, B, C. Under the letter A have been classed accessory considerations, such as those of safety and police. These are of special importance in towns. But their relative importance varies somewhat with the habits of the people as well as with the requirements of the authorities ; for instance, in one locality or country conditions are not objected to, which in another locality are considered entirely prohibitory. The conditions under this head are:

1. Absence of steam. 2. Absence of smoke and cinders. 3. Absence, more or less complete, of noise. · 4. Elegance of aspect. 5. The facility with which the motor can be separated from the carriage itself. 6. Capacity of the brake for acting upon the greatest possible number of wheels of the vehicle or vehicles. 7. The degree to which the outside covering of the motor conceals the machinery from the public, while allowing it to be visible and accessible in all parts to the engineer. 8. Facility of communication between the engineer and the conductor of the train. In deciding upon the relative merits of the several motors, so far as the eight points included under this heading are concerned, it is clear that, except possibly as regards absence of noise, the electrical car surpassed all the others. The compressed air car followed, in its superiority in respect of the first three points, viz., absence of steam, absence of smoke, and absence of noise ; but the Rowan was considered superior in respect of the other points included in this class.

Under letter B have been classed the considerations of maintenance and construction. 9. Protection, more or less complete, of the machinery against the action of dust and mud. 10. Regularity and smoothness of motion. 11. Capacity for passing over curves of small radius. 12. The simplest and most rational construction. 13. Facility for inspecting and cleaning the interior of the boilers. 14. Dead weight of the train compared with the number of seats. 15. Effective power of traction when the carriages are completely full. 16. Rapidity with which the motor can be taken out of the shed and made ready for running. 17. The longest daily service without stops other than those compatible with the requirements of the service. 18. Cost of maintenance per kilometre. (It was assumed, for the purpose of this subheading, that the motor or carriage which gave the best results under the conditions relating to paragraphs 9, 10, 12, and 13, would be least costly for repairs.)

As regards the first of these, viz., protection of the machinery against dirt, the machinery of the electrical car had no protection. It was not found in the experiments at Antwerp that inconvenience resulted from this ; but it is a question whether in very dusty localities, and especially in a locality where there is metallic dust, the absence of protection might not entail serious difficulties, and even cause the destruction of parts of the machinery.

In respect of the smoothness of motion and facility of passing curves, the cars did not present very material differences, except that the cars in which the motor formed part of the car had the preference.

In the case of simplicity of construction, it is evident that the simplest and most rational construction is that of a car which depends on itself for its movement, which can move in either direction with equal facility, which can be applied to any existing tramway without expense for altering the road, and the use of which will not throw out of employment vehicles already used on the lines ; the electric car fulfilled this condition best, as also the condition numbered 13, as it possessed no boiler.

In respect to No. 14, viz., the ratio of the dead weight of the train to passengers, if we assume 154 pounds as the average weight per passenger, the following is the result in respect of the three cars in which the power formed part of the car:

Electric car, ... $\dfrac{9{,}350 \text{ lbs.}}{154 \times 34} = 1.78.$

Rowan, ... $\dfrac{15{,}950 \text{ lbs.}}{154 \times 45} = 2.30.$

Compressed air, ... $\dfrac{22{,}000 \text{ lbs.}}{154 \times 56} = 2.55.$

The detached engine gave, of course, less favorable results under this head.

Under head No. 15 the tractive power of all the motors was sufficient during the trials, but the line was practically level, therefore this question could only be resolved theoretically, so far as these trials were concerned, and the table before given affords all the necessary data for the theoretical calculation.

As regards the rapidity with which the motors could be brought into use from standing empty in the shed, the electric car could receive its accumulators more rapidly than could the boiler be brought into use for heating the exhaust of the compressed air car.

Under letter C are classed considerations of economy in the consumption of materials used for generating the power necessary for working.

TABLE I.   TABLE II.

Description of motor.	Total number of miles run	Total consumption of fuel.	No. of lbs. per train mile.	No. of places indicated on the cars per mile run	Consumption of fuel.	No. of lbs. of fuel consumed per places indicated per mile run.
		Lbs.			Lbs.	
Electric	2.358.9	14,786	6.16	80,203.5	14,786	.18
Rowan	2,616.9	14,498	5.42	148,399.6	14,498	.09
Wilkinson	2,473.3	22,000	8.82	119,085.1	22,000	.18
Krauss	2,157.8	22,726	9.10	108,983.9	22,726	.20
Compressed air	2,259.1	90,420	39.48	126,189.3	90,420	.69

TABLE III.   TABLE IV.

Description of motor.	No. of seats per mile.	Consumption of fuel.	No. of lbs. of fuel consumed per seat per mile run.	Total number of miles run.	Total consumption of oil, tallow, etc.	Consumption of oil, tallow, etc., per train mile run.
		Lbs.			Lbs.	
Electric	61,591.2	14,786	.23	2,358.9	99.0	.038
Rowan	135,928.8	14,498	.10	2,616.9	106.7	.038
Wilkinson	93,065.6	22,000	.23	2,457.8	188.5	.073
Krauss	86,039.9	22,726	.25	2,473.3	255.4	.101
Compressed air	132,732.7	90,420	.06	2,259.1	585.2	.255

As regards the figures in these tables, it is to be observed that the consumption of fuel for the electric car is, to a certain extent, an estimate; because the engine which furnished the electricity to the motor also supplied electricity

for electric lights, as well as for an experimental electric motor which was running on the lines of tramway, but was not brought into competition. Capt. Galton summarized his views as follows:

"The general conclusion to which these experiments lead is that, undoubtedly, if it could certainly be relied upon, the electric car would be the preferable form of tramway motor in towns, because it is simply a self-contained ordinary tram-car, and in a town the service requires a number of separate cars, occupying as small a space each as is compatible with accommodating the passengers, and which follow each other at rapid intervals. But the practicability and the economy of a system of electric tram-cars has yet to be proved; for the experiments at Antwerp, while they show the perfection of the electric car as a means of conveyance, have not yet finally determined all the questions which arise in the consideration of the subject. For instance, with regard to economy, the engine employed to generate the electricity was not in thoroughly good order, and from its being used to do other work than charging the accumulators of the tram-car, the consumption of fuel had to be to some extent estimated. In the next place, the durability of the accumulators is still to be ascertained; upon this much of the economy would depend. And in addition to this question, there is also that of the durability of parts of the machinery if exposed to dust and mud."

As electricity was thus awarded the first prize, we cannot, therefore, complain, but we might show that the conditions were very unfavorable for good results, and this is evident when glancing at the consumption of fuel. We find the electric railway requiring nearly twice as much coal as one of the steam locomotives, and about an equal quantity with another steam motor. The report itself explains how this is to be accounted for—other electrical service being rendered—and we may add that the test, if more minutely carried out, would have shown a far greater efficiency for the electric system than appears from the report. This assertion is based upon several facts. In the first place, the engine which was employed in driving the dynamo which charged the accumulators, was a portable one, the economy of which was not determined. A good stationary engine would in all probability have yielded better results.

Starting with an indifferent engine, the charging dynamo is found to have an efficiency of *only fifty-seven per cent.*, a figure far below that attained every day in practical work by all leading types of American dynamo electric machines. What the efficiency of the motor upon the car was, is not stated, but it is evident that if its efficiency was low, another factor against the system is introduced. If, therefore, the consumption of fuel was high, it cannot be wondered at, as the causes are apparent. It would not be asserting too much to say that with a good engine, dynamo, and motor the coal consumption could have been reduced one-half, so that even in this respect the electric propulsion would be equal, if not superior, to steam direct. What makes this trial all the more interesting is the fact that accumulators have been found to give satisfactory service in a trying position, for none can deny that the constant handling and necessary rough usage are far from conducive to the good standing and long life of the storage battery. It appears from the report that the accumulators employed were constructed in August, 1884. If they were in use all that time and yet gave the service they did during the test, it is again evident that the storage battery has entered upon its commercial sphere of usefulness. Instances of this kind, substantiating the durability of the battery, are cropping up almost daily, and the fact is dawning upon the world that the storage battery is not a name but a reality.

In this connection the work of Mr. A. H. Bauer, of the Electric Storage Company, of Baltimore, deserves notice. During 1885 a very successful experiment was made by him on one of the Daft motor cars, for the purpose of demonstrating the practicability of secondary batteries for street car propulsion. Since then Mr. Bauer has devised a novel system that can be applied to existing cars at a very small expense. An experimental car equipped with the system has been running for some time on an eighth-of-a-mile track at the Viaduct Manufacturing Company's works.

Unlike other experimenters, instead of using light-weight cars for his test, Mr. Bauer has attacked the problem from the opposite side; that is to say, he has begun with larger weights than would appear in practice.

The car used is an old one loaned by the Union Railway Company, is twenty feet in length and weighs 5,400 pounds. The equipment consists of two beams extending from one axle to the other. These beams carry a motor, the armature shaft being extended and having pinions on each end which mesh into counter gears. The countershaft carries a pinion, which in turn meshes into a gear on the car axle. The motor is wound in three sections in multiple arc, and is connected with a double switch located on the platform for throwing one, two, or three of the sections in circuit with the battery, as desired, depending, of course, upon the amount of power the motor is required to develop.

To accommodate the batteries, which are placed under the seats and are entirely out of sight, the body of the car is raised three inches, so as to bring the wheels below the floor. The batteries are set on trays upon rollers, and when necessary to make changes they can be run out on to platforms through doors in the sides of the car, and freshly charged ones run in. This can be done within the time required to change horses. Access to the motor and gearing is had through a trap-door in the floor, or they can be got at from underneath the car.

The weights of the different parts of the experimental car are as follows :

	Lbs.
Car,	5,400
60 cells battery,	5,400
Motor,	923
Gearing,	900
Total,	12,623

or about 6¼ tons.

In practice the above total weight will be reduced to about 7,500 pounds.

The whole car, internally and externally, has nothing whatever strange in its appearance ; it looks indeed similar to an ordinary street car propelled without horses or other visible motive power.

The track is one-eighth of a mile in length, beginning at the car-house at the foot of a one in twenty grade, 200 feet in length, on a curve of forty-five feet radius. As already stated, this car has been running almost daily for about two months. With the exception of a bolt or collar, working loose, not a single fault has developed, the car running smoothly and satisfactorily during every trip.

When running at a speed of six miles per hour, the armature makes 800 revolutions per minute. With 5,400 pounds of battery the car will continue to run without cessation for six hours at a speed of six miles per hour; or a total of thirty-six miles. Tests have also been made with half horse-power cells having a total

FIG. 103.—GEARING OF ELIESON CAR.

weight of 2,580 pounds, running continuously for three hours, or eighteen miles, before requiring to be changed.

Calculations based on the efficiency of the above-mentioned and other trials, show that the cost of running a line of street cars equipped with the Bauer system should not exceed $2.21 per car per day. The average cost of horsing is understood to be $4 per car per day. Given a line running say twenty-four cars it will be seen that by substituting this system or any analogous, a saving of $1.79 per day per car should be effected. This for twenty-four cars amounts to $15,680 per annum, a sum that at six per cent. represents the interest on $261,340.

In Figs. 103, 104, and 105 is illustrated the method of Mr. Elieson, put in operation by the Electric Locomotive and Power Company, London. The novelty in the apparatus consists in the arrangement of the gearing by which the motor can be driven at a very high velocity, and thus work under favorable conditions for economy.

The mechanical connection of the motor is very ingenious. Mr. Elieson has applied a lever between the electro-motor and the axle of

14

the locomotive in such a way that the motor, which must necessarily run at a high rate of speed in order to develop the greatest efficiency, acts through the lever by a method analogous to the case of a man using a crowbar for the purpose of lifting a heavy weight. By this contrivance the *vis inertia* of the loaded tram-car is easily overcome, and it is evident, even to non-scientific readers, that speed is then easily attained until the natural speed of the electromotor is approached by the rate of speed of the driving wheels.

Instead of the electro-motor being a fixture, and having motion transmitted from it through belt or crank gearing to the wheels of the car, the motor itself revolves, the motion being transmitted through bevel gearing. It has, as will be seen, Figs. 100 and 101, a vertical shaft through its centre, to which a motion lever projecting horizontally about two feet, and carrying at its outer end a spur-wheel gearing into a fixed circular rack, is secured. This vertical shaft carries at its lower end a bevel wheel, which gears into one or other of two similar wheels on the driving axle of the engine. The mitre gearing is equipped with a mechanical

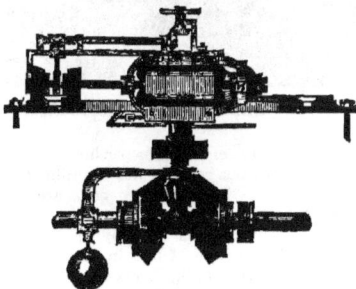

FIG. 104.—GEARING OF ELIESON CAR.

clutch, by means of which the locomotive may be made to run either backward or forward, a lever, acting mechanically, throwing the motor in or out of gear, or adjusting the clutch by a simple movement of the hand. This suggests itself as being a very good arrangement indeed. The electrical details are very simple. The motor is fitted with collecting brushes travelling on two fixed circular rings of copper, separated from each other by a flange. The speed of the motor is varied by inserting resistances in the ordinary way, and it is evident that great care has been exercised to avoid anything like complexity. The locomotive, Fig. 102, in appearance resembles a small car, and weighs four tons seventeen cwt. The motor, which is of four horse power, consumes about forty ampères per hour, so that it carries power sufficient for six or seven hours of motion, and makes about 600 revolutions per minute when in full swing, or a maximum of 1,000 on a level road. The speed obtained is eight miles per hour, it being for certain obvious reasons not desirable to exceed this rate. Fifty storage cells are used, giving 280 ampère hours.

Mr. Reckenzaun has recently applied electricity to the haulage of coal in the Trafalgar collieries of Drybrook, Gloucestershire, England, and some recent tests made with the motor are of considerable interest, as they show a remarkable uniformity of action under various loads, together with a high efficiency. A view of the locomotive now in use is given in Fig. 106. The construction of the motor and driving gear is similar to that adopted by the inventor in his electric street cars, but the conditions to be satisfied were widely different from and more difficult than those obtaining in an ordinary tramway. The space is very limited, and since both sharp curves and heavy gradi-

ents occur at frequent intervals, it was somewhat difficult to stow away the necessary power in so limited a space. Within the narrow gauge of 2 feet 7 inches, and an extremely short wheel base, there had to be arranged an electric motor of eight horse power, with suitable gearing, brakes, and attendant details. There is a foot-board which runs all round the locomotive, and there is a brake lever at each end. The box forming the body of the car serves to receive the accumulators, and there is a compound switch at each end by which the motor can be started, stopped, and reversed by the attendant who stands at one end or the other of the foot-board, according to the direction in which the locomotive is travelling. The switches are inclosed in a box to protect them from accidental injury.

At the test, electrical energy was supplied to the motor from a number of "E. P. S." storage cells, and the mechanical work was ascertained by means of a balanced Prony brake. The following table gives the results of the test:

FIG. 105.—ELIESON CAR WITH ACCUMULATORS.

| No. of test. | Revolutions per minute. | Electrical measurements. | | Prony brake. | | Electrical energy. | Mechanical work. | Percentage of return. |
		Volts.	Current in ampères.	Length of lever in feet.	Weight on scale in lbs.	Horse power A×C / 746	Horse power measured.	
1	1,020	105	35.5	2.625	7	4.08	3.568	71.65
2	1,013	107	40.0	"	8	5.73	4.027	70.50
3	982	104.75	39.5	"	8	5.55	3.930	70.8
4	800	95.5	43.25	"	9	5.53	3.870	70.0
5	970	106.5	43.0	"	9	6.11	4.385	71.4
6	900	101.78	46.75	"	10	6.38	4.500	70.5
7	1,022	113.0	47.0	"	10	7.15	4.110	71.4
8	1,048	114.88	47.75	"	10	7.35	5.240	71.3
9	1,047	121.88	53.0	"	12	8.66	6.282	72.5
10	1,070	122.64	54.0	"	12	8.87	6.420	72.3
11	950	118.87	62.0	"	14	9.46	6.650	70.3
12	1,010	127.7	72.0	"	17	12.32	8.85	71.8

In the motor tested there are only two brushes used (one pair), which were not shifted or adjusted during the tests. The motor is so designed that the brushes remain fixed in position, and the direction of rotation of the armature is controlled by merely reversing the direction of the current. Electricity has for some time been successfully applied in these mines for pumping water and for ventilating, and it is now intended to supplant the horses used in the haulage of the coal.

There has now entered upon active duty at Hamburg, Germany, a tram-car which obtains its power from accumulators carried by it. Herr Huber, the engineer in charge, was one of the members of the board which awarded the tram-car run by the Julien accumulators at the Antwerp Exhibition, the first prize, in competition with several other forms of locomotors. He is evidently willing to practise what he believes.

The Hamburg car, which will soon be supplemented by others, weighs, fully equipped, and can easily be drawn out by opening two long traps in the side of the car. In the car house the vehicle is drawn between two tables, on which the charging takes place, and the accumulators are slid from the car on to the tables. The shunting of the boxes, both in the car and on the charging table, takes place automatically by a contact apparatus, both simple and sure, constructed by Herr Huber.

Four double conductors lead from the accumulators in the car, which are shunted in four groups, to the Julien commutators, of

FIG. 106.—RECKENZAUN MIXING LOCOMOTIVE.

4,830 kilogrammes. Of this weight, 1,200 kilogrammes is that of the accumulators. The accumulator consists of ninety-six cells, of which every three are united in a single three-cell holder. The cells are formed out of a new material, something like hard gutta percha, but rather more flexible. Each cell contains fifteen plates, seven positive and eight negative. The plates have a surface of only 134 by 147 millimetres, and are about four millimetres thick. The charging requires about eight hours.

The accumulators are distributed in eight low wooden boxes, of which four are stowed away on each side of the car in the space under the seats. The boxes move along greased slides, which one is placed on each platform. By turning a handle which forms part of the key of the commutator, six different positions can be given to it, viz.:

(1) So that there is no connection between the accumulators and the motor. (2) The four groups of accumulators are connected in parallel arc and placed in connection with the motor. (3) The groups are connected two and two in parallel arc, and the two pairs in series and in connection with motor. (4) Two in parallel arc behind the other two in series. (5) All four groups in series. (6) All four groups in parallel arc, but unconnected with the motor. The commutator stands in this position with the key up during the periods of stoppage.

A great advantage of this arrangement is that different velocities are obtained without the application of any current regulator or resistances.

The positions (2), (3), (4), (5) of the key correspond to the electromotive forces 48, 96, 144, mits its power by means of hemp cords to a loose axle between the two wheel axles, and from hence the power is given out by means of chains to the wheel axles. The hempen ropes are protected as much as possible against the influences of weather by a special preparation.

FIG. 107.—CHARGING STATION, HAMBURG ELECTRIC STREET RAILWAY.

and 192 volts on the motor, and its velocity again corresponds to these electromotive forces. The normal current amounts to about eighteen ampères, while on inclines and curves the current may sometimes reach eighty ampères. The motor is a Siemens series machine, model D, with about 0.6 ohm resistance. It is hung under the car and trans-

The reversal of the direction of rotation of the machine is brought about by changing the position of the brushes; there are two pairs of brushes about ninety degrees apart, of which only one pair rubs at one time.

One charging is sufficient to drive the car fifty kilometres. Since such a car has to traverse about 100 kilometres daily, one change of the

FIG. 108.—CAR, HAMBURG ELECTRIC STREET RAILWAY.

accumulators is enough. Fig. 107 shows the charging station, Fig. 108 the car.

The installation for charging the batteries occupies but a very small space. A small vertical steam engine is mounted on the wall and belts to a countershaft which drives a dynamo of the Schwerd pattern. In addition, there are a Buss speed indicator and the necessary voltmeters and ammeters.

The use of electric motors with storage batteries, for marine and aerial navigation, is treated in another chapter.

CHAPTER VIII.

THE INDUSTRIAL APPLICATION OF ELECTRIC MOTORS IN EUROPE.

THE Paris Electrical Exposition of 1881 was marked by a revival of interest in electric motors, and many of the new types produced were of great merit, though the rapid advances in this field may have relegated some to obscurity. One of the best known is the machine of M. Paul Jablochkoff, which he calls the "ecliptic," Fig. 109. The construction of the "ecliptic" is

FIG. 109.—JABLOCHKOFF MOTOR.

said to realize an improvement by reducing the mass of magnetic metal which is subjected to changes of polarity, whereby the magnetic inertia is reduced to the lowest possible limit.

This motor is composed essentially of two coils, one of which is stationary and disposed in a vertical plane, while the other is movable and fastened to a horizontal axis in an inclined position. It is from this inclination, resembling that of the ecliptic to the equator, that the name given to the machine by the inventor is derived. The stationary coil, while fixed in

the vertical plane, is not in a plane perpendicular to that of the axis of rotation, but forms with that plane a certain angle, determined by experiment, and which depends on the working conditions of the apparatus.

The stationary coil is wrapped around a copper framework ; the movable one is fixed upon an iron core, which, when a current is passed through the coil, becomes an electro-magnet, the poles of which are formed by two circular discs. A commutator is placed on the revolving shaft, against which four brushes bear. This commutator is so arranged that during the revolution of the shaft the movable coil is

FIG. 110.

traversed by a current always in one direction, and which maintains a constant polarity in the discs of the electro-magnet, but at each half-revolution the current is reversed in the stationary coil, which has no soft iron core.

The motor, then, is operated by the reciprocal attractions and repulsions between a movable, constant electro-magnet, and a fixed solenoid, traversed by currents alternately in opposite directions. These reciprocal actions tend to produce a rotation of the movable electro-magnet placed in the interior of the fixed solenoid. The object of the commutator is to make these actions co-operate in the same direction, thereby producing a continuous movement.

M. Jablochkoff's motor is reversible in the true sense ; that is, it can not only convert electricity into mechanical work. but can also convert mechanical work into electricity.

The work of M. Marcel Deprez with electric motors, large and small, has always been full of interest and instruction. Deprez has aimed directly at higher efficiency while avoiding complicated construction. He observed the fact that in ordinary motors of the old Siemens type about one-third of the energy expended is consumed in energizing the magnetic field and in maintaining its power. He therefore conceived the idea of using permanent magnets for the magnetic field. He made comparative tests of two machines, in one of which the field magnets were permanent, and by measuring the power obtained and the amount of zinc consumed in the battery he discovered that the efficiency of the motor having permanent magnets was about sixty per cent. higher than that of the other. M. Deprez found that on using this motor as a dynamo-electric machine that its electrical equivalent for the same power, as used on the other machine, was very much higher. He constructed a small generator on this plan, having an internal resistance of .5 ohm and producing an electromotive force of nearly twenty volts, with a total weight of only twenty-five kilogrammes (fifty pounds). Such a generator would give a carbon pencil four millimetres in diameter and fifty millimetres (two inches) long, a bright cherry red glow. M. Deprez found that the use of such a magnetic field was attended with difficulties, how-

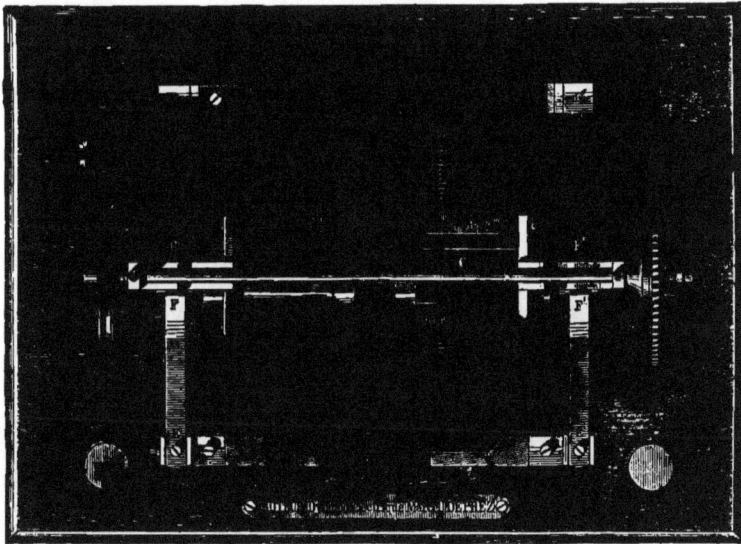

FIG. 111.—DEPREZ'S SMALL MOTOR.

coived the idea of using permanent magnets for the magnetic field. He made comparative tests of two machines, in one of which the field magnets were permanent, and by measuring the power obtained and the amount of zinc consumed in the battery he discovered that the efficiency of the motor having permanent magnets was about sixty per cent. higher than that of the other. M. Deprez found that on using this motor as a dynamo-electric machine that its electrical equivalent for the same power, as used on the other machine, was very much higher. He constructed a small generator on ever. Just as the poles of an armature A B (Fig. 110) move into line with the poles of the magnetic field N S, the current is reversed in the armature, and instead of attraction repulsion results. Now, if the current in the armature is too powerful, the magnetism of the armature will be sufficient to neutralize the polarity of the permanent magnets, even when the best magnets are used. The result was that, although the efficiency of the motor was very great at first, the power of the motor soon dwindled down by the weakening of the magnet. Another disadvantage of permanent magnets was that for the same power they must be much larger than electro-magnets. M. Deprez, therefore, concluded to sacrifice economy in favor of convenience to some extent, and returned to electro-magnets, striving to use them to the best advantage.

Figs. 111 and 112 are *full size* illustrations of a small motor constructed by M. Deprez, which will run a small sewing machine with two Bun-

brushes E F E' F^v are supported by small wooden or vulcanite arms attached to the backs of the electro-magnets, which are secured to the base board, as shown, and form the supports for the whole apparatus. These brushes are made long and flexible, so as to provide a light yet smooth and perfect contact with the commutator segments. The speed of the motor is reduced by means of a pinion gearing with a

FIG. 112.—DEPREZ'S SMALL MOTOR.

sen cells. Two armatures C C' are mounted on the same shaft, each of which revolves in the magnetic field between the two opposite poles A B' and B A' of two U electro-magnets placed opposite each other. These armatures are sixteen millimetres in diameter and twenty millimetres long, and are fastened to the shaft with their poles at right angles to each other, so that the dead centre of one corresponds to the active period of the other. The poles of the electro-magnets are joined at each side by a brass framework, which extends outward, and has bearings on which the shaft turns, as will be readily understood from the figures. The

toothed wheel which revolves at one-tenth the speed of the pinion.

By changing the connections of the electro-magnets and armatures from series to multiple, three different variations of power may be obtained. The change may be made without trouble. The motor is very compact and light. It has no dead centre and the magnets are disposed so as to secure a powerful magnetic field, which must, of necessity, be exactly equal at both armatures.

The electric motor of M. Estève (Fig. 113) brought before the public in France contemporaneously with that of Deprez, possesses original features of great interest. In this motor the ordinary H Siemens armature is materially modified in form, and the magnetic field in which it rotates is also different. In all other small electric motors of this old Siemens type the original H form of armature has always been adhered to with magnetic field pieces of varied form. M. Deprez preferred in his earlier

Fig. 113.—Estève Motor.

forms to give to the armature poles A B a considerable expansion, and to restrict the field pieces in size, so that they do not surround the armature so completely. This is the disposition adopted in his motor just described, the relative proportion of armature and field-pole expansions being indicated by the section shown in Fig. 114.

In his motor M. Estève evidently follows a different theory, as the magnetic field is expanded so as to surround the armature as completely as in ordinary dynamo-electric machines, while the polar expansions of the armature are entirely suppressed, and it assumes the sectional appearance of the letter I, as shown in Fig. 114, which is a sectional plan of this armature and its magnetic field. The armature is, in fact, made of a flat plate of iron revolving on its longer axis, as if the polar expansions of the H armature had been filed down to a level with the central part. The difference will be readily understood on comparing Figs. 110 and 114.

Better results are obtained when the armature core is made up of layers of thin sheet iron separated from each other by means of paper, but M. Estève prefers to make the core with a hollow centre, using insulated plates, which

15

must be sufficiently thick, otherwise the magnetic reaction of the armature on the magnetic field is lessened materially, just as when no iron cores are used, and the efficiency of the motor is considerably reduced. M. Estève has also tried the use of armature cores constructed of insulated iron wire, but the efficiency obtained was not greater than with the hollow armature just mentioned. He finds that the coercive force of this hollow armature is extremely small, consequently that the polarity may be reversed with extreme rapidity.

The construction of this motor is quite simple and will be readily understood from Fig. 113. The cores of the field magnet are made of cast iron forming one piece with the base, though the winding is more convenient when they are cast separate. The wire is wound on the armature in two equal sections C C' D D' (Fig. 114), leaving a small space E E' between them. These sections may be connected either in series or in multiple circuit, according to circumstances. The ends go to the two commutator segments C, which M. Estève prefers to make of iron. This is a new departure, because the use of copper in commutators has always been regarded as well-nigh indispensable. M. Estève says, in favor of iron commutators, that they wear out much less and do not spark as much as copper ones, while their resistance of contact is not sensibly different. At any rate, he does not find any loss in efficiency in consequence of using them. The brushes F F are both held by an oscillating lever, which can be

Fig. 114.

secured at any convenient angle by the set screw V, the arrangement being particularly useful in permitting the adjustment of the brushes to the point of least sparking.

The field magnet and the armature are connected in series, and in this condition there is little or no sparking at the brushes, even when using powerful currents. M. Estève experimented with one of his motors placed in a branch circuit from a Gramme machine, and found that its operation was entirely satisfactory, the motor remaining cool and free from sparks at the commutator. Experiments have also been tried with this motor when its field circuit was derived from the armature circuit, as in "shunt" dynamos, but the results were not as satisfactory.

It is plain that when constructed in this form, the motor is open to the objection of having a dead centre, but M. Estève suggests that by making the magnetic field somewhat wider, two or three armatures may be placed on the same shaft, just as in the Deprez motors.

The design of this little motor both electrically and mechanically is very good, considering the limitations which convenience, cheapness, and durability impose. The form of armature appears to be an improvement in the right direction. It has been shown by M. Trouvé that with the ordinary Siemens H armature the magnetic attraction which causes the rotation is not effective for more than a small portion of the revolution. A little reflection will make this clear. In a motor of this type we have practically two magnets, one of which tends to move constantly so as to place itself axially in the magnetic field of the other with like poles near each other. If the poles of the moving magnet are broad and expanded it does not require to move so much before a certain part of the pole arrives at the axial position. The magnetic attraction becomes concentrated at these points, and there is little or no tendency to the further motion of the armature so as to bring the rest of its mass into the axial magnetic position. The result is that the rotative impulse ceases at a certain distance before the armature reaches the axial position shown in the figure, and the armature must depend on its momentum to carry it as far as the point where the current is reversed and where repulsion will begin. M. Trouvé, in his motors, sought to remedy this difficulty in two ways: First by making the face of the polar expansion of the armature curve on a shorter radius, and second by making the field more open at certain points, or elliptical in shape. The object in either case was to provide for a more gradual approach between the pole of the armature and the iron of the magnetic field poles, so that the motion would not cease until the whole mass of the armature was in the magnetic axis. By these means the efficiency of the motor was greatly increased and the dead centre, which before that comprised a certain period of the rotation, was now reduced to a mere point.

In the motor of M. Estève the armature necessarily attains the object more readily and surely. It is more certain to reach the position shown in the figure, because no portion of it reaches the axial position sooner than the rest. However, the magnetic attraction must undoubtedly become partially satisfied as soon as the poles A B (Fig. 110) approach the magnetic field pole, and this must tend to weaken the rotative impulse. By making the field elliptical the approach would be still more gradual, and the result would be a more equable rotative impulse. The conchoidal field of M. Trouvé may also be recommended. An idea of this form of field will be obtained by supposing the field piece N (Fig. 114) to be depressed so that the upper edge of the pole is nearest to the armature and the lower edge most distant from the armature, while the pole S is elevated so that its lower edge is nearest and its upper the furthest from the armature.

The field magnet, of which the iron base forms a part, is comparatively massive, and its point of magnetic saturation is not so soon reached as when there is less iron. This is an important quality, especially when the motor is worked to its highest capacity, as well as when the motor is used as a generator of current. When once the point of saturation is reached, then it avails nothing to increase the strength (magnetizing power) of the current, because the iron cores are "full," and will not receive any more magnetism. Any current beyond the amount necessary to produce saturation is wasted. Another argument in favor of a good mass of iron is that the nearer the point of saturation a magnetic metal is, the more current is required to cause a proportional increase in magnetism. When there is plenty of iron, the "margin" between the non-magnetized and the saturation point is wider.

One of the most indefatigable and successful inventors of electric motors has been Mr. Anthony Reckenzaun, C. E., of London. The accompanying illustration (Fig. 115) is a perspective view of his motor made in 1884, and exhibited at the International Electrical Exhibition at Philadelphia that year, by Mr. Fred. Reckenzaun, brother of the inventor. The magnets are, in appearance, somewhat similar to those employed in the Siemens dynamo, except that, as will be seen from the cut, the cores are in an inclined position, the upper and lower core-ends meeting at a rather acute angle. This arrangement saves space, reduces the weight, and renders the frame rigid. The armature consists of a ring, made up of a series of rings, each of which is again composed of a number of links provided with holes at their ends to receive the bolts which hold the links as well as the rings together. The links, overlap-

shaft inside the armature. These inside collars are in metallic connection with a pair of similar collars at the commutator, where another pair of brushes rests on them, picking up a small current for the internal magnet. This internal circuit forms a shunt to the main circuit. The internal magnet, on being excited, offers two poles, each facing a like-named external field-magnet pole. Hence the passing armature bobbins are exposed to strongly magnetized pole pieces inside as well as outside,

FIG. 115.—RECKENZAUN MOTOR.

ping one another, are insulated from each other in order to avoid Foucault currents. From twelve to thirty-six bobbins surround the ring thus formed and connect with a commutator made up of a corresponding number of sections. A pair of brush-holders carry two brushes, movable within a certain range to adjust the speed of the motor. Inside the armature is a magnet, resting loosely on the shaft by means of rollers. This internal magnet is, in cross-section, H-shaped, having two pole pieces, between which a quantity of fine wire is wound lengthwise, the ends of which are connected to copper brushes which, in running, rub against two brass collars fitted upon the

thereby utilizing also the inner parts of the wire bobbins. The internal magnet is made for larger sized motors, and may be taken out and the motor run without it. On top of the machine are two binding posts mounted on a block of wood, to which the mains are connected. All the iron in this motor is best soft wrought iron, no cast iron being employed. All parts are carefully proportioned for light weight, high efficiency, and strength. In case the armature should require repairing, the bobbins need not be unwound as in some other machines, but any one may be slipped off its section after taking out the nearest bolt, thus saving time, labor, and material.

FIG. 116.—GRAMME MOTOR WITH TAVERDON DRILL.

The motor exhibited in Philadelphia was of one and one-half actual horse power. It will strike many of our readers as noteworthy that this motor weighed no more than 106 pounds, which gives a co-efficient of 467, or, in other words, 467 foot-pounds of work per minute to every pound of its own weight. Its bulk was likewise exceedingly small. The motor measured in height nine and one-half inches, width sixteen and one-half inches, and length of shaft twenty and one-half inches—other sizes in proportion. These facts have amply justified its application in England, not only for stationary purposes, but also for various kinds of service where light weight and small bulk, combined with high efficiency, are of great importance, such as in connection with electric launches, telpher lines, mining work, and other purposes.

One of the most important departments of mining operations is drilling and tunnelling, and naturally, as the beds and strata of rocks and minerals near the surface of the earth become exhausted, shafts and galleries are carried deeper and deeper, with a corresponding increase in the number and extent of the difficulties that tend to hinder the successful exploitation of rich deposits. But, thanks to the progress in electric lighting and in the transmission of power, work can be carried on at a greater depth below the surface than ever before, with a decrease of danger and expense. Electricity is now applied to the most varied work in mines, and is found equally available in illuminating the galleries about which float gases of noxious character, in piercing rocks, in hoisting and pumping, and in ventilating; and its use enables the apparatus to be made in small and compact form, the generator of current or prime source of power being at a distance.

The accompanying illustration, Fig. 116, shows the Taverdon drill, with Gramme motor, used in boring subterranean galleries. This invention is a striking example of the transmission of power and the application of electricity to a new and difficult kind of work. Numerous systems of rock drills have been in favor from time to time, driven by steam or compressed air, but it is asserted on behalf of this new plan or device that it is far less cumbersome and far more easy to control than any of its predecessors. M. Taver-

don applies electricity to a rotary drill, which is worked by a motor in the manner indicated. In his system, the drills carry at their striking end black diamonds, capable of penetrating the hardest rocks. In order to fix the diamond solidly, so as to keep all its facets properly at work, M. Taverdon employs a hard metallic solder that fills all the cavities. As he could not apply the solder directly to the stone, he first covers the latter by electrolysis with a thin coating of copper. This allows the application of the solder, but does not interfere at all with the parts of the diamond presented to the rock. A special carriage is provided both for the perforator and for the motor. The perforator is fixed upon an upright column adjusted by a spiral spring to the roof and floor of the gallery in such a way as to keep the platform of the car on which it is mounted perfectly steady. It is capable of movement on vertical and horizontal axes, and thus can be set in any desired position. The butt end can be fitted with an ingeniously constructed motor, with a view to the use of steam, compressed air, or hydraulic pressure, indifferently, for driving, but in the electrical arrangement a box replaces the motor and a simple pulley receives and turns with the belt or cable connecting with the electric motor on the rear car. The motor consists of a Gramme octagonal machine, similar to that used in the famous plowing experiments at Sermaize, and its strong cast-iron frame evidently fits it for rough mining work. The pulley at the end of the axis of the motor transmits the power to the box at the end of the drill by the cable, which passes over two other pulleys as shown, one of the latter being adjustable. The generating machine is, of course, outside the gallery at any convenient distance.

On the car carrying the motor is a water tank, from which water is forced to the perforator and is then used to wash away the sand as quickly as it is formed and accumulates. We ought to say here that in another drilling machine of M. Taverdon, the drill or boring tool is fitted direct to the axis of the motor, which is driven in the usual way and is carried on a car. This plan is simpler than the other, and apparently more economical of power, but M. Taverdon speaks highly of the apparatus shown in the illustration and reports that from it he has obtained results equal to those of the best steam drills and better than those with compressed air. The advantages of the use of electricity for this work are easy to see. The little gallery is not cramped and choked up with steam, air, or water pipes, which, besides occupying valuable space, are liable to leakage and sometimes stop the work while their de-

FIG. 117.—LEE-CHASTER MOTOR.

fects are being remedied. A noteworthy feature of the scene is the use of the incandescent lamp; and if necessary the blasting charge in the drill holes can be fired by an electric current.

A fair idea of the extent to which motors of moderate size have already been introduced in France, may be formed from the summary of installations made up to April 1, 1886, by the Compagnie Electrique. This single company had effected forty-two installations, using one hundred and ninety Gramme machines, with a total of 310 horse power. Of these machines, thirty-three are employed for cranes and elevators, forty in driving machinery and tools, fifty-two with ventilators, eleven in pumping, etc., and fifty-four for miscellaneous purposes.

The small motor illustrated in the engraving, Fig. 117, and known in England as the Lee-Chaster, was lately introduced to public notice. It occupies a space of 8 in. by 8 in., and is said to be capable of developing energy equal to nearly three-fourths horse power, and can be started, stopped, or reversed by the simple movement of a switch. It is driven by a battery, which is said to involve the minimum of trouble, which will run for about twenty hours

without recharging, and can then be restarted with fresh solution in a very short time. The cells are charged with Lee's new double bichromate, soluble in its own weight of cold water, and which does not deposit crystals in the pores of the carbons or in the cells. Moulded corrugated carbons are used, and the zincs are so cut as to economize the metal. The cells rest on a tray in a box, and the tray can be raised by means of a treadle until the elements are fully immersed in the solution. Both zincs and carbons are suspended from a board, which forms a lid, so to speak, and the connections are made by means of brass plates, thus avoiding the use of wires. Two wing nuts hold this board to the mechanism of levers, and by removing them, the whole battery of elements can be lifted out, leaving the cells exposed. The zincs and carbons are automatically removed from the solutions when the treadle is released, and the amount of immersion can be regulated to a nicety.

Professors Ayrton and Perry have devoted much attention to the study of electric motors, and as a result they have promulgated the theory,—which we have already drawn attention to in a preceding chapter,—that whereas in the dynamo the field should be of great magnetic strength and the armature a weak one magnetically, the reverse should be observed in the mo-

FIG. 118.—AYRTON-PERRY MOTOR.

tor; i. e., the field should be a weak magnet and the armature a powerful magnet. They have embodied their ideas in a form of motor which differs from those of ordinary construction in that the armature is kept stationary while the field magnet revolves within it.

Fig. 118 shows the Ayrton-Perry motor in perspective; Fig. 119 shows the construction of the motor more in detail. The stationary armature, as will be seen, consists of a laminated cylinder built up of toothed rings of sheet iron, and resembles very much the Pacinotti toothed ring armature. The wires are wound on in sections, joined in series, and at each joint are

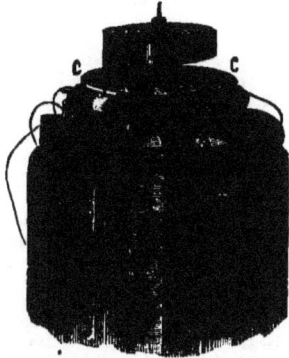

FIG. 119.—DETAILS OF AYRTON-PERRY MOTOR.

connected to a segment of the stationary commutator C C. The spindle of the revolving field magnet carries the brushes which revolve with it.

In explanation of the operation of the motor, Professor Ayrton says that wherever the brushes B happen to be at any particular moment, there two opposite magnetic poles at N and S are produced on the armature, as shown in Fig. 119. As the brushes revolve so do these poles, and the brushes, which are carried by the field magnets, are so set that the magnetic poles in the armature are always a little in front of those in the field magnet. The latter, therefore, are, as it were, perpetually running after the former, but never catching them.

Professors Ayrton and Perry have also devoted considerable attention to the regulation or governing of electric motors, and have devised several methods of accomplishing this result. One of their oldest forms, known as their "spasmodic governor," consisted of a trough of mercury, which revolved with the field magnet, and had a wire dipping into the mercury through which the current passed. As the speed increased, the mercury would take the character-

istic parabolic contour of revolving liquids enclosed in vessels, and at a certain speed the wire would be left out of contact with the mercury. The circuit would then be broken, the current cease, and the speed of the motor would fall again. The great objection to this form of governor is that it either supplies full power when the motor is running too slowly, or no power when the motor is running too fast, and hence is incapable of maintaining constant speed.

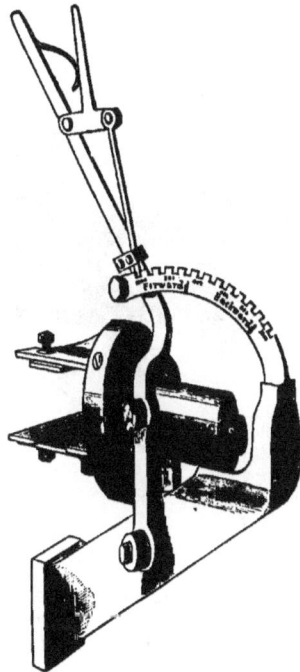

FIG. 120.—REGULATOR OF AYRTON-PERRY MOTOR.

Professors Ayrton and Perry have, however, designed several other forms and experimented with other and more perfect methods of governing electric motors, one among them consisting in winding the motor with two distinct circuits in such a way that the current passing through one of them magnetizes the iron, causes the machine to act as a motor, and consequently

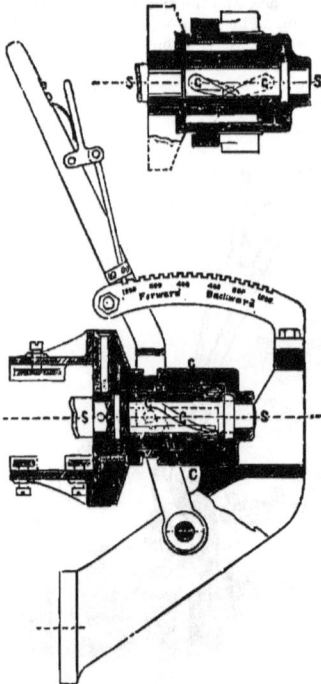

Fig. 121.—Regulator of Ayrton-Perry Motor.

is itself resisted, whereas the current passing through the other circuit tends to demagnetize the iron and stop the motion. This evidently is

equivalent to a differential winding and does away with all mechanical governing.

Where it is required to change frequently the speed and direction of rotation of an electric motor, such as upon a tram-car, Professors Ayrton and Perry have applied the method of varying the lead of the brushes. Figs. 120 and 121 illustrate the manner in which this is accomplished. By pushing the handle fully forward the motor revolves rapidly in one direction; when pulled back in the other direction the motor reverses. At intermediate position corresponding lower speeds are obtained. The action of this lead adjuster is as follows:

Attached to the rotating field magnet is the spindle SS, which is itself attached to and rotates with the outer collar CC. On pushing the handle forward or backward, this collar is moved along the spindle, and the effect of this is to cause a pin to move along the groove GG and so cause the inner collar PP, which usually rotates along with CC and the field magnet, to move a little forward or backward relatively to CC. Since the collar PP is screwed to the brush-holder, it is possible, even when the motor is running, to shift the brushes relatively to the field magnet together with which they are rotating, and consequently with only *one pair* of brushes to give any desired lead forward or backward. In other cases the lead is altered by means of a wheel and screw which permits of very accurate adjustment.

From the peculiar construction of the Ayrton and Perry motor, it may be operated without any wire at all upon the revolving field magnets. This arises from the fact that the magnetism in the stationary armature induces opposite magnetism in the iron of the field magnets, and, as pointed out before, the brushes are so placed that the magnetic poles in the armature are always just in front of those in the iron, which latter are always running round after those in the former but never catch up with them.

CHAPTER IX.

THE INDUSTRIAL APPLICATION OF ELECTRIC MOTORS IN AMERICA.

THERE are, as the previous chapter exemplifies, a thousand and one places to-day where small electric motors are greatly needed and can be used. In large cities, and manufacturing towns of any importance, where hundreds of small steam engines have been in use, each requiring to be fired and attended, the electric motor is of the greatest utility, its chief features of recommendation being that it generates no heat, smoke, or smell, requires scarcely any care the year round, can be entrusted to unskilled hands, makes little noise, is ready to start or stop at a turn of the switch, keeps up a steady motion, is under perfect control, and where a central power or lighting station exists, is cheap to operate. The success that has attended the central power stations already in existence in America attests the appreciation in which such a convenient source of supply is held, and the record is made almost daily of new installations. There can be no doubt that as regards convenience and applicability, electric transmission of power, for the reasons above given, stands without rival. It is often the case, too, that power users crowd into buildings where they pay high rents, solely because in other places they cannot get, or are not allowed to use, steam or gas. The electric motor is highly economic of space, and the wires leading to it can be run out of sight, through the smallest cracks and holes, around corners of the sharpest angle, and to heights or depths at which no one dreams of placing ordinary power-generating machinery.

We have already spoken briefly of the work now being done in Europe, industrially and otherwise, by small motors. The application has not been neglected in America. One of the most successful of recent inventors has been Mr. Griscom, of the Electro-Dynamic Company, of Philadelphia, whose motor during the last four or five years has come into very extensive application, not only here, but in England as well. Its principal excellence is to be found in the neat and compact design given to it—a feature which recommends it to many uses where other machines would be rejected, especially in domestic work. The motor is of the same type as the Deprez and Trouvé motors, inasmuch as its armature is of the old Siemens form; but the magnetic field in which the armature revolves is entirely different in shape. The motor has received the name of "double induction motor," from a peculiar phenomenon which was noticed by its inventor while experimenting with it. Ordinarily, the armature is included in the same circuit as the coils on the field magnets. Mr. Griscom once happened to pass a current through the armature circuit alone, the field circuit being disconnected. In this condition the armature, being powerfully magnetized, acted on the iron of the field magnets and tended to move so as to bring its poles in a line with the poles of the field magnet, and there, the magnetism of the armature being reversed, a mutually repulsive effect between this pole and the residual magnetism of the field magnets arose and tended to cause the armature to move away toward the other poles, so that once started the armature would continue to turn somewhat slowly. It was found, however, that if the field circuit, while still detached from the battery circuit, was simply short-circuited upon itself, the motor would begin to revolve very rapidly. Mr. Griscom concluded that by the motion of the armature its lines of force are cut by the field-magnet coils, and thus give rise to a current in the latter which helps to magnetize the field, and he ascribed the phenomenon to the peculiar conformation of the field magnet, which is such as to bring its wire close to the revolving armature. This phenomenon attracted much attention and caused considerable discussion, especially in Europe.

One sufficient theory, as opposed to the above, is that the closed circuit simply prevents the induced magnetism from diminishing on account of the "Lenz effects" which arise in the closed circuit, and that as the magnetism is slightly increased at each turn by the induction of the armature, the magnetic intensity of the field soon reaches a maximum. Whatever induction is produced by the lines of force of the armature must necessarily be of a nature to oppose the motion instead of helping it. This inference is indeed a valid consequence of Lenz's law. When the field magnet is in the same circuit as the armature, and fed from a battery current, the phenomenon does not oc-

FIG. 122.—THE GRISCOM MOTOR.

cur after the magnets have reached saturation, which takes place almost instantly; so that the term "double induction" is a misnomer. However, the motor can sustain its reputation quite well, even without the supposition of double induction, for it certainly attains a remarkable efficiency. It is said that it can lift 2,000 times its weight (forty ounces) in one minute, when working with the full battery power. This gives it a capacity of nearly one-sixth of a horse-power.

The motor is remarkable for the small space it occupies, due to its neat and compact design, shown in Fig. 122, which is nearly full size. The armature is entirely encased by the cylindrical electro-magnet within which it revolves, and by the metallic caps or discs fitted to this cylinder at each end. The cylindrical field magnet is composed of a cylinder of soft iron wired in two large coils, each of which covers nearly one-half of the cylinder, the space left

between the two coils at opposite sides of the cylinder constituting the magnetic poles of this cylindrical electro-magnet. The current which passes through the wire on this magnet circulates in opposite directions in each coil or section, so that both coils combine to produce a north pole in one of the open spaces and a south pole at the other. The result is practically the same as if two U electro-magnets were brought together with like poles in opposition, these forming a circular magnet with two consequent or combined poles, one at each junction. The iron of the cylindrical magnet projects laterally at each pole, and to these projections an ornamental brass disc is screwed firmly at one end, as shown in the figure. This disc forms one of the bearings of the armature shaft, which passes through it, and at the same time serves to protect the armature from injury. At the other end of the motor another brass plate is similarly fastened to the lateral projections of the poles of the cylindrical magnet. The shaft of the armature has a bearing in this plate also. The binding posts which receive the current from the battery pass through this plate. In the figure one is shown at the top and the other a little to the side of it. The binding post shown at the top is prolonged on the other side of the metallic cap, and carries one of the brass springs or brushes which serve to convey the current to the armature by pressing on the commutator. The other brush, touching on the opposite side of the commutator, is held in place by a special screw device attached to the metallic cap. The armature and the field magnet are connected in series, as may be readily seen from the figure. The current, entering the armature by the upper commutator spring, leaves it by the lower, from which it passes to the field magnet, whence it goes to the second binding post.

The Griscom motor weighs only forty ounces, and, as said above, can develop a power of 5,000 foot-pounds per minute (that is, nearly one-sixth of a horse power) without difficulty. It has been in great demand for working sewing machines, and is also being used very extensively for many other industrial purposes. It has proved of great utility and convenience to surgeons, and especially to dentists, in driving various surgical instruments. The well known dental engines used for rotating the excavating drills used to remove the decayed portions of teeth are all operated by a treadle, which not only obliges the operator to remain in fatiguing positions for several hours, but requires him to keep up a monotonous and tiresome movement. By a clever adaptation of the electro-dynamic motor to the flexible shaft of a dental engine, these disadvantages are obviated. The apparatus is suspended either by balanced cords from the ceiling, or from an adjustable arm, which allows it to remain balanced at any height or desired angle, thus relieving the operator of the weight of the apparatus and permitting him to manipulate the drills as delicately as a pen.

nets are divided so that there are two or more circuits around the core. By suitable devices these are so related that they can be thrown into series or into multiple arc, or into other combinations when there are more than two circuits, for the purpose of changing the strength of the magnetic field, to suit the electromotive force and strength of current supplied to the motors.

The armatures are modelled in principle after the Gramme, but their construction is much improved, especially in respect to the manner of mounting them on their shafts. Instead of

FIG. 123.—DAFT GENERATOR OF 1884.

New forms of the Griscom motor are, we understand, now being designed and constructed for more general use.

To this department of electricity, as well as to the use of motors on railways and street-car lines, Mr. Leo Daft has paid considerable attention. We illustrate here some of the machines made by him in 1884,—it being unnecessary to go further back. Fig. 123 shows a typical Daft generator, already referred to in the preceding chapter on American electric railways, and Fig. 125 a motor. The field magnets are made after what is called the Siemens plan. That is, they lie horizontally, have consequent poles, one above and the other below the armature. They are series wound, but the coils of the field mag-

wood, suitably insulated metallic spiders and connections are used for the purpose. Thus there is left a space within, around the shaft, through which air can circulate for ventilating and cooling purposes.

As high speed is favorable to the efficiency of electric motors, they are provided with gearing so that the armatures may be run at high speed, but communicate to the driving pulley only a moderate rapidity of rotation. With that end in view, the armature shaft is lengthened on the end opposite the collector; two bearings are there provided, and between them there is a steel worm gearing with a phosphor-bronze wheel on the shaft carrying the driving pulley. This makes a good wearing combina-

FIG. 124.—NEW DAFT GENERATOR.

FIG. 125.—DAFT MOTOR WITH GEARING.

tion. The bearings of the armature shaft are of phosphor-bronze, and an end plate and adjusting screw are provided to receive the thrust of the shaft due to the gearing. The whole makes a very practicable combination, taken in connection with the means of varying the field strength of the magnets of both generator and motor.

Two of the motors have been in use in Spruce street, New York city, one since January, 1884, and the other since April, 1884, giving great satisfaction in the operation of freight elevators which have a capacity of 2,000 pounds with a speed of from thirty to thirty-five feet a minute. One of them was also put to work as long ago as 1883, in the steam mills, Newburgh, N. Y., to operate an elevator raising a load of 1,800 pounds. The New York elevator motors run continually during working hours. When not engaged in raising the elevators they run faster than when doing work. Consequently their counter-electromotive force cuts down the current supplied to them to the point of supplying the energy necessary to overcome the friction of the motors and their gearing. The average speed when at work is 1,200 revolutions per minute. The difference of potential at the binding posts is about ninety volts. The current varies with the load, but averages about twenty-five ampères to each machine. The actual power recovered is said to be sixty-six per centum.

The armature of the generator has a resistance of 0.23 ohm. Its speed is 1,100 turns per minute. Its electromotive force is ninety volts, and its extreme practical current capacity seventy ampères. Hence it can deliver 6,300 voltampères of electrical energy, or $\frac{6300}{746} = 8.44$ horse power. Fig. 124 shows the latest type of Daft generator.

With a Daft motor of this type, and of one and one-half horse power, *The Electrical World* gave an interesting exhibition of printing by electricity at the International Electrical Exhi-

bition in Philadelphia, September and October, 1884. For six weeks, the regular and special editions of the paper were printed from electrotypes on a 31x46 Cottrell press made specially by Messrs. C. B. Cottrell & Sons for the occasion. Although this was not the first time printing by electricity had been accomplished, the idea was quite new to a great many visitors to the exhibition and attracted unusual notice. The motor and press worked without the least

FIG. 126.—DAFT MOTOR WITH BLOWER.

trouble, under the supervision of Mr. Clarence E. Stump, the business manager of the paper, who had charge of the exhibit and who found the printing to compare very favorably with that done on a press directly actuated by steam. It may be mentioned here, as of interest, that the Ilion (N. Y.) *Citizen* was printed by a Parker motor, March 14, 1884, through a break-down of its steam engine; and that the Lawrence (Mass.) *American* has been printed daily since July 6, 1884, by an electric motor. In a letter written to one of the present authors, in October, 1884, Mr. George S. Merrill, proprietor of the *American*, said: "We formerly used a ten horse-power engine, necessitating the employment of an engineer, but the employment of the Edison Company's power gives a saving in expense of more than thirty-three and one-third per cent. The speed is uniform and the power satisfactory in every respect." The motor is used to run several cylinder and job presses.

Fig. 126 shows a Daft motor attached to a No. 4 Sturtevant blower, the two together forming practically one machine. In the blower illustrated, which requires two horse power, three speeds are obtained, of 800, 2,000, and 2,700 revolutions per minute, by changing the resistance of the field, thus doing away with outside resistances and entailing no loss of work due to their employment.

FIG. 127.—NEW DAFT MOTOR.

For some time past, Mr. Daft has devoted his energies to meeting the requirements of the various central power stations established to operate under his system. The success experienced by these power stations, especially in Boston and Worcester, Mass., has induced him to enter upon the manufacture of a more extended series of sizes and to remodel his machines to meet the requirements of more varied work.

The form which he now employs is shown in our engraving, Fig. 127. It will be seen that the field magnets are of the simple horseshoe form, and that the armature is of the Gramme type, as in Mr. Daft's previous models. The machine is designed to deliver normally six horse power, but upon test it has been driven

to as high as eleven horse power without injurious effect.

Constant speed at all loads is naturally the first requisite in an electric motor designed for stationary power plants, and hence accurate regulation must be provided for. The present machine, according to Mr. Daft, maintains its speed within two per cent., between maximum load and no load. To obtain this result, the machine has its field wound compound with three different windings. Of these one is a "series" and the other two are shunts.

The coils are wound on spools which are slipped over the wrought-iron cores, and can readily be removed for examination when necessary. The principal data regarding the machine are as follows:

Electromotive force designed for . . .	100 volts.
Resistance of armature,	0.15 ohm.
" " series coil in field, . . .	.024 "
Resistance of first shunt in field, . . .	32.73 ohms.
Resistance of second shunt in field, . .	7.50 ohms.
Power, ...	6 h. p. nom.
Number of revolutions per minute, . .	1,300.
Weight, . . .	875 lbs.

It is evident that in a system of electric distribution where power is furnished and sold to various consumers, it is necessary to provide some means of controlling the maximum amount of power each customer may use, as well as to prevent injury to the machine by unskilled persons starting it with the full force of the current before the inertia of the armature is overcome and it has attained speed enough to develop a suitable working resistance or counter-electromotive force.

· Mr. Daft has provided for these necessities, and places upon the premises of each power consumer an apparatus designed for this purpose. It is shown in Fig. 128. Figs. 129 and 130 are details which are shown in outline in Fig. 128.

It will be seen that mounted on a shaft supported in journals is a gradually increasing cam A of insulating material, and this cam is rotated by a toothed wheel B on the same shaft, and a worm C meshing with the wheel and operated by a crank from the outside. Upon the periphery of the cam are arranged strips or sections D D of copper, and supported in standards and insulated from each other are two contact pieces E E, which bear upon the strips on the cam. At the highest part of the cam is

so that the operator may properly manipulate the switch and know the position of the parts inside by the location of the indicator with reference to the words "off" and "on."

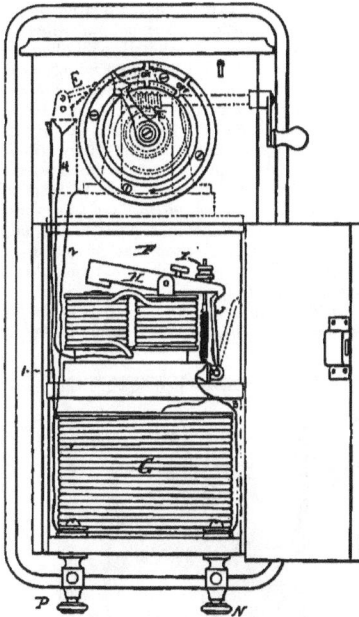

FIG. 129.—DETAILS OF LIMIT SWITCH.

The circuits are traced as follows: Entering at the post P the current passes, by wire 1, to one of the spring contact pieces E, which, in its normal or "off" condition, rests upon the insulating material of the cam, and no current can pass. As the cam is slowly rotated the contact brushes bear upon the conducting-strips, one of which is electrically connected to the frame, and thus by the wire 2 through the resistance coil to wire 3 and out by post N. When both contact pieces bear upon the cross-strip M, the current passes through the same and the contact piece, by wire 4, through the cut-out and to post N, short-circuiting the resistance.

The operation of the apparatus will now be readily understood. Supposing the indicator to point to the word "off," the spring contacts

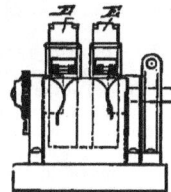

FIG. 128.—DAFT MOTOR—LIMIT SWITCH.

placed a transverse conducting strip M insulated from the other strips and adapted to complete the circuit direct between the spring contacts. In the bottom of the box is a resistance coil G and a cut-out switch F, the armature H of which is adjusted by a regulating screw I to withstand the desired degree of attraction before moving and to retain the snap switch J in position.

An indicator K is attached to the cam-shaft, which shows the relative position of the cam,

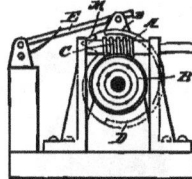

FIG. 130.—DETAILS OF LIMIT SWITCH.

will rest upon the smallest part of the cam, and as the strips on the periphery do not extend to this part, the contacts will rest upon the insulating material of the same and no current will pass. If, now, the handle is turned, the cam will be slowly rotated, bringing the conducting strips under the spring contacts, and as these strips are connected to the resistance coil, the current will first flow through the coil, and the armature of the motor will not be endangered; and as it requires a number of turns of the worm to complete the rotation of the varying-cam, some little time will elapse after the first

FIG. 131.—SMALL VAN DEPOELE MOTOR.

contact of the strips with the spring contacts before they will reach the transverse conducting-strip, when the resistance will be cut out and the direct circuit be completed through the strip and spring contacts, permitting the motor to have the full force of the current, and the indicator will point at "on." It will be seen that, during this operation, the armature of the motor will have attained considerable velocity, thus developing a suitable working-resistance to prevent injury to the brushes or other parts.

As stated before, there is a cut-out added, which prevents the use of more power than is contracted for. The coil of the cut-out is placed in the main circuit, and the armature may be adjusted so as to allow a current of a certain specified strength to pass without operating it; but any abnormal increase due to overloading the motor or otherwise would cause it to be attracted toward the magnet core, thereby releasing the snap-switch and breaking the circuit. The adjusting screws permit of regulation for a wide range of current, and can be adjusted for any desired consumption of power. The cut-out box is kept locked and under the control of the parties at the central station, and the consumer is limited, therefore, to the use of the amount of power contracted for. In the

event of an attempt to take more power than contracted for, the speed of the armature is reduced, the internal resistance of the machine being thereby decreased, the flow of the current will quickly reach the point at which the adjustable cut-out has been set, when the circuit is severed and the consumer is obliged to notify the company before the machinery can be placed in working order again. The use of the gradually increasing cam is also valuable in preventing the operation of the cut-out when the machine is started, as otherwise an abnormal flow of current is likely to occur which would operate the cut-out. It also furnishes a safe and effective stop-switch, for as soon as the increasing cam in its rotation causes its highest part to pass the spring contacts, they will instantly fall upon the smallest insulated part of the cam, thus severing the circuit without the possibility of forming an arc for any appreciable time. If at any time there should be a sudden abnormal increase in the current in the line from any cause, the cut-out operates to prevent injury to the machine.

We illustrate next two industrial types that have been made by the Van Depoele Company.

FIG. 132.—SMALL VAN DEPOELE MOTOR.

Fig. 131 shows the small motor for light work. The principle of construction will be easily seen from the cut. The ring armature has numerous sections, insuring steady motion, and the pole pieces are of special form. The design of the whole is simple and symmetric; there are no parts to get out of order, and with a few drops of oil occasionally, these motors will run for years without the slightest irregularity, and without perceptible wear. The battery furnished with them is a bichromate, of improved pattern, which can, of course, be connected with or disconnected from the motor at will by means of a switch, and can be put away in any convenient place in a box.

Fig. 132 is a motor for running large horizontal fans, which can be coupled directly with the vertical shaft. When it is necessary to run several fans at once, instead of using several

The base carries the upper core and pole of the field magnet attached permanently to it. The under core and pole is hinged at the rear of the motor just above the binding post there shown. It is so hinged for the purpose of changing the speed of the motor to suit requirements by moving the under pole to or from the armature by means of a connecting rod and treadle not shown.

When the pole is moved away from the armature its attractive influence is not so strong, and therefore the power and speed are decreased.

Fig. 133.—Diehl Motor.

motors, one motor of sufficient power for the lot can be used, with the intervention of belting in the ordinary way.

At the Singer Manufacturing Company's exhibit in the International Electrical Exhibition at Philadelphia in 1884 were seen several sewing machines run by various electric motors invented by Mr. Philip Diehl, the inventor engaged by the sewing machine company, and one whose other work in the practical application of electricity has been marked by great originality.

One form of motor is made part of the flywheel of the sewing machine, and the one we illustrate by Fig. 133 shows the motor at about two-thirds size.

If the pole be swung downward until its lowest limit is reached, the electric circuit is broken entirely at the point where the button on the under pole and the spring projecting from the upper pole meet. The post with the two jam nuts on it is for the purpose of fixing the position of the upper pole.

The armature core axis and pulley are cast as one piece of iron, and the armature is of the Siemens H type. One pole of the armature core is extended and bent so that the axis at the commutator end is in its proper place, and there is a space between this extension and the other pole. At the other end of the other pole a like extension for the pulley axis is provided. Thus there are longitudinal and transverse spaces

17

lengthwise around the core to receive the coil; and yet the poles are not connected by iron except within the coil.

The commutator is of the kind usually attached to bipolar armatures—that is to say, it has two sections.

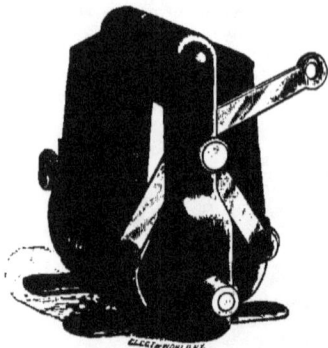

Fig. 134.—Diehl Motor.

One of the most novel and ingenious of recent motors is that shown in Fig. 134. This motor is also the invention of Mr. Diehl. The present motor, though built on the same principle as the one exhibited by the inventor at Philadelphia, 1884, and illustrated above, differs con-

Fig. 135.—Keegan Motor.

siderably from it in construction and general appearance.

By referring to the engraving, it will be seen that the field magnets are placed vertically and hinged at the top, being supported by two side rods, cast solid with the base. The lower ends of the field magnets encircle the armature,

which is also carried by journal bearings in the side rods.

The method of regulation of the motor consists in separating the pole pieces from the armature. This is accomplished by means of two connecting rods fixed to the lower ends of the magnets and joined together by a pin which lides in a slot on the upright. A rod connected to the pin serves to raise and lower the upper ends of the two connecting rods, and in doing so the field magnets are separated or brought together, as the case may be.

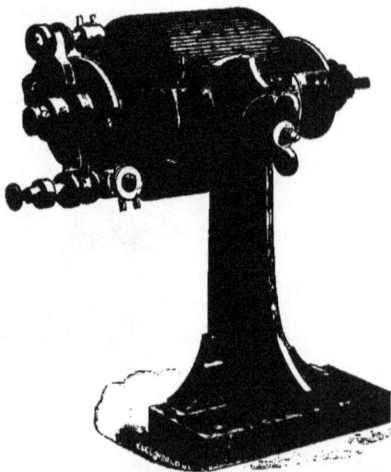

Fig. 136.—Pendleton Motor.

When used in connection with a sewing machine, the motor is secured to the under side of the table in an inverted position, and the regulating lever connected to the treadle. In this position the field magnets fall apart of their own weight and the machine does not work. It is only when the treadle is pressed and the magnets are brought together that motion is obtained. It is evident that by varying the distance between the armature and the magnets any desired speed can be obtained for fast or slow work. The motor is finished in a very ornamental style, and runs very smoothly. The armature shaft is provided with a pulley, and its end is bored so that the power can be transmitted by belt or applied directly, as when driving a fan.

FIG. 137.—ARMATURE OF PENDLETON MOTOR.

Another good motor recently devised for small work is that of Dr. V. E. Keegan, of Boston. The main objects sought have been those

five volts for the standard size, according to the power required. The motor, which is made by Mr. Wm. J. Keenan, of Boston, is neatly built. The commutator is platinized to prevent corrosion.

At a meeting of the electrical section of the American Institute, held in New York, in July of the present year, Mr. John M. Pendleton exhibited to the society a small electric motor of his own design which embodies several novel features. Mr. Pendleton remarked that the general introduction of electricity had drawn considerable attention to electric motors, from the recognized fact that power is not only more

FIG. 138.—PENDLETON MOTOR.

of simplicity of apparatus and economy in running. As the illustration shows—Fig. 135—the motor consists of two horseshoe electro-magnets, the one acting as field, the other as the armature. The pole pieces are extended inwards until they come within a quarter of an inch of each other, and at this point, where the magnetism is the strongest, each takes the form of a semicircle; thus, instead of an alternate action of attraction and repulsion, the two forces are always acting together, and hence the effect of the induction current, according to the inventor, is largely neutralized. In order to make the motor economical of battery material, it is wound for high resistance, requiring a current of only two ampères at from ten to twenty-

readily transmitted by electricity, but also on account of the facility with which it may be subdivided and distributed without loss—a point in which neither steam nor gas engines can compete with it.

The motor, which is shown in perspective in Fig. 136, weighs forty ounces, and is capable of developing power sufficient to run a sewing machine or other light running apparatus, such as a dental drill, mallet, or a fan.

The armature of the motor, which is shown detached in Fig. 137, is of the three-pole type, and each of the three segments of the commutator is consecutively cut out in such a manner as to reverse the polarity of the pole of the armature so that for one-half the distance of the pole-piece attraction takes place, and during the other half repulsion.

This construction obviates any dead point, and the motor starts instantly at any position of the commutator. This is evidently an indispensable quality in a motor designed to operate intermittently, stopping and starting at frequent intervals. Such work also frequently requires a change in speed, so as to run fast or slow, and this has also been provided for, by making the brush-holder adjustable. The latter is controlled by a spring which normally maintains it at the position of maximum speed. By ro-

tating the brush-holder, however, the position of contact, and hence the speed, can be varied at will from the maximum down. This shifting of the brushes does not require any manipulation by hand, but is accomplished by a cord attached to a treadle, thus leaving the operator's hands free to guide the work.

The engraving, Fig. 138, shows a larger size of the same type. The latter weighs twenty pounds, and is said to be able to develop as high as one-quarter horse power.

(The latest American motors of the industrial class will be found in Chapter XII.)

CHAPTER X.

ELECTRIC MOTORS IN MARINE AND AERIAL NAVIGATION.

THE use of electric motors in marine and aerial navigation has been chiefly studied with a view to obtaining the necessary current from storage batteries. It is true that bichromate of potash has been employed, but storage is regarded by almost all who have investigated the subject, as the ultimate means to be adopted in any practical work on a large scale.

The experiments on the Neva, fifty years ago, have already been noticed. There is nothing to record in the present chapter from the efforts of Jacobi until we come to those of the ingenious and versatile Trouvé of Paris, who put a small electric boat on the lake at the exhibition in 1881. This boat, which had previously been shown in operation on the Seine, was equipped with a double motor, or, in other words, with two bobbins put close together fixed on the rudder-head. The current was furnished by a bichromate of potash battery placed in the middle of the boat. Motion was communicated by means of an endless chain to a small screw fitted in the rudder itself. A speed of about three and one-half miles was obtainable, with a load of four or five passengers, and the battery was only active when wanted.

The launch "Electricity," operated on the Thames in 1882, is said to have been the third boat propelled by an electric motor. It was twenty-five feet in length and about five feet in the beam, drawing one foot nine inches forward and two feet six inches aft, and was fitted with a twenty-two inch propeller screw. On the trial trip on the Thames there were stowed under the flooring and seats forty-five electric accumulators of the Sellon-Volckmar type, which had been charged by wires leading from dynamos, and were calculated to supply power for six hours at the rate of four horse power. These storage cells were placed in electrical connection with two Siemens dynamos, furnished with special reversing gear and regulators, to serve as motors to drive the screw-

propeller, the arrangement being such that either or both of the motors could be switched into circuit at will. The party on board consisted of four persons, Mr. Volckmar being one of the number. The launch would carry twelve passengers. The ability of the boat to go forward, slacken, or go astern, at the pleasure of the commander, was satisfactorily tested, and a speed of eight knots an hour was made against the tide. The return trip from London, Bridge to Millwall, coming down with the ebb, was made in twenty-four minutes, the mean speed of the vessel being nine miles per hour. The actual expenditure of electric energy was calculated to be at the rate of three and one-eleventh horse power.

During 1883, a launch built by Messrs. Yarrow, of England, and shown at the Vienna Electrical Exhibition, attracted considerable attention. The boat was forty-six feet in length, and was capable of accommodating some forty-nine or fifty passengers—an extraordinary number, considering the carrying powers of any steam launch of corresponding dimensions. The whole of the boat, with the trivial exception of a small space at the stern—hardly more than is sufficient for the "man at the wheel"—was available for use instead of having, as is the case of the best constructed steam launches, a large portion of the centre of it occupied by the machinery. Comfortable seats extended through the entire length of the launch on each side, and there was nothing to interrupt a promenade from end to end of it.

The motive power lay *perdu* in seventy boxes, each of one horse power, stowed away under the floor of the launch, and at the end there was a Siemens D 2 type of motor, the spindle of which was continued so as to form the shaft of the screw. There was no gearing whatever between the dynamo and the screw, to which 600 to 800 revolutions per minute could be imparted without the slightest noise, and a speed of from eight to nine miles an hour kept up with far less than the usual amount of wash. There was no noise nor heat, nor smell of machinery, nor smoke, and, as we have said, the whole of the boat was practically available for use, without any obstruction of boilers and engines. The advantage of such a motive power is thus in many ways quite obvious, and the cost of the launch complete in every respect was, it is said, only about $3,000.

Since 1883 various other trials have been made, and experiments tried. One of the most successful workers along this line has been Mr. Reckenzaun, who at the present time has a launch running successfully on the Thames fitted with his motor and secondary batteries. In June, 1885, Mr. Reckenzaun took the Duke of Bedford for a cruise in the electric launch "Australia," on the Thames. The Duke was so pleased with the performance of the "Australia" that he decided to order a boat of similar design, but of more elegant appearance, and the Electrical Power Storage Company was intrusted with the construction of the propelling apparatus of this new vessel, which is some three feet longer than the "Australia"; the internal arrangements, however, are very similar. Twenty-nine E. P. S. accumulators are placed in a box in the centre of the boat, this box serving as a seat for passengers; the cells actuate a Reckenzaun motor, and the speed obtained is of the average rate of six knots per hour for four and a half hours. The accumulators of this boat serve also for lighting the yacht when the electric launch is suspended from the davits, and the cells are charged from the dynamo which usually lights the "Northumbria." The official trial took place at Westminster, in the presence of numerous spectators. Mr. Reckenzaun has had several designs of electric boats in progress, for some time past, embodying further substantial improvements. One of these is being executed to the order of the Italian government, and a second for an Indian prince; the former is for war purposes and the latter for pleasure. The prince's launch is to be fitted most luxuriously, and electrically lighted, even the fans being actuated by electricity.

During September of the present year, the launch "Volta," fitted with two Reckenzaun motors and a set of accumulators, made the trip from Dover to Calais and back, with ease and safety, the batteries being charged but once for the whole journey. The "Volta" is 37 feet long, has 7 feet of beam and is 3½ feet deep. She is built of galvanized steel plates. Her propelling power consists of sixty-one accumulators, each eight inches square, placed as ballast under the floor with the motors. The accumulators were charged over night from a dynamo worked by a small steam engine in a carpenter's shop facing Dover harbor, the connection to the boat being by short sections of a cable. Seven passengers were carried and a speed of over six miles an hour was maintained, while over twelve miles was reached.

As in marine navigation, so with aerial—the use of electric motors has been of an experimental character, and yet its results are most significant and encouraging. Up to 1881, one of the greatest desideratums in ballooning was a light motor that would not require fire and would not be subject to loss of weight in operating. Clearly, the electric motor was the thing wanted, and M. Gaston Tissandier applied himself to the problem of adapting the means to the end. In a note to the French Academy of Sciences, read August 1, 1881, he said: "The recent improvements made in dynamo-electric machines have given me the idea of employing them for the directing of balloons, combined with secondary batteries, which although of relatively light weight, store up a large amount of energy. Such a motor, connected by a propelling screw, offers advantages over all others, from an aerostatic standpoint. It operates without any fire, and thus prevents all danger from that element under a mass of hydrogen. It has a constant weight, and does not give out products of combustion which continuously unballast the balloon and tend to make it rise in the air. It is easily set running by the simple contact of a commutator." M. Tissandier carried out these ideas in a model with which he experimented publicly and successfully at Paris, during the electrical exhibition of 1881. He then went to work with a balloon equipped with a light Siemens machine and a bichromate of potash battery, and resolved to try the principle of screw propulsion. Finally, in October, 1883, he made a notable experiment near Paris with a balloon having a total weight of 1,240 kilogrammes. Allowing 10 kilogrammes, the lifting force was 1,250 kilogrammes. The bichromate

FIG. 139.—VIEW OF THE TISSANDIER BALLOON.

of potassium batteries were composed of four troughs with six compartments, making twenty-four elements in circuit. By means of a mercury commutator, 6, 12, 18, or 24 elements could be used, thus giving four different speeds of the screw, varying from 60 to 180 revolutions per minute. The results of this experiment were summarized by M. Tissandier as follows:

"We have concluded from this first trial that:—1, electricity furnishes a balloon with the most convenient power, the management of which in the car is remarkably easy; 2, in our own case, when our screw, 2.8 metres in diameter, made 180 revolutions per minute, we were able to keep head to wind, moving three metres per second, and, when proceeding with the current, to deviate from the line of the wind with great ease; 3, the mode of the suspension of a car from an elongated balloon by means of bands running obliquely and supported by flexible side-shafts, insures perfect stability to

the whole." Our illustrations, Figs. 139 and 140, give an excellent idea of the appearance of the balloon and of its motive mechanism. The length of the balloon was 28 metres, and its diameter at the centre 9.2 metres. The Siemens motor, which weighed only 54 kilogrammes, had an armature very long in proportion to its diameter, and which made 1,800 revolutions while the screw to which it was geared made 180.

The speed obtained in 1883 was three metres per second. During a trip made by the Tissandier brothers in 1884 a speed of nearly four metres was obtained, and it was also found that the balloon could be brought back to its starting point even in calm weather. The next noteworthy experiments in this direction were those of Capt. Renard and Capt. Krebs, who on August 9, 1884, made a highly successful demonstration with their directible balloon, the outcome of six years' quiet work, and of a grant of 100,000 francs from the French government. The shape of the balloon was not unlike that of a cigar pointed at both ends. The car suspended by network contained seats for two aeronauts, the motive power, and the steering apparatus. Capt. Renard invented for this trial a secondary battery of unusual lightness, and Capt. Krebs devised the screw and rudder, and the motor gearing. The dimensions and weights were these:

Length of the inflated ellipsoid, . . 50 m. 42 cent.
Central diameter, ... 8 m. 40 cent.
Volume, ... 1,864 cub. m.
Length of car—Nacelle, ... 33 m.

Weights:
Balloon and ballonet, . . . 869 kilos.
Silk covering and net, . . . 127 kilos.
Car complete with rigging, etc., . . 152 kilos.
Rudder, ... 46 kilos.
Screw-propeller, ... 41 kilos.
Motor, ... 98 kilos.
Wheelwork, ... 47 kilos.
Shaft, ... 30 kilos, 500 grams.
Battery complete, ... 435 kilos, 500 grams.
Average velocity per second, ... 5 m. 50 cent.
Diameter of the propeller, ... 7 m.
Number of revolutions per minute, ... 30 to 40
Number of elements employed, ... 32

The electric motor was constructed to develop 8.5 horse power upon the shaft, and it transmitted its motion thereto by means of a pinion gearing with a large wheel. The battery was divided into four sections, that could be connected either for quantity or for potential, and was calculated to deliver 12 horse power—8,952 watts—to the motor for four consecutive hours. The trip was made in the neighborhood of Paris. In his official report to the Academy of Sciences, M. Hervé Magnon said: "The balloon rose to an elevation of fifty metres above the ground, at which elevation it was kept per-

the balloon descended gradually, obliqued right and left, forward and backward, at the pleasure of its pilots, and finally landed exactly at the point indicated." The time occupied in making the entire circuit of 7,600 metres (about five and one-half miles) was only twenty-three minutes. The maximum velocity obtained was nineteen kilometres per hour. In later trips an average velocity of twenty-five kilometres was shown as the result of the various improvements in details. Here, then, was the

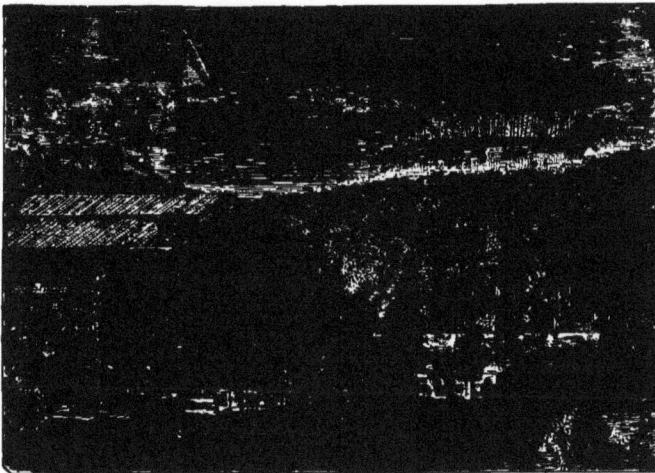

FIG. 141.—THE KREBS-RENARD BALLOON.

manently by Capt. Renard, Capt. Krebs manoeuvring the rudder. As soon as the propeller was given a rotary movement the aerostat took its course toward the Hermitage of Villebon, which, previous to the ascension, had been designated as its objective. The wind at this moment moved with a velocity of five metres per second, and the balloon moved against it. So soon as arrived at its destination the officer who held the tiller waved a flag, the signal of return, upon which we saw the aerostat luff, describe majestically a half circle of a radius of about 300 metres, and sail back to Meudon. Upon reaching the lawn, whence it had started,

attainment of practical ballooning. As Col. Fred Burnaby, an enthusiastic aeronaut, had said but a short time before in discussing the availability of electricity, "to put the case in a nutshell, aerial navigation is a mere question of lightness and force," and the two French officers had undoubtedly succeeded in putting weight and strength in their right proportions to be effective. M. Gaston Tissandier, whose ample experience qualified him to speak authoritatively on the subject remarked, not only with generosity but with truth: "These new experiments are decisive. Navigation of the air by means of long balloons provided with

18

screws, is demonstrated. We will repeat what we have already said many times, that to be practicable and useful, aerial ships must be very long, of very large dimensions, which shall carry very large machines capable of giving a speed of from twelve to fifteen metres a second, allowing their working at almost any time. When the wind is high, or there is a squall or tempest, aerial ships must remain in port, as other vessels do. It becomes now only a question of capital."

A view of the Renard-Krebs balloon described in the above passages is given in Fig. 141, which shows also its starting place.

CHAPTER XI.

TELPHERAGE.

WHILE in the electric railway, as in electric lighting, the tendency of inventors has been to preserve old forms and methods, for the purpose of more easily adapting their devices to public use, in what is known as "telpherage" a decidedly new departure is taken. Mr. Herbert Spencer, if we remember aright, once drew attention to the survival of conventional curved lines in the bodies of the English railway cars, which thus present the aspect of the old and obsolete stage coaches; and we might instance the more recent case of incandescent lighting, in the introduction of which to general notice and use, Mr. Edison sought as far as possible to adhere to methods that had become familiar in the employment of gas. His mains, branches, meters, brackets, "electroliers," and switches are, practically, so many like parts of a gas-lighting system, and may be safely left to the handling of the most inexpert; only the generating apparatus requires technical skill and knowledge on the part of those who deal with it. Some may say that telpherage is after all simply an old idea, plus electricity, but we believe that to the vast majority of people, the transmission of freight or passengers, along a wire road, is a surprising innovation, an application for which their information or experience can find no parallel.

The word "telpherage" was coined by the late Professor Fleeming Jenkin, who conceived the invention now being developed by Professors Ayrton and Perry. In a lecture before the University of Edinburgh, Professor Jenkin said: "The transmission of vehicles by electricity to a distance, independently of any control exercised from the vehicle, I will call 'telpherage.' The word should, by the ordinary rules of derivation be 'telphorage,' but as this word sounds badly to my ear, I ventured to adopt such a modified form as constant usage in England for a few centuries might have produced; and I was the more

ready to trust to my ear in the matter, because the word 'telpher' relieves us from the confusion which might arise between 'telephore' and 'telephone' when written."

Generically considered, a telpher line system consists of a rod or rail track of considerable length, suspended several feet from the ground, connected with a source of electricity placed at some suitable and convenient place at or near the course of the track, and

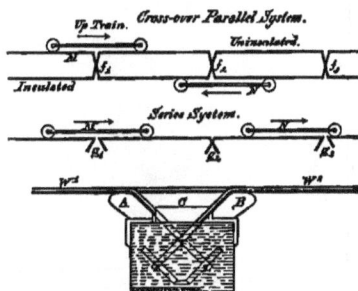

FIG. 142.—DIAGRAM OF TELPHERAGE TRACKS.

traversed by an electro-locomotive which derives its motive power electrically from the said track, draws a number of small holders of freight or passengers, and is controlled, as to its motion, from a place or places other than itself.

A telpher line can be built on either the "series" or the "cross-over parallel" system. Figs. 142 and 143 show the "series" system as put into experimental operation at Weston, Hertfordshire, England, about two and a half years ago. M and N are two trains of cars running on the line upon which the wheels bear. The line has make-and-break mechanism at g_1 g_2 g_3, points about 120 feet apart.

These make-and-breaks are normally closed, so that a current of electricity may flow from end to end. But when an electric train is started over the line, say from left to right, as the forward wheel of the motor strikes the make-and-break g_1, it opens the circuit by moving the latter; and then the current takes the course through to the rear wheel of the train, the motor thus receiving the current that energizes it. The train is made a little longer than the train receives no current. It is clear that any desired number of trains may be run upon the line at one time, it being necessary only to have the electromotive force adapted to the number of trains operated. In the larger illustration of the series system, the rough posts carry cross-arms securely bolted to them. On the overhanging ends of these arms, are the junction blocks for the ends of the sections. These junction blocks are placed only on alternate

FIG. 143.—VIEW OF "SERIES" TELPHER ROAD, WESTON, ENGLAND.

120 feet between breaks, so that when the forward wheel strikes break g_2, though it opens the circuit there, the current still flows through the train, spanning the section of the line from g_1 to g_2. But when the rear wheel of the train strikes the break g_1, it closes it, so that current may still flow through the line and train. As the train moves onward, it successively opens the line circuit by the "break" under its foremost wheel, and closes the line circuit by the "make" under its rearmost wheel, so that the current for the motor is derived from the sections of the track under the forward and rear wheels of the train, at which time the track immediately under the posts, because the posts are 60 feet apart, and the sections of the line are 120 feet in length. The intermediate posts carry only suitable supports for the line. Fig. 142 represents one of the junction blocks. Cast-steel supports A and B are bolted down on a wooden block, which is in turn bolted to the ends of the cross-arms; or this block may be the end of the cross-arm itself. The upper surfaces of A and B are channelled to receive the conductors W^1 and W^2. These conductors pass one on each side of the cast-steel piece C, and go through holes in the wooden block, being secured in the latter by nuts, as shown. The piece C is bolted on the wooden block in a position intermediate

between A and B, but is insulated from them. This piece serves as a continuation between the rods W^1 and W^2 so that the wheels of the locomotive and skips can ride from W^1 to W^2 with regularity and smoothness. The circuit is completed either by ground, a conductor for the purpose, or, preferably by a return line over which the locomotive may run. The wire in this line is five-eighths of an inch in diameter. The load is carried in seven skips, the first being seen in Fig. 143. About half a ton can be put into each skip and a speed obtained of six miles an hour.

is supported by what is practically one long, continuous steel rod; but, in reality, at the tops of the posts the rods are electrically subdivided into sections and joined across by insulated wires, one of which can be seen on the post in the foreground of Fig. 144, which gives a good idea of the line in actual operation. To prevent the metallic wheels of the skips from short-circuiting the two sections as they cross the tops of the posts, there are insulated gap-pieces, also to be seen in Fig. 144, at the tops of the posts where the steel rod is electrically divided. It is found that for moderate inclines,

FIG. 144.—VIEW OF "CROSS-OVER PARALLEL," TELPHER ROAD, GLYNDE, ENGLAND.

The principle of the cross-over parallel system of telpherage is best shown in a commercial line—Fig. 144—recently put in operation at Glynde, England, for the New Haven Cement Company, to carry clay from a pit to the Glynde railway siding, whence it is delivered into trucks and taken by rail to the cement works. Fig. 145 illustrates the construction of the track for two trains. D is the dynamo furnishing current to the circuit A_1 and B_1, respectively, positive and negative. The wheels, L and P of one train and L_1 and P_1 of the other are insulated from their trucks and connected in the case of each pair by a wire on the motor. Consequently as the trains move, a current is always passing from a positive section of the line to a negative section through each motor. Mechanically, each train

direct driving with pitch chains, of two wheels with india rubber treads, gives a gravitation grip sufficiently strong for haulage purposes. In the earlier lines, Ayrton and Perry motors were used; in this, the Reckenzaun has been tried.

The automatic governing of the speed of the train is effected in two ways,—first, there is a governor attached to each motor, which interrupts the electric circuit, and cuts off the power when the speed becomes too high; secondly, there is a brake which is brought into action should the speed attain a still higher value. To avoid the formation of a permanent electric arc when the circuit is broken, the governor (Fig. 146) is so arranged that the diverging weights are in *unstable* equilibrium between two stops: they fly out at about 1,700 revolutions per minute of the motor, and fly back at about 1,600. When the circuit is closed, the current is conveyed across the metallic contact at C. When the weights $W\,W$ fly out, this contact is first broken, but no spark occurs,

In this line which has now been working for nearly a year, the steel rods are three-fourths of an inch in diameter and are supported on wooden posts about eighteen feet high, at either end of the cross-piece, which is eight feet long.

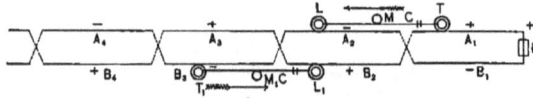

FIG. 145.—DIAGRAM OF GLYNDE TRACK.

because a connection of small resistance is continued at B between the piece of carbon and a piece of steel, which, being pressed out by a spring, follows the carbon for a short distance as the arm A begins to fly out. This contact is next broken, producing an electric arc; which,

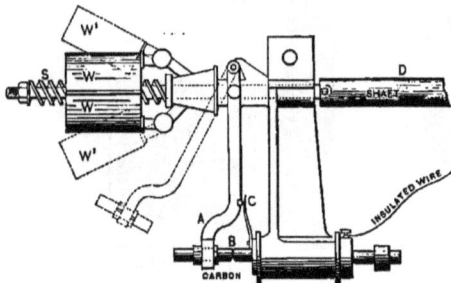

FIG. 146.—TELPHER GOVERNOR.

however, is instantly extinguished by the lever A flying out to the dotted position. The brake is shown on Fig. 147, and consists simply of a pair of weights, $W\,W$, which, at a limiting speed greater than 1,700 revolutions per minute of the motor, press the brake-blocks $B\,B$ against the rim $C\,C$, and introduce the necessary amount of retarding friction. In practice, however, with the gradients such as exist at Glynde, and which do not exceed one in thirteen, the economic method of cutting off the power automatically with the governor is all that is necessary to control the speed of the train, the brake rarely coming into action. With steeper gradients, the brake would be of more service.

The skips are trough-shaped. Each holds about two hundred weight, and is suspended from the line by a light iron frame, at the upper end of which is a pair of grooved wheels running on the line of rods. Ten of these skips, five each side of the motor, make up a train. At the charging end of the telpher line, the skips are loaded each with about two hundred weight of clay, the train thus carrying one ton. A laborer, by touching a key, starts the train, which travels at a speed of from four to five miles an hour along the overhead line to the Glynde station. Arrived there, another laborer upsets each skip as it passes over a railway truck, into which the clay is thus loaded. This upsetting, however, will eventually be performed automatically by means of a lever on

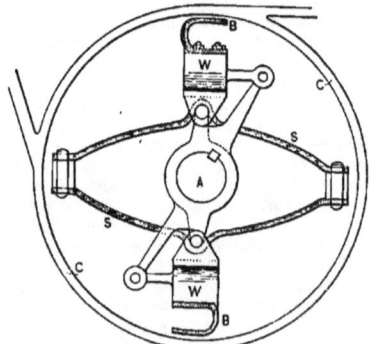

FIG. 147.—TELPHER BRAKE.

each skip, which will come in contact with a projecting arm as it passes over the truck.

The laborer at the discharging end of the line has full control over the train, and can stop, start, and reverse it at will, as can also the man at the other or loading end. There are two trains at Glynde, but only one is at present $6,000, that sum including outlay for an equipment to consist of stationary steam engine, generating dynamo, and five trains with electro-locomotives, with a capacity to carry over a hundred tons daily. The total cost of operation is put at six cents per ton of material carried. The figures of the Telpherage Company

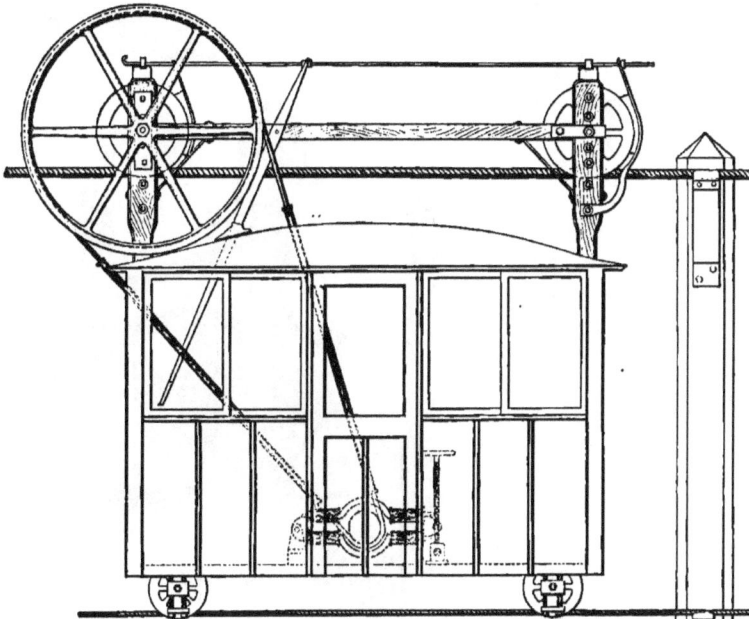

Fig. 148.—Chandler System of Suspension Transportation—Elevation of Locomotive Car.

used, that being found sufficient to deliver 150 tons of clay per week at the station. The trains need no attention when running, as they are governed to run at the same speed both on rising and falling gradients. An automatic block system is provided, so that as many as twenty trains can be run on the line without the possibility of collision.

As a few figures in regard to expense will be interesting, it may be stated, that such a telpher line as that at Glynde can be put up for of London show a cost of about $50,000 for a line ten miles long, to carry 30,000 tons of freight yearly. It need hardly be pointed out that such lines can be made important feeders for main lines of railway. To quote once more from the modest but brilliant electrician, the late Professor Jenkin, whose ideas have been carried out by his associates, Professors Ayrton and Perry:

"Mineral traffic is only one small part of the work which these lines can do. Where railways and canals do not exist, telpher lines will provide the cheapest mode of inland conveyance for all goods—such as corn, coal, root crops, herrings, salt, bricks, hides, and so forth —which can be conveniently subdivided into parcels of one, two, or three hundred weight. In new colonies the lines will often be cheaper to make than roads, and will convey goods far more cheaply. Surely I am not too sanguine in expecting that great changes will be produced in agriculture by these new facilities for trans-

FIG. 150.—SMALL FREIGHT CAR, CHANDLER SYSTEM.

port, coupled with the delivery of power at will from any point of the telpher road. It must not be supposed that I look on the new telpher lines as likely to compete with railways or injure their traffic. On the contrary, my feeling is that they will act as feeders of great value to the railways, extending into the districts which could not support the cost, even of the lightest railway. It is idle to endeavor to foretell the future of any new idea; but this much is certain—a novel mode of transport, offering some exceptional advantages, will be publicly shown on a practical scale to-day."

A system of this nature, for transporting freight and passengers, is now being introduced by the Suspension Transportation Company, of Boston, under various patents, covering

FIG. 149.—CHANDLER SYSTEM OF SUSPENSION TRANSPORTATION—END VIEW OF CAR.

FIG. 151.—MOTOR ON CHANDLER WIRE ROAD.   FIG. 152.—PASSING A POLE ON CHANDLER WIRE ROAD.

FIG. 153.—DESIGN FOR MAIL OR COAST SERVICE.

the use of electricity, steam, or other motive power for aerial transportation.

The general *modus operandi* of this company's system for an electric wire road is well

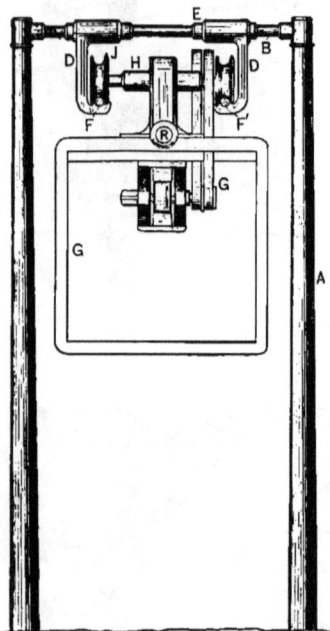

FIG. 154.—VAN DEPOELE TELPHER SYSTEM.

illustrated in the accompanying engravings. It is evident that such a system, while not interfering with the cultivation of the soil or with pasturage, is free from danger to man and beast, and that a mere right of way is sufficient without the cost of the fee. Again, it requires no cuttings or fillings, and is thus adapted to uneven, rocky, or uncleared land, and the turning of sharp curves presents no obstacles. In crossing streams no bridges need be used. The opportunities for the application of this system are vast, and the variety of uses to which it can be put is very great, both in cities and in the country.

An experimental line of this character, with a capacity for transporting several tons in weight upon the cables, is now in successful operation at the works of Mr. Leo Daft, and the details of the system will prove interesting.

The posts that carry the cables are placed thirty feet apart, and the cables are supported upon wrought-iron brackets bolted to them. The cables rest upon rubber strips placed in clamps at the end of the braces. This is clearly shown in Figs. 148 and 149, which represent the locomotive in elevation and in section as it appears suspended between the cables. The latter are of steel, the upper being two inches in diameter and the lower one inch. These are placed in the same vertical plane and are seven feet six inches apart. The current passes from the upper to the lower cable, the motors being in parallel or multiple arc between them. The motor within the car is geared to a large wheel mounted upon the same shaft with the forward grooved traction wheel, and is pivoted on one end, so that all slack in the belt may be readily taken up. The braking will be accomplished by electrical means, but hand brakes are also provided as shown, which can be applied through the medium of the long lever entering the car, the other end of the lever being attached to the rod connecting the brake-shoes on the two wheels. Two safety catches are attached to each car and are placed

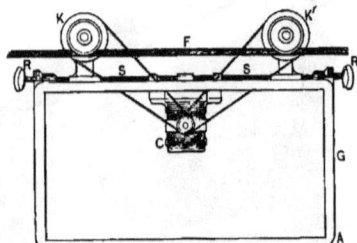

FIG. 155.—VAN DEPOELE TELPHER SYSTEM.

alongside the upper wheels. They prevent the car from leaving the cable in case of a running off of the wheel.

In all places where the grade is much above ground it is proposed to stretch a safety wire from bracket to bracket, thus insuring against accident.

The system is evidently adapted not only for the transportation of goods but for passenger traffic. The rate of speed will depend on the service to be rendered; as high as twenty-five miles an hour, and even more, is spoken of.

Fig. 150 shows the small "freight express" car operated successfully for several weeks at the Novelties Exhibition in Philadelphia last year. Figs. 151 and 152 illustrate the section of track now in use experimentally at the Daft Works, Greenville, N. J. The illustrations are made from photographs, and show accurately the slight sag in the wire caused by the passage of the car, weighing one ton. Fig. 152 shows the manner in which the car is enabled to pass the poles on the line. The car seen carries a small Daft motor taking its current from the wire cable.

Fig. 153 is an illustrative design embodying a plan proposed by the company to be adopted for mail service or for life-saving service along the coast. There can be no question that the use of such a device would immensely expedite the mail delivery between New York and Brooklyn, for example, or could be applied in the transportation of mail bags from the general post-office to the Grand Central Depot. The saving in time alone would be worth a great many thousand dollars annually.

The "telpher" system devised and patented by Mr. C. J. Van Depoele, has been in operation for some time at the factory of his company in Chicago. The method adopted by Mr. Van Depoele consists in suspending the car upon two cables supported by pillars and cross-bars, as shown in the accompanying illustrations, Figs. 154 and 155, which represent respectively a front and a side view of the arrangements. The hangers D, which support the cables $F F'$, are insulated, and the cables themselves form the positive and negative terminals of the motor. The band wheels $K K'$, are connected with the driving wheels of the electric motor C. Buffers $R R$ are also provided at the ends of the rods $S S$, the inner ends of which are provided with buffer springs.

CHAPTER XII.

LATEST AMERICAN MOTORS AND MOTOR SYSTEMS.

SINCE the foregoing chapters were written and prepared for press—almost entirely as they now appear—phenomenal activity has been displayed in America in the production of new motors and motor systems. It has therefore been thought well to bring the work down to

FIG. 156.—THE STOCKWELL MOTOR.

date by including all the latest developments in the motor field. This chapter should be read in connection with, and as supplementary to, Chapters VII. and IX.

At the meeting of the National Electric Light Association, held at Detroit, in August, 1886, the interesting fact was brought out that more than 5,000 electric motors of all sizes were in operation in this country at the present time. Among those mentioned as being largely in use was the Stockwell motor, which has for

some time been employed in the running of light machinery, especially sewing machines, and it is said that not less than a thousand of these are in actual service to-day.

The motor, which we illustrate in the engraving, Fig. 156, is enclosed within a case, one end of which is removed so as to expose the interior. The magnets are of the converging, consequent pole type, and form an integral part with the top and bottom of the casing. The two sides are cast separate and held together by screws.

The armature, or more correctly the armatures, for there are two of them, are shown in Fig. 157. As will be seen, they are of the Siemens shuttle-wound type, and are placed at right angles to each other. The commutator has four segments and the terminals of the wire on each armature are connected to opposite segments. The latter are not made parallel with the spindle, but are helical in shape, so that there is no break in the circuit at that point, since the brush passes the current to one armature before leaving the other. By this arrangement only one armature is in action at one time. Taking the one to the right, for example, it is at its maximum effect during the quarter revolution, when the polar faces of the armature are approaching the pole-pieces, and until they come directly opposite each other. During the next quarter revolution the armature is cut out of the circuit entirely; on the third quarter it again comes into the circuit until occupying the same relative position as in the first quarter; and, finally, in the fourth quarter it is again cut out. But it is evident that during each of these idle periods of the armature to the right, that to the left comes into circuit and goes through relatively the same cycle of operations. The action is quite analogous to that in two steam engines coupled with their cranks at right angles to each other. While one is passing over the centre, and practically doing no effective work, the other is in the position of

FIG. 157.—ARMATURE OF STOCKWELL MOTOR.

maximum power, with the crank at right angles to the line of stroke. In both cases, there can be no dead point, and the motion is smooth and continuous.

Where motors are applied to machinery required to be run at different speeds, some method of regulation becomes necessary, and in the present instance this has been worked out in a very simple manner. Where the mo-

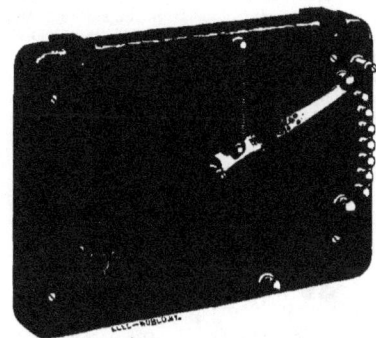

FIG. 158.—PERSPECTIVE OF RESISTANCE BOARD.

tors, as usual, are connected in series with each other, an adjustable resistance is provided which is contained within a box, such as that shown in perspective in Fig. 158, and in section in Fig. 159. This adjustable resistance is placed in a shunt to the motor and consists of a series of carbon bars of gradually decreasing conductivity. As the switch lever is passed over

the successive contacts, increasing resistances are introduced in the shunt, which consequently allows more current to pass into the motor and increases its speed correspondingly. The spring attached to the switch lever keeps the latter in the position of "no current" in the motor, and by attaching a cord or other device connected to a treadle, the operator on a machine has both hands free to work with.

The carbon resistance bars are copper-plated at their ends and firmly clamped, and by making them of gradually decreasing cross-section, a relatively greater increased resistance is thrown into the circuit as the switch lever passes from one contact to another. A wide range of control is therefore afforded with a comparatively limited movement of the

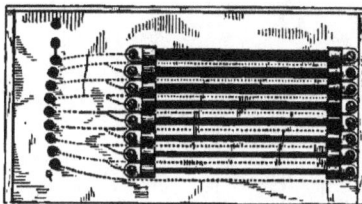

FIG. 159.—DIAGRAM OF RESISTANCE BARS.

lever. The motor is provided with a clamp, so that it can be readily attached to a table or work bench.

The varying loads which in practice are thrown upon an electric motor driving a number of machines, require that some provision be made for keeping the speed constant under each change of condition. Mr. John Beattie, Jr., of Westport, Mass., in solving the problem employs a motor whose field magnets are provided with several independent coils upon each leg, as shown at C, Fig. 160. Geared to the motor there is a governor, D, which oscillates a lever K' having a circular rack M at its extremity. The latter swings a lever K, which touches both the terminals L of the field-magnet coils, and those of a corresponding number of resistance coils I, at L'. The lever K, as shown, is in series with one or more of the field-magnet coils which are in parallel circuit and with one or more of the resistance coils, which are also in parallel circuit.

It will now be readily understood that any increase of speed in the motor operates so as to

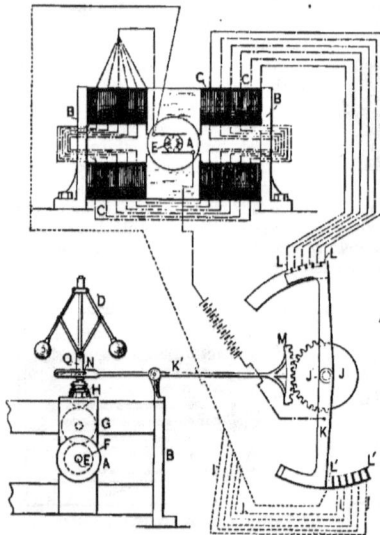

FIG. 160.—BEATTIE MOTOR.

cause the substitution of one or more of the resistance coils for a like number of field-magnet coils. This of course reduces the strength of the field and reduces the speed of the motor to its normal amount.

The armature of the motor employed by Mr. Beattie is provided with two sets of grooves, Fig. 161, $P\,P'$, parallel with the armature shaft, and at right angles to these run a series of deep annular grooves Q. The wires are wound according to the method of Siemens.

Recognizing the growing importance of the electric motor in the field of applied electricity, Mr. C. F. Brush has for some time past devoted his attention to the construction of a motor which should fulfil the conditions required in a successful prime mover. Steadiness of power and constancy of speed under all loads are two of the principal objects to be sought

for, and in the new motor these have been provided in a very ingenious way.

The motor, which is illustrated in the engraving, Fig. 162, closely resembles the Brush dynamo, which is too well known to require extended notice; but the devices added to the machine for the purpose of securing the advantages above mentioned are decidedly interesting, and merit a detailed description.

It will be seen that, mounted on the shaft between the commutator and the journal bearing, there is a cylindrical shell. The shell contains the governor by which the speed of the motor is maintained constant. The mode of regulation adopted by Mr. Brush consists in causing the governor to adjust the commutator automatically with relation to the brushes. To this end the commutator segments are mounted upon a sleeve on the shaft, so that they can be revolved to any desired extent under the influence of the governor.

The illustrations, Figs. 163 and 164, show the governor in detail. As will be seen, the commutator brushes $C\,C$ remain fixed, and loosely mounted on the shaft E is the commutator sleeve a, which turns freely. The commutator sections d are insulated from the sleeve a, and are connected to the armature bobbins by flexible wires, so as not to interfere with the rotary adjustment of the commutator. To the inner periphery of the cylindrical shell G, which is bolted to the shaft, the governor arms $H\,H$ are

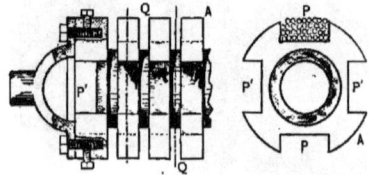

FIG. 161.—DETAILS OF BEATTIE ARMATURE.

pivoted. The inner free ends of the arms are connected to the opposite arms by means of spiral springs $I\,I$. In addition, the arms carry each an adjustable weight K. The links $L\,L$, attached to the arms $H\,H$, are connected to a disc upon the commutator sleeve. Hence, it will be readily understood that as the governor shell rotates with the pivoted weights $K\,K$, the latter, by centrifugal force, will be removed toward the periphery of the shell, and, through the medium of the connecting links L L, will impart a rotary movement to the commutator, varying its position on the armature shaft.

The action of the governor is precisely analogous to that in a steam engine. When in a state of rest, the springs draw the weights toward each other and maintain the commutator segments at the maximum point of effect with relation to the brushes. When current is switched on to the motor, the governor weights in their revolution are thrown outward and rotate the commutator, carrying the maximum points away from the contact points of the brushes and in the direction of rotation of the armature. This action decreases the effect of the driving current until a point is reached where the effect of the driving current is balanced by the load on the motor, and the speed of the latter remains constant. Now, should the speed of the motor be retarded by a decrease of current-strength with no corresponding diminution of load, or by an increase of load with no increase of current-strength, the governor balls will be retracted and drawn toward each other by the spiral springs, and thereby rotate the commutator in a direction opposite to the motion of the armature shaft, the effect of which is to move the maximum points on the commutator nearer to the brushes, and thereby increase the speed of the motor. On the other hand, should the speed of the motor be increased above the normal rate, owing to an increase of current-strength or to a decrease of load, the governor balls will be caused to recede from each other and rotate the commutator in the same direction as that of the armature shaft, and cause the maximum points on the commutator sections to be moved away from the brushes, and thereby decrease the speed of the motor. In this manner provision is made for all contingencies affecting the working of an electric motor. The parts constituting the governor are few and simple.

Another of the more recent systems designed for the purpose of running light machinery is exemplified in a neat combination of electric motor and battery, designed by, and named after, Messrs. Curtis and Crocker of New York.

The little motor, shown in Fig. 165, is series wound, having an internal resistance of .12 ohm, and is capable of carrying a current of 16 amperes with safety. At 2,000 revolutions it generates a counter electromotive force of six volts, with the current of sixteen amperes, and is said to exert a pull of five pounds on the circumference of the pulley, which is one and one-half inch in diameter. The magnet cores and

FIG 162.—BRUSH MOTOR.

pole-pieces are continuous, and are wrought-iron drop-forgings. The armature of the motor is of a novel construction, on the Gramme principle, and is completely enclosed, so as to exclude all dust and keep it from accidental injury.

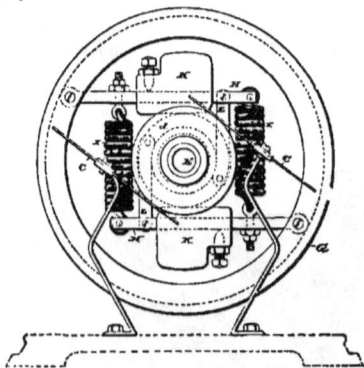

FIG. 163.—DETAILS OF GOVERNOR, BRUSH MOTOR.

In connection with the motor there is made a battery consisting of two cells, giving an electromotive force of nearly four volts, with a current of from eight and one-half to ten ampères. The chemicals used in the battery are made up in the form of bricks, which are dropped into

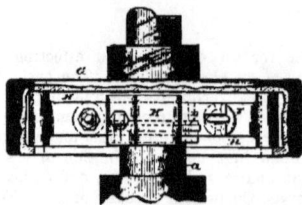

FIG. 164.—DETAILS OF GOVERNOR, BRUSH MOTOR.

the cells filled with water and soon bring the battery up to its full work. This makes the handling of the battery very convenient for those inexperienced.

Another feature is the method by which the speed of the motor can be regulated for fast and slow working. This is accomplished by sus-

pending the zincs of the battery upon a lever so that they can be immersed to any extent or raised entirely out of the solution. The lever is operated by hand, and falls into different notches. The strength of the current, depending upon the extent of the immersion of the zincs, can thus be regulated to any extent, and with it the speed of the motor.

Since the days of Pacinotti it has been known that dynamos and motors are reversible, and it is now known that it is almost impossible to put any assemblage of copper and iron together which, when a current is passed through it while in a magnetic field, will not show some evidence of motion. For some years past dynamo construction has been carried to a very high

FIG. 165.—THE "C. & C." MOTOR.

degree of efficiency, and machines have been built which would convert ninety-six per cent. of the mechanical energy delivered to them into electricity. Yet, notwithstanding this fact and the assertion of Prof. Henry A. Rowland, that the best dynamo must be the best motor, in other words, that the best apparatus for converting mechanical power into electricity must be the best apparatus for converting electricity into mechanical power, a statement which on its face carries the elements of truth, electric motors have remained for a long time a subject concerning which most literature was sadly at fault, both the theory of the motor, a knowledge of its action, and its practical application being remarkably limited. One of the causes, perhaps, of the difficulty of handling this subject, has been that most motor experiments have generally been conducted with crudely made dynamo machines, and without any definite idea of the relations that exist in the different parts of the circuit. The terms electromotive force, potential, current, and resistance, in their relation to what is commonly called the counter-electromotive force and to each other, have not been at all generally understood, nor even the law of the electro-magnet until Deprez and Hopkinson began their researches on the saturation of iron. Even to-day well-known scientific men differ on this latter law.

Deprez in Paris, Ayrton and Perry in London, and Sprague, among others, in the United States, have been the most active in developing the true theories of motors. Of the latter little was known in this line until the fall of 1884, when the Electrical Exhibition was held in Philadelphia. Mr. Frank J. Sprague had for some time before this been pushing his researches with energy, and at the Philadelphia Exhibition exhibited a number of machines which were the first of the kind ever shown. These machines were run on an Edison constant potential circuit. They were thrown into circuit gradually with a very strong rotary effort or torque, ran at constant speed with brushes at fixed points, and without any evidence of sparking, under all loads from the minimum up to the maximum allowed. In addition to those machines, Mr. Sprague showed others, one of which, starting under a heavy load, could be made to run forward or backward, fast or slow, at will, the reversal being made with an ease and rapidity and freedom from sparking which were remarkable. Another machine could be made to run in either direction, and was provided with adjustments so that it could be made to run at different determined constant speeds under varying loads. Since the exhibition of these machines, all of which were experimental and many of which are now in practical use, Mr. Sprague's progress in this work has been remarkably rapid. Some idea may be obtained of the operation of some of the different classes of machines built under the Sprague system from the following general explanations:

A motor when running may be looked upon as a dynamo machine propelled by a current; it has a field magnet like any other dynamo; it has an armature situated in that field which, either because of the attraction and repulsion of the lines of force, or of the double attraction and repulsion of the poles which are set up in

the armature acting on the poles of the field magnet, is caused to rotate. This armature rotating in the magnetic field has an electromotive force developed in it which is precisely of the same kind as would be developed were the motor driven by a belt instead of by a current. The strength of this electromotive force depends upon the resulting strength of field and the speed of the armature. This electromotive force, which may be termed a *motor* electromotive force, is ordinarily called the counter-electromotive force, because it is opposed to that of the line current which is flowing into the motor. The difference between this line electromotive force and that of the motor is what may be called the effective electromotive force, and determines, in combination with the resistance of the circuit, the strength of current which will flow in a circuit into which these elements enter. In any case of a single transmission from a dynamo to a motor the combination of these two determines the differences of potential which exist in the different parts of the circuit, which difference of potential determines the strength of current which will flow in any derived circuit. This counter-electromotive force likewise determines the efficiency of a motor or a system of transmission of power.

Motors may be described as belonging to one of three different systems. First, those in which the field magnet is excited by a coil in parallel circuit with the armature, that is, in shunt relation thereto; second, those in which the field magnet is in series with the armature circuit; and third, those in which there is a combination of these two circuits. There are in addition a very large variety of each of these classes, different conditions demanding different performance. Furthermore, similar machines may be placed upon three different kinds of circuits, their performances varying widely in each case. These three conditions are: First, the case of special transmission with varying potential and current; second, constant current circuits in which the main current is kept at a constant quantity, and third, constant potential circuits. The special transmission of power, unless carried out under certain well-defined laws based either upon a constancy of current or constancy of potential at some definite part of the circuit, is unsatisfactory, but if made according to a law no difficulties present themselves.

20

We will briefly consider the action of these different kinds of machines on two classes of circuits only. First, on the constant current circuit. If a series wound machine be placed upon such a circuit, the same current passing through the field magnet, it will develop a constant torque, which torque is directly proportional to the strength of the field magnet and to the current in the armature. If the mass of iron is sufficiently great this torque will be directly proportional to the effective ampère-turns in the field magnet, and the work done will be directly proportional to the speed. If the machine be at rest there will exist a difference of potential at the terminals of the machine equal to the product of the current and the resistance of the machine. When running, however, an electromotive force will be developed in the machine and the potential at the terminals of the machine will rise by the same increment. The work done may be likewise expressed by the product of this counter-electromotive force and the current, or $e\,C$, and is independent of the resistance of the machine. The resistance, however, determines, in combination with the other elements, the total efficiency of the motor. The total energy expended is the product of the difference of potential existing at the terminals of the motor and the current flowing, or $E\,C$. The efficiency then is $\dfrac{e\,C}{E\,C}$, or $\dfrac{e}{E}$, and the heat wasted $(E - e)\,C$.

When running at any particular speed the work will be increased directly as the field magnet strength is increased. So also will be the economy. The heat wasted with any given resistance in a machine under these conditions is a constant. The direction of rotation of such a machine can be reversed by reversing either the armature circuit or the field circuit; if both circuits are reversed, then the machine will run in the same direction. For many classes of work this kind of machine is exceedingly useful, because it admits of a great range of hand control. If such a machine, however, be put on ordinary work, and this work be lightened up, the machine will run faster and faster, and unless the field be weakened or the brushes shifted to check it, the speed will practically increase without limit. Every change of speed and every change of

load is accompanied by a corresponding change in the potential which exists at the terminals of the machine. Moreover, on a constant current, the motors being in series with each other and with lamps, this continual variation of potential is apt to cause trouble on the circuit, especially if the machines are not automatic, since, as already stated, with any fixed field the torque is constant, the work done is directly proportional to the speed. The machine has the highest efficiency when running at the highest speed.

With shunt machines, however, the action on the constant current circuit is much different. Here the current is divided in two circuits, such division, when the motor is at rest, being inversely proportional to the resistances of the two parts of the circuit. With such a motor, the field is weakest when the machine is at rest, and its torque or rotary effort is also very weak. If the load be not too great, as the speed of the machine increases a counter-electromotive force is set up, the potential at the terminals of the armature and field magnets rises, the current in the armature diminishes, and that in the field magnet increases. Provided there is sufficient iron in the field magnets, the torque or rotary effort will vary in a decreasing ratio until one-half of the current is flowing through the field magnet. At this moment the machine will be doing its maximum amount of work, and at less than fifty per cent. total efficiency. If the work be lightened, the machine will increase its speed until, when the work is entirely removed, there will be practically no current through the armature; all will have been shunted through the field magnets, and the potential at the terminals of the machine will be at the maximum. Such a machine will do the same total work at two different speeds and efficiencies.

If a machine be wound with a double set of coils it will behave very much the same as a shunt machine does, its field magnet being strengthened in a more or less rapid ratio, or being kept constant, depending upon whether the series coil is cumulative or differential.

Because of the fact that constant current circuits in ordinary use deal with small currents and very high electromotive forces, and do not admit of such perfect regulation, Mr. Sprague has preferred ordinarily to work the constant potential circuits, although some of his machines are running on the constant current circuits for other than automatic work; that is, for work where the speed is under control and where the work done for a given speed is constant.

There are two ways by means of which a constant current motor can be governed. One consists in automatically changing the counter-electromotive force by changing the position of the brushes on the commutator to positions more or less removed from their normal one. To this objection is offered because the proper position for the brushes of any machine is at the points of least sparking. The other method consists in varying the counter-electromotive force by automatically weakening the field as the load is diminished, or strengthening it as the load is increased. Several methods have been proposed for doing this, generally by the action of a centrifugal governor. To this also some objections are raised. Mr. Sprague, desiring to get rid of all such rapidly moving adjuncts to motors, is now engaged on a totally new system, which promises, he thinks, entire freedom from these defects.

On constant potential circuits the behavior of these different classes of motors is entirely different.

A plain series wound motor when there is sufficient iron in the field has a torque proportional to the square of the current flowing through it. It is capable of exerting a great rotary effort and doing a large amount of work at a slow speed. The range of speed for different loads is, however, great, and the motor is unfitted for ordinary work where steadiness of speed is an object; as the load is diminished, the speed increases and, if thrown off entirely, the motor will run faster and faster, the field continually growing weaker and the armature all the time accelerating its speed in a vain attempt to generate an electromotive force equal to the initial. For some classes of work, this kind of a machine, with some essential modifications, is exceedingly useful.

On the other hand, the shunt-wound machine will run fairly well on a constant potential circuit. The field, being excited independently of the armature, is constant, and since the load varies with the motor electromotive force, and the field is constant, it follows that the speed must vary with c. The torque is proportional to the current in the armature, and the

speed will be slowest with the greatest load and fastest with the lightest, that is, when $c - E$. The lower the resistance of the armature, the less the variation in speed.

It is with the third class of motors, when used on constant potential circuits, that the difficulties which are involved in the governing of a motor entirely disappear, and, without the use of any such apparatus as centrifugal governors or movable contacts, it becomes possible to satisfy the most exacting conditions, both as regards efficiency, steadiness of running, power to start under very heavy loads, and freedom from sparking.

When Mr. Sprague first proposed his constant speed machines for constant potential circuits, he enunciated the following seemingly paradoxical proposition:

In a motor with the armature and field magnet independently supplied, the work which the motor will do in a given time, its economy and efficiency, are all independent of the strength of the field magnet, provided the translating devices intermediate between the motor and whatever is the recipient of its energy are not limited as to the rate of transmission of the motor speed; and that in all cases where a motor is working on a constant potential circuit and not up to its maximum capacity, in order to increase the mechanical effect of speed or power, or both, or to compensate for any falling off of the potential on a line, it is necessary to weaken the field magnets, instead of strengthening them, and *vice versa*.

The strength of the field determines the speed at which a motor must run to get a required efficiency. With a given initial potential at the armature terminals, no matter how the load varies from the maximum allowed, the speed may be maintained constant by changing the strength of the field; such strength being diminished as the load is increased, and, *vice versa*, increased as the load is diminished.

These facts may be demonstrated as follows:

Let us consider the motor current as derived from mains having a fixed difference of potential, and the motor with its field and armature in shunt relation. In this case the armature runs with a velocity dependent upon the strength of field, the initial potential, the number of turns, resistance, etc., of the armature, and the load, and a counter-electromotive force is set up which regulates the armature current. The higher the speed the greater this counter-electromotive force. Let E be the initial and e the counter-electromotive force, and r the resistance of the armature. The current flowing in the armature is then $\dfrac{E-e}{r}$. With a given armature and given field e varies with the speed. The power at any given speed and strength of field varies with the current, and with any given current varies with the strength of field. The total work done is the product of the speed by the work per turn, and since the speed is as e and the work per turn as the current $\left(\dfrac{E-e}{r}\right)$, the total work done is expressed by $\dfrac{e\,(E-e)}{r}$. The efficiency is the ratio $\dfrac{e}{E}$. It will be seen that both these expressions—the total work done and the efficiency—are independent of any function of the field, but depend only on the initial and counter-electromotive forces and the resistance of the armature, and any given value of e can be attained with any strength of field by attaining proper speed.

Considering the speed of machine constant, its field alone being varied, and differentiating the expression for work done, $\dfrac{e\,(E-e)}{r}$, we have $\dfrac{de}{r}\left(E-2e\right)$ as the rate of variation of work.

It follows then that to maintain the speed constant with a current of constant potential under varying loads, when the load increases so that the speed would naturally decline, the field is weakened, the counter-electromotive force diminished and armature current increased, the tendency to reduced speed is counteracted, and there is an increase in the mechanical effect—power. For a decreased load the field is strengthened, the counter-electromotive force increases, the current decreases, the speed remains the same, and the power is decreased.

To maintain speed or power constant under varying initial potential, if the potential at the motor terminals increases, these mechanical effects increase or tend to increase. By strengthening the field an increased counter-electromotive force is produced, so that the increased

power or speed, or the tendency thereto, is counteracted, and this counteraction may evidently be itself considered a decrease in mechanical effect, whether the regulation is performed simultaneously with the increase of potential or before or after such increase. If the regulation is performed simultaneously, with a gradual change of potential, there may be less change in counter-electromotive force or armature current; but there is still the counteracting of the tendency to increased mechanical effect, which counteracting is itself a decrease of mechanical effect. For a decreased or decreasing initial potential, the field is weakened to counteract the decrease in mechanical effect which would otherwise occur, and therefore to produce an increased mechanical effect.

Hence to change the speed or power of a motor on a circuit of constant potential, the speed or power is increased by weakening the field, which produces a decreased counter-electromotive force and an increased armature current, and consequently the increased mechanical effect desired; and such mechanical effect is decreased by strengthening the field, and thus increasing the counter-electromotive force.

In brief, then, Mr. Sprague's method of regulation consists in strengthening the magnetizing effect of the field-magnet coils of the motor to decrease the mechanical effects, such as speed or power, or both, and *vice versa*, weakening such magnetizing effect to increase the mechanical effects, and under varying loads the speed is maintained constant by an inverse varying of the strength of the field magnets.

This may be accomplished in two ways, one by varying the field circuits by a mechanical governor which responds to any variation in the speed of the motor. This, however, is not satisfactory, and Mr. Sprague's ordinary method of working is to make use of certain coils in series with the armature and dependent upon it, which coils have a resultant magnet action which is opposed to that of the main coils of the machine. While the main principle is the same, Mr. Sprague has a number of different methods of applying it. The first has a series coil in series with the armature, and its action in the above laws will be understood from the following description. All these machines, it should be said, could be used as constant speed machines on constant current circuits, provided the field coils are properly proportioned for the current which they would have to carry, but with certain disadvantages, as will be shown.

The magnetic moment of a coil may be defined as the product of the ampères flowing therein by the number of turns, and if the main and governing coils are practically similarly situated with regard to the field-magnet cores, the magnetic field may be considered as proportional to the effective magnetic moment; that is, to the difference of the magnetic moments of the shunt and series field coils, so long as we are working on a straight or nearly straight line characteristic. This characteristic can be determined for any particular cores in any of the well known ways; for instance, by running the motor as a dynamo at a constant speed, passing variable known currents through the field coils, and noting the potential existing at the free armature terminals.

For a properly constructed motor the field magnet must at no time be too highly saturated, that is, it must be worked with a characteristic which is a straight or very nearly a straight line.

Let f denote the resistance of the main or shunt field coils; m the number of turns therein; r the resistance of the differential or series field coils, and n the number of turns; E, the difference of potential at the shunt terminals; e the counter-electromotive force set up in the armature; and R the resistance of the armature.

The work done $= e \dfrac{E - e}{r}$; that is, it depends

upon e, a variable quantity, and upon the constants E and r.

Now e varies with the speed and field, or the effective magnetic moment of the field, but the conditions are that the speed remains constant, hence e must vary with the field alone.

Current in shunt field $= \dfrac{E}{f}$;

Magnetic moment of same $= m \dfrac{E}{f}$;

Current in series field $= \dfrac{E - e}{R + r}$;

Magnetic moment of same $= n \dfrac{E - e}{R + r}$.

The effective magnetic moment must then be

$m \dfrac{E}{f} - n \dfrac{E - e}{R + r}$; and the conditions are such

that (for two different counter-electromotive forces or two different loads)

$$\frac{e}{e^1} = \frac{m \dfrac{E}{f} - n \dfrac{E - e}{R + r}}{m \dfrac{E}{f} - n \dfrac{E - e^1}{R + r}};$$

or, $\dfrac{e}{e^1} = \dfrac{m E (R + r) - n f (E - e)}{m E (R + r) - n f (E - e^1)}$;

or, $\dfrac{e}{e^1} = \dfrac{m E (R + r) - n f E + n e f}{m E (R + r) - n f E + n e^1 f}$;

or, $e\, m\, E\, (R + r) - e\, n f E + e\, n e^1 f = e^1 m E (R + r) - e^1 n f E + e^1 n e f$.

Cancelling we have $e\, m\, (R + r) - e\, n f = e^1 m (R + r) - e^1 n f$,

or, $m (R + r)(e - e^1) = n f (e - e^1)$,

or, $\dfrac{m}{n} = \dfrac{f}{R + r}$.

That is to say, the number of turns in the shunt coil must bear the same ratio to the number in the series coil as the resistance of the shunt coil bears to the sum of the resistance of the series coil and the armature.

This is the Sprague law of winding for a machine of the kind mentioned, and so wound it will be self-regulating for any constant potential up to the maximum allowed by the construction of the machine, and from no load up to the maximum.

There is a feature of motors so wound which may be here noticed.

The ratio of the magnetic moments of the shunt and series fields is

$$\frac{m \dfrac{E}{f}}{n \dfrac{E - e}{R + r}} \quad \text{or,} \quad \frac{m E (R + r)}{n f (E - e)}.$$

But $\dfrac{R + r}{f} = \dfrac{n}{m}$.

Hence the above ratio $= \dfrac{m\,E\,n}{m\,n\,(E-e)}$ or, $\dfrac{E}{E-e}$.

That is, the ratio of the effective electromotive force is the same as the ratio of the moments of the shunt and series coil.

When $e = 0$ this ratio becomes $\dfrac{E}{E} = 1$; that is, the moments are equal, and this means that, in a perfect machine, if both coils be closed and in their normal position, for any potential or current, a zero field, or practically so, will be formed, and the motor will either not start at all, or, if it does start, will run at a very great speed, take the maximum current at any given potential, and do little or no work at all.

How to obviate the bad effects of this peculiarity, yet to take advantage of it, will be shown later.

What has already been pointed out may be again stated, that the motor will regulate itself perfectly for all potentials so long as we work with a straight line characteristic, but it must be with a theoretical efficiency of not less than fifty per cent., for if we go below this, the governing coil works in the wrong direction.

Referring to the equation $\dfrac{m}{n} = \dfrac{f}{R+r}$, it will be seen that m and n can be increased in the same ratio. That is, if means are provided for varying the effective magnetic moments of shunt and series coils the motor can be set to run at different determined speeds. It is evident that f and r can also be varied to change the speed.

Let us now consider the same class of motors with constant speed, varying load, and *constant current*.

Let the resistances and turns be designated as before. Let K be the constant current. Let E be the variable potential at the terminals of the motor and e the variable counter-electromotive force.

The work done $= \dfrac{e\,(E-e)}{R+r}$.

We must eliminate E, making it dependent upon e and the constants R, r, f, and K, and hence the work can be expressed in terms of R, r, f, K, and e, of which e is the only variable quantity; e depends upon speed and field, but speed is constant. Hence our conditions require that with the same current we make e, and hence the work, variable, but by changes in the field alone.

Field current $= \dfrac{E}{f}$;

Armature current $= \dfrac{E-e}{R+r}$.

But $K = \dfrac{E}{f} + \dfrac{E-e}{R+r}$;

or, $f\,(R+r)\,K = E\,(R+r) + f\,E - f\,e$;

or, $f\,(R+r)\,K + f\,e = E\,(R+r) + f\,E$;

or, $\dfrac{E}{f} = \dfrac{(R+r)\,K+e}{R+r+f}$, and $K - \dfrac{E}{f} = \dfrac{f\,K-e}{R+r+f}$.

Moment of shunt field $= m\,\dfrac{(R+r)\,K+e}{R+r+f}$

Moment of series field $= n\,\dfrac{f\,K-e}{R+r+f}$.

Effective moment $=$
$$\dfrac{m\,(R+r)\,K + m\,e - n\,(f\,K-e)}{R+r+f}.$$

Our conditions are such that
$$\dfrac{e}{e^1} = \dfrac{m\,(R+r)\,K + m\,e - n\,(f\,K-e)}{m\,(R+r)\,K + m\,e^1 - n\,(f\,K-e^1)};$$

or, $e\,m\,(R+r)\,K + m\,e\,e^1 - n\,f\,K\,e + n\,e\,e^1 = e^1\,m\,(R+r)\,K + m\,e\,e^1 - n\,f\,K\,e^1 + n\,e\,e^1$.

Cancelling and transferring, $m\,(R+r)\,(e-e^1) = n\,f\,(e-e^1)$;

or, $\dfrac{m}{n} = \dfrac{f}{R+r}$,

which is the same law as found for constant potential.

The ratio of moments is $\dfrac{m\,(R+r)\,K+m\,e}{n\,f\,K-n\,e}$.

When $e = 0$ this becomes $\dfrac{m\,(R+r)}{n\,f}$.

But $\dfrac{R+r}{f} = \dfrac{n}{m}$;

hence substituting we have $\dfrac{m\,n}{n\,m} = 1$.

That is, if the motor is at rest and any current is sent through it a zero field will be produced. This of course follows from what has been already said about the constant potential motor.

The potential E which will exist if $e = E'$ and no current is passing in the armature is $f K$, and the maximum work is done when $e = \dfrac{f K}{2}$.

To be self-regulating, the motor can be worked up to this point, but not beyond it, for then the regulating coil works in the wrong direction.

In another variety of motor this series coil is placed outside the terminals of the shunt coil. The laws governing the action of this machine on a constant potential circuit may be described as follows:

Let the same letters of reference be used.

Then the potential existing at the shunt terminals will be $E − r C$.

$$\frac{E − r C}{f} = \text{shunt current;}$$

$$\frac{E − r C − e}{R} = \text{armature current;}$$

$$\frac{E − r C}{f} + \frac{E − r C − e}{R} = C;$$

$$E R − r C R + f E − r f C − e f = C R f;$$

or, $C R f + r f C + r R C = f E − e f + E R.$

Whence $C = \dfrac{f (E − e) + E R}{f R + (f + R) r}.$

Work done $= e\ \dfrac{E − r C − e}{R}.$

But since C can be expressed in terms of e and constants, the work can be also expressed in terms of e and constants.

$$m \frac{E − r C}{f} = \text{shunt current,}$$

$$n C = \text{series current;}$$

$$m \frac{E − r C}{f} − n C = \text{effective current,}$$

$$= m \frac{E − r \dfrac{f (E − e) + E R}{f R + (f + R) r}}{f} − n \frac{f (E − e) + E R}{f R + (f + R) r}.$$

But our conditions are such that

$$\frac{m \dfrac{E − r \dfrac{f(E − e) + E R}{f R + (f + R) r}}{f} − n \dfrac{f (E − e) + E R}{f R + (f + R) r}}{m \dfrac{E − r \dfrac{f(E − e^1) + E R}{f R + (f + R) r}}{f} − n \dfrac{f(E − e^1) + E R}{f R + (f + R) r}},$$

$$\frac{e}{e^1} = \frac{m E [f R + (f + R) r] − m r [f (E − e) + E R] − n f [f (E − e) + E R]}{m E [f R + (f + R) r] − m r [f (E − e^1) + E R] − n f [f (E − e^1) + E R]};$$

or, $m E e f R + m E e f r + m E e r R − m e r f E + m e r f e^1 − m e r E R − e n f^2 E + e n f^2 e^1 − e n f E R = m E e^1 f R + m E e^1 f r + m E e^1 r R − m e^1 r f E + m e^1 r f e − m e^1 r E R − e^1 n f^2 E + e^1 n f^2 e − e^1 n f E R.$

Cancelling we have—

$$m f R (e − e^1) = n f^2 (e − e^1) + n R f (e − e^1);$$

or, $\dfrac{m}{n} = \dfrac{f + R}{R}.$

That is, the number of turns in the shunt main field bears the same ratio to the number of turns in the series differential field, as the sum of the resistances of the shunt field and the armature bears to the resistance of the armature.

This is the Sprague law of winding for a machine of this character, and so wound the machine will be self-regulating for any constant potential and for any load up to the maximum allowed, and even with a resistance in circuit and with varying potential.

The same peculiarity exists in those motors which has been pointed out in connection with the first class of differentially wound motors, and this will now be described.

The ratio of the magnetic moments of the shunt and series fields is,

$$\cfrac{m E - m r \; \dfrac{f(E-e) + E R}{f R + (f+R) r}}{\dfrac{n f(E-e) + n E R}{f R + (f+R) r}};$$

or,

$$\frac{m E[fR + (f+R)r] - m r[f(E-e) + E R]}{f[nf(E-e) + n E R]}.$$

If $e = 0$, this becomes

$$\frac{m E f R + m E f r + m E R r - m r f E - m r E R}{f^2 n E + f n E R};$$

or,

$$\frac{m R}{n(f+R)}.$$

But $\dfrac{m}{n} = \dfrac{f+R}{R}$.

Hence the ratio becomes $\dfrac{m\,n}{n\,m} = 1$.

That is to say, if a motor of this character is at rest and the series coil in its normal governing position, and the circuit be closed to the motor, a zero field, or nearly so, will be produced; for under these circumstances the magnetic moments are equal, and either the motor will not start at all, or, if it does start, will run at a very great speed, take the maximum current at any given potential, and do little work or none at all.

Referring to the equation $\dfrac{m}{n} - \dfrac{f+R}{R}$ it will be seen that m and $f + R$ can be increased in the same ratio. This means that the determined constant speed of the motor can be varied for any given potential. Also m and n can be increased in the same ratio—that is, if means are provided for varying the effective magnetic moments of shunt and series coils, the motor can be set to run at different determined speeds.

This motor with constant speed, varying load, and *constant current* will now be considered.

Let the turns, resistance, etc., be designated as before. E is the variable potential at the terminals of the shunt field, and e the corresponding counter-electromotive force.

We must eliminate E and express the work in terms of e and constants; e depends on speed and strength of field, but since speed is constant e depends on the field alone.

$$\frac{E}{f} = \text{current in shunt field,}$$

$$\frac{E - e}{R} = \text{current in armature,}$$

$$K = \text{current in series field;}$$

and therefore $K = \dfrac{E}{f} + \dfrac{E-e}{R}$;

whence $f R K = E R + (E - e) f$;

or, $\dfrac{E}{f} = \dfrac{R K + e}{f + R}$.

The conditions are $\dfrac{e}{e^1} = \cfrac{\dfrac{m R K + m e}{f + R} - n K}{\dfrac{m R K + m e^1}{f + R} - n K}$;

or, $\dfrac{e}{e^1} = \dfrac{m R K + m e - f n K - R n K}{m R K + m e^1 - f n K - R n K}$;

or, $e m R K + m e e^1 - f n K e - R n K e = e^1 m R K + m e e^1 - f n K e^1 - R n K e^1$;

or, $(e - e^1) m R K = (e - e^1) f n K + (e - e^1) R n K$;

or, $\dfrac{m}{n} = \dfrac{f + R}{R}$.

This is the same law of winding that holds when a machine of the same class is used for constant potential; and the same remarks in regard to the zero field apply as in the former case.

Also, as in the former case, the speed for any given current can be varied by varying the resistance and turns or the effective turns.

From the foregoing demonstrations, it follows that a motor of either class depending for its regulation on this differential winding will regulate with a constant current only when working at less than fifty per cent. armature efficiency; and that the same machine with the same winding will regulate on a constant potential circuit only when working at over fifty per cent. armature efficiency.

The laws above set forth are for pure electrodynamic motors; if there is any permanent magnetism, as in hard cast iron, or where permanent steel magnets are used, the law of winding is modified in so far as the residual or permanent magnetism is the equivalent of an electro-magnetic moment; but in this case, too, there should exist a zero field if the governing coil is normally closed when the motor is at rest.

The fact already pointed out, that in the best self-regulating motor there is a zero or very weak field when the motor is started, necessitates in both classes of motors, especially when it is desired to start at a speed not greater than the normal, or when there is any load on the motor, in which case there is danger of burning out, the use of devices whereby the action of the governing coil may be modified.

This may be done by the introduction of a resistance, by shunting the coil with a resistance or by the variable shunting of the armature upon the main field. Mr. Sprague, however, prefers to use a switch to short-circuit the governing coil or to short-circuit and reverse it. If it is reversed, then the first rush of current makes a very strong field instead of reducing it to zero or nearly so, increases the rotary effort and prevents the burning out of the machine.

As an instance, if a constant potential motor has the series coil reversed when the full circuit is closed, if there is margin enough on the field characteristic we shall have a field twice as strong as the strongest normal field, four times the strength when the motor is doing its maximum work per unit of time, and a momentary rotary effort eight times that existing when the maximum work is on. As soon as the speed comes up, the governing coil is short circuited and then reversed, and then the motor is self-regulating.

Having obtained a machine which was thus automatic, Mr. Sprague made another step in overcoming the distortion or, rather, counteracting the distortion set up by the armature, by producing a distortion in the field magnets, which is dependent on precisely the same current that flowed through the armature, and he uses two methods, of which one only will be described.

21

Main field-magnet coils are employed in shunt relation to the armature, differential field-magnet coils in series with the armature, and additional accumulative field-magnet coils, also in series with the armature. The main field coils may be shunted upon the armature alone, or upon the armature and both the cumulative and differential series coils, or upon the armature and either of the series coils, the other series coil remaining outside the terminal of the main field shunt.

The object sought is to maintain the non-sparking points of the commutator cylinder constant by opposing the distortion of the magnetic field due to variations in the armature current by a counter distortion dependent upon such variations, whereby the magnetic resultant due to the armature and field magnet is unchanged, and the line of parallel cutting of the lines of force or point of least sparking is maintained in the same position.

In accomplishing the counter distortion of the field, the motor used is one in which the field-magnet cores extend in different directions from the field of force in which the armature revolves. The differential series coils are wound or arranged so that their greatest effect is produced on diagonally opposite parts of the magnetic field; and the cumulative series coils, so that their greatest effect is produced on the other diagonally opposite parts. The differential coils are arranged to have a greater magnetizing effect than the cumulative coils. A decrease of load, causing a decreased armature-current, tends to shift the magnetic resultant of the armature and field magnet; but this also decreases the magnetizing effect of all the series coils, and therefore the parts of the field principally affected by the cumulative coils are weakened, and those principally affected by the differential coils are strengthened, whereby a distortion of field is produced opposed to that produced by the decrease of armature current, and hence the magnetic resultant—the line of parallel cutting and the points of least sparking — remains unchanged. Thus no shifting of the commutator brushes is ever required, except on account of wear.

The arrangement of two sets of series field coils—one differential, the other cumulative—may be employed simply as a means of field regulation where it is not desired to produce the counter distortion. In such case the coils may be evenly wound on all the legs of the field magnet and used only to regulate the motor, being wound in the proportions above stated. The differential and cumulative series coils have a differential effect, which, as the differential coils predominate over the cumulative coils, produces a weakening of the total strength of the field magnet when the armature current increases, and a strengthening of the field magnet when the armature current decreases, and so maintains constant the speed of the motor.

FIG. 166.—SPRAGUE ELECTRIC MOTOR.

One of Mr. Sprague's methods for varying the speed and power is to wind the field magnets with a series of coils of different cross section and resistance. These coils are all in series with each other, and the bights of the coils are brought to a commutator.

In the simplest form of this motor, one end of the armature circuit is connected with a contact arm arranged to travel over a contact range, thereby making electrical connection with different sections of the field coils. The other end of the armature circuit is connected with one end of the series of field coils, preferably at the junction of such series with the supplying circuit. As the arm moves over the successive contacts the armature is shunted around a greater or less number of the sections of the field coils, and the difference of potential between the terminals of the armature cir-

cuit is varied between the maximum and zero; but in this arrangement it is not reversed unless the connections are reversed. The contact arm is provided with an adjustable contact piece, to allow for wear. In another method, where a single set of field coils is broken up into sections, the wires from the bights are connected in a special manner to a circular range of contact pieces, and a double arm, the two parts insulated from each other and bearing on opposite sections, is used. To each of the arms is connected an end of the armature circuit, and as the arms are made to travel around the contact range the difference of potentials at the arms or the brush terminals is reduced from the maximum to zero, is then reversed, and increases; and if the arms continue in the same direction another half revolution it is diminished to zero, changed again, and increased to the original maximum. The connections of the field sections are made thus: the first and last to single and opposite blocks of the range; the next adjacent sections have double connections to the next adjacent blocks on either side of the first ones connected; the next sections to the next adjacent pairs of blocks on either side, and so on till the blocks meet. In a third method two series of field-coil sections are used, and the bights connected to two ranges of contact pieces arranged in one or two circular or partly circular sets. Here, also, two arms insulated from each other are used, and the two arms are connected to the two terminals of the armature circuit—that is, the armature circuit becomes what corresponds to the galvanometer circuit in the Wheatstone bridge. As the arms are made to travel over the successive contact surfaces, the difference of potential existing at those arms or at the terminals of the armature circuit decreases from the maximum to zero, changes, and increases again to the maximum reverse potential.

In this last arrangement the speed, torque, and direction of rotation can be varied as rapidly as desired without any sparking at the moment of reversal.

In types 1 and 3 as referred to above and illustrated in Fig. 166, the standard machines are wound in the sectional method described and in addition are arranged so that when the motor is started the governing coil is in series with the armature and works accumulatively, while the potential at the armature terminals is progressively raised by moving along the field sections by means of the commutator at the top, part of the field sections being in series and the rest in shunt with the armature; this arrangement gives a very strong rotary torque; when full potential has been reached, the coarse coils are short circuited and then re-

up to the maximum, and promptly recover their normal speed under sudden and marked changes in load. There is no change necessary in the lead of the commutator brushes.

In the larger type of machines, however, two forms of which are shown in Figs. 167, 168, and 169, Mr. Sprague prefers to use a rheostat for throwing the machines into circuit, instead of winding the field coils in sections, because it is a much cheaper process of working, and as

FIG. 167.--SPRAGUE ELECTRIC MOTOR.

versed and the machine becomes an automatic machine, having the following qualities:

It can be thrown into a circuit at a dead rest or slow speed, without any disturbance of potential and consequent flickering of light. It can be started gradually, whether free or under full load, without burning of brushes or flickering of light, the potential being raised progressively from zero to maximum. If the load is such as to prevent starting until the full difference of potential exists at the brushes, the motor then starts with a rotary effort, or torque, very much in excess of what exists under the condition of maximum work. These motors are perfectly automatic, running at nearly the same speed for all loads

in case a heavy machine should be damaged in the sectional winding, it would be far more costly to make repairs to it than in the case where a rheostat is used. Of course this rheostat carries no current, except at the moment of starting the motor. These motors will lower a varying weight at the same speed that they will pick it up, and with the same freedom from sparking.

In another form, a variable speed machine, such as is now in use in the Western Union operating room, the rheostat is of peculiar construction. By a single movement of the switch the machine is thrown into circuit with a very strong field, the potential at the armature terminals is gradually raised, and after full potential has been reached a resistance is then thrown into the field magnet, and the field thus weakened so that the speed of the machine is increased. This method of working allows of the finest gradations of speed.

Another machine which is just being brought out for use on constant potential circuits permits of nine or ten variations of speed from a single switch movement without the use of any external rheostat, and another type permits of a like variation of speed and entire reversal of movement, also with the single switch. This latter type is designed for operation on street cars.

FIG. 168.—SPRAGUE ELECTRIC MOTOR.

For all ordinary work motors are built for constant potential circuits of about 100 volts. But the demand has come for 220 volt machines to go on the Edison three-wire circuits. Before long the Sprague Company expect to undertake some special cases of transmission of power in connection with mining work which involve the transmission of very large powers over long distances and under high pressure, such as 200 horse power sixteen miles with 1,000 volts at the motors, the entire electrical conditions being perfectly automatic, both at the generating and receiving end. A very large number of special problems are now being considered and motors are about to be used where none but the most enthusiastic believers in their adaptability will believe it possible.

Among the very interesting facts which have been brought out by experience is this, that on all ordinary classes of work motors do not

average over thirty-five to forty per cent. of the maximum capacity which may be safely demanded of them. A central station hence can take advantage of this falling off of work, and since there is an actual recovery of about sixty-five per cent., and this sixty-five per cent. is only about forty per cent. of the capacity of the motors which are in operation, it follows that, for every 100 horse power in a steam engine at a central station, including ten per cent. loss on distribution where the work is widely distributed, about 170 horse power can actually be contracted for.

We come next to Mr. Sprague's work in connection with electric railways. In December of 1885, a paper was read before the Society of Arts, Boston, by Mr. Sprague, on the subject of the application of electricity to the propulsion of motors on the elevated railroads of New York. This paper was an elaborate technical article which made a thorough investigation of the power used and its distribution, and indicated some of the methods which the writer proposed to carry out in the system that he had devised. The Third Avenue Elevated Road was taken as an example. It was shown that on this road the work was expended in three different ways, viz.:

1. In overcoming the inertia of the train, which was fifty-nine per cent. of the total.

2. In lifting the train on up grades, which amounted to twenty-four per cent.

3. In traction, seventeen per cent.

It was pointed out that because of the great frequency of stoppages and the necessity of high speeds on this road, most of the energy of the train which was put into it on getting under way and lifting it on up grades was of little value for traction. At the time the paper was written, there were at commission hours sixty-three trains in operation at one time on the up and down tracks. This was on a double track line of only eight and one-half miles length. The aggregate power that the engines on this road were capable of exerting was nearly 11,700 horses, the engines being of about 185 horse-power capacity. The average power exerted during the time trains were in motion was 4,640 horses, or, for the entire time on a trip, including stoppages, about seventy-four horses for each train of four cars.

The problem of how to handle this tremendous power on grades running up as high as

105 feet to the mile, with trains stopping every third of a mile, and sometimes not half a station apart, and sometimes reaching a speed of twenty to twenty-two miles an hour, is no mean one. It is true that small roads have been operated, among them one six miles long at Portrush, Ireland, but the conditions are totally different, and the demands which would be made upon an electrical system by the conditions of service on the elevated railroads present a new problem, and this problem not alone an electrical but also a mechanical one.

The great amount of power which is used on the elevated railroads and the distance over which the trains are hauled necessitate in the electric circuit large currents of a high potential. The potential decided on was 600 volts, and the experiments on the Thirty-fourth street section have been carried on with that pressure. No such electrical potential has ever been used in practice for this kind of work. The occasion for it has not existed, and motors which might be used with small powers over short distances and with low potentials would not avail there.

FIG. 169.—SPRAGUE ELECTRIC MOTOR.

The elevated system presents the result of a great many years of careful thought in engineering study. It is the culmination of a great many improvements. It has been carried to a degree of efficiency far higher than its most earnest supporter thought possible. It has been taxed to its uttermost. Recognizing these difficulties, Mr. Sprague has been for a long time engaged in the elaboration of a system which for some months past has been in experimental operation on the north track of the Thirty-fourth street branch of the Manhattan Railroad. Car No. 293, a full-sized standard passenger car of the elevated railroad, was placed at his disposal by the officers of the Manhattan Company, and this has been equipped and is now a thorough-going experimental car in which a great many problems are being worked out.

Some idea of the current and potential necessary for operating the Third avenue line may be easily gathered from the following facts.

As mentioned above, there are at one time 4,640 horse power actually being developed. With an efficiency of eighty per cent. for the motors, this would mean a current of ·43,291 ampères if one hundred volts were maintained at the terminals of the motors. With 600 volts this would be reduced to 7,215 ampères. The handling of a current of 7,215 ampères and of from 600 to 665 volts electromotive force is a somewhat difficult matter. A conductor to carry this amount of energy without a very large loss under ordinary conditions must be large, but with the stations properly put in and with the rails properly reinforced, together with the methods of working which will be described more in particular below, it will be seen that the difficulty of handling this has been very largely reduced. We will now enter into a somewhat detailed description of this system, both with regard to what is being done and what will probably be done in the future.

The first subject to be considered is the generating station. The system preferred by Mr. Sprague is the operation of a number of dynamos wound so as to generate at their normal speed and with a full load an electromotive force of about 670 volts at their terminals. These dynamos are wound for constant potential circuits. They are of very low armature resistance, and have high-resistance shunt fields. The dynamos may be built so as to maintain a constant potential under all loads at the junction of the mains with the track. There is one disadvantage, however, about this, and that is, if the electromotive force of the dynamos rises automatically, and there should be any very serious cross on the line, the machines might be burned out. Where they are wound with the field magnets in a simple shunt circuit, and no cumulative coil in series with the armature, any very bad cross on a line will lower the potential at the terminals of the machines, and while a very heavy load will come upon them for a brief interval of time, the drop of potential at the terminals will be sufficient to so far demagnetize the field magnets that the machines cannot be burnt out. In addition, however, to the ordinary shunt coil, Mr. Sprague employs a special winding, one which now appears in his railroad motors. This special winding is a coil in series with the armature, whose polarity is exactly at right angles to the polarity set up by the shunt coils, and is so proportioned that it automatically maintains the point of non-sparking coincident with the line of contact with the brushes on the commutators. This series coil would not have the effect of an ordinary cumulative coil. It would not raise the potential of the dynamos, but simply makes them non-sparking with fixed brushes under all loads.

Considering the length of road and the amount of power used it would be better to have two central stations instead of one. These stations would be of the most improved possible mechanical construction. The engines would be compound, condensing, and placed near the water. By this means the coal con-

sumption could be reduced at the central station to as low as two pounds of coal per indicated horse power. By having two stations, each removed about a quarter of the distance of the length of the road from either end, the size of conductor which is necessary for the middle rail is only one-fourth that which would be required were there only one station in the middle. Furthermore, the points of supply of current from each station should be maintained at the same differences of potential, to obtain which Mr. Sprague runs an independent line wire from station to station, with suitable indicators in it, showing whenever there is any inequality of potential existing at the supply points. This is done because the highest possible economy requires perfectly equal differences of potential at all points of supply, no matter how many the trains, nor where they may be situated on the track. The combined capacity of the two stations would be something more than equal to the highest total horse power appearing at any one time on the road, which, as we have seen, is about 4,700 horse power. It will be noted that there are no losses allowed for here. Why this is so will be explained in describing the system of braking which is used. This, then, would give a capacity at each station of about 2,500 horse power. On account of the rise and fall of the work done on the line, it being light at night, somewhat heavier during the middle of the day, and at its maximum during the morning and evening, this 2,500 horse power would be divided up into about four units, and to allow for any break-down of an engine these units would be of about 800 horse power each. The travel on the road is so perfectly known, and follows such a well-defined law of increase and decrease, that there would be no difficulty whatsoever in starting the engines at the proper time and throwing the dynamos into circuit. This system of power generation would be the most economical possible. With improved boilers and improved methods of burning cheap fuels, and with high grades of engines, compounding and condensing, results would be obtained which would be very gratifying.

We come now to the system of distribution, and this will be described more particularly with regard to the demands of the elevated railroads, leaving out particular reference to street roads, which form a department by themselves. There are many things which pertain to this system of distribution which would, of course, appertain to street work. The main rails are grounded, and form one side of the circuit, being connected to the structure of the road at suitable intervals. · Four single rails, together with the superstructure and the ground connections, form a path of very low resistance, and there would probably be no need of any reinforcement at the fish-plates. Should such reinforcement be found to be advisable, a short connecting piece would be made from one rail to another, very much in the same manner as is now done where the track is used for electric signals, or, as with the middle rail, a main conductor would be used. The other part of the circuit consists of a very light rail of special construction, thoroughly well insulated in a simple manner, and raised so that its top is from three to four inches above the plane of the ordinary traffic rails. This rail is not continuous, being of necessity broken at all switches, turn-outs, sidings, and cross-overs. The ends terminate about eighteen inches from all crossing traffic rails, and instead of ending abruptly, they are bent down slightly, so that when the collecting wheels, running on the central rail, leave or enter them, they do so without any shock or jar to the spring mechanism which carries them. This middle rail is further divided up into sections of any convenient length desired, say at intervals of 500 or 600 feet.

In addition to this middle rail, there extends along the entire length of the line a heavy, continuous conductor, thoroughly insulated. This is connected to both ends of each section by fusible plugs or cut-outs and a short branch circuit. The branch circuits of the cut-outs form a Y connection, the main conductor being secured to the stem of the Y and one end of each section to the arms of the Y. It will be seen now that in the normal condition of affairs if current is flowing from one part of the road to another part and there is no train between these two parts, that this current is carried over a double ladder-like circuit. The main conductor carries the major part of the current and the sectional working conductors a smaller part. So long as there is no train on the sections adjacent to any connection, it is evident that there is no difference of potential existing at the two opposite ends of the connecting branch, and no current will flow over it, al-

though very powerful currents are flowing past each end of it. These currents will, of course, be in the same direction. When, however, a train enters a section it does not make any contact whatever with the main continuous conductors, but only with the working conductor, and current is supplied to this working conductor from both ends, partially, it may be, through the working conductors next adjacent, but mainly through the branches connecting it to the main conductor; that is, there is a difference of potential set up in the different parts of this circuit, and parts which were inert before become active the moment a train passes on to a section, no matter whether the train be taking current from the line or giving it to it. The current that flows through these branches may be made to actuate any kind of special device which is necessary, and thus forms a perfect block system of signalling, which operates by the presence of a train upon a section, since this train automatically sets signals at both ends of its section. These signals are of a variety of kinds, visual or audible, or both. Some are day and some night signals; and the incandescent lamp, preferably two or more in multiple circuit with each other, are used for the night signals. Since the current on a motor is under perfect control, it follows that even if the train is at rest on the section, the engineer is able to set his signals.

One of the great advantages of this system of main and working conductors is this: If there is any bad cross or accident on the line the section will be cut out. The rest of the road will not be interfered with in the slightest, but the whole circuit will remain intact with the exception of the one particular branch of 500 or 600 feet, which has been affected. The signals may be made of that automatic character such that when a cross does occur sufficient to break the safety catches of that particular section a signal is set and cannot be replaced until the section is repaired. We have here a perfect safeguard against any extended disabling of the line. Furthermore, if it becomes desirable at any time to operate a signal at only one end of the section, the other end of the section can be cut out. If repairs are made, a section of the road being taken out or replaced, the track foreman can at once cut that particular section out of circuit, and after his repairs are made put it in again without interfering with the main line. In addition to these devices, the main working conductor can be divided up into sections and switches inserted, so that if it be desirable to cut out any extended portion of the track in case of any accident which makes the passage over a section of the track inadvisable, as in case of a fire, that portion can be cut out without the necessity of disconnecting each individual section.

In addition to these arrangements, the conductors of like potentials on different tracks and switches are connected by cross circuits which tend to equalize the potentials on the

We now come to the question of motor construction. The elevated railroad presents a special problem, as the strength of the superstructure is limited. At present the trains are drawn by locomotives which aggregate about twenty-two and a half tons in weight. Of this weight only fifteen tons is available for traction, this being the weight on the drivers. The weight of twenty-two and a half tons is centred in a very small space. Immediately behind the locomotive is the forward truck of a car with a proportionate weight of nearly nine tons. There is then a total weight of over

FIG. 170.—SPRAGUE ELECTRIC RAILWAY SYSTEM—ONE-CAR TRAIN.

line, especially where there are any bad joints in the rail, and also when one track is more heavily loaded than the other. Another great advantage of these cross connections is that the current generated by trains running on down grades and stopping, can not only be sent back to the conductor on its own particular track and circulate through the system, but it can take a shorter and more direct path to the opposite track where a train may be moving on the up grade or just starting. It should be further stated that both tracks are supplied from the same source, forming one complete circulating system. All motors are run in parallel circuit with each other, the current in each being independent of the current in all others, and the motors on the one track are in parallel circuit with the motors on the other.

thirty-one tons in a space of about thirty-six feet, which is less than the distance between two columns. The consequence is that the strains, both tensile and shearing, are very great; but these strains are not the only source of danger. The vibration set up by a moving train, both vertical, due to the weight, and longitudinal, due to the motion of the train, has a shattering effect which is very great. It tends to loosen the bolts and badly strains the whole structure. There is an additional vibration due to the reciprocal strokes of the steam locomotive and its consequent unevenness of pull. If an electric locomotive were applied to handle a train, and it were made of fifteen tons weight, it would pull more than a steam locomotive of equal weight, since all of it could be put upon the driving wheels, and there would bo no necessity of additional truck wheels. But a fifteen-ton electric locomotive properly constructed and handled would pull even more than a twenty-two and a half ton steam locomotive with fifteen tons on its drivers. If the weight was distributed on four wheels, the wheels being on two perfectly independent axles, there would be absolutely equal pressure on each. This, however, is not the case with a steam locomotive. In addition to this, the strain could be simultaneously brought on all the wheels of an electric locomotive with such a perfect progression that they would adhere to the rail more firmly than an equal weight

This is the manner in which a single car is now being operated on the Thirty-fourth street branch of the Third Avenue Elevated Railroad in New York city. The accompanying illustrations, Figs. 170, 171, and 172, show the car in one and two unit combination, and in end view, as it appears upon the track. The truck upon which the car is mounted is shown in perspective in Fig. 173, and in detail plan and elevation in Figs. 174 and 174 a.

As will be seen, the latter represent a standard iron truck such as is in use on all the new cars of the elevated railroad, with, of course, some omissions and changes which were neces-

FIG. 171.—SPRAGUE ELECTRIC RAILWAY SYSTEM—TWO-CAR TRAIN.

where the motion is derived from a reciprocal movement. Furthermore, there is a certain amount of increased adhesion of the wheels, just how much it is impossible to say, because each in the space between the axle and the it varies under different conditions, and this is probably due to the heating effect of the current passing from the rail into the wheel.

Another method of handling the cars, and this is the most logical, although it may be a somewhat more costly method of working when dealing with old rolling stock, is the placing of the motors underneath the cars on the trucks which carry them. In this way at least one-half of the weight of the car and the passengers, as well as the motors, is available for traction.

If the motors are thus placed under the cars, each can be made an independent unit, or a dozen cars can be operated in a single train by a small regulating truck placed ahead of them.

sary for attaching the motors. The principal omission is that of all braking apparatus. There are two motors carried on this truck, each in the space between the axle and the centre cross-piece. The field magnets, which are made of the finest selected scrap wrought iron, are built up of four segments, all forming parts of circles. Two of these form the pole pieces and to these are attached heavy bronze hangers. The latter carry the armature, which is wound on a special modification of the Siemens system, and has at each end forged steel pinions of three inches face, and 3.7 inches diameter on the pitch line. There are thirteen teeth only. The hangers are extended and embrace the axle, which is turned off to a perfectly smooth surface, leaving a small shoulder at each side. Part of the hangers extending from the magnet pole pieces embrace one-half of the

22

axle, and the opposite half is embraced by heavy bronze caps, and inside each there are split liners to take up the wear. The armature shaft, as it passes through the hanger, is carried by two curved self-concentrating sleeves.

On the axles, close to the hub of the wheels on each side, are two split gears. These differ in character. One is keyed and bolted directly on to the axle, which is first turned off, and is

FIG. 172.—END VIEW OF CAR.

a fixture. The other is composed of four parts, two being inner webs which are keyed on to the axle; the two outer ones form the geared section and are bolted together and have corresponding webs projecting inwardly, and fit snugly both on the outer edge and on the face of the webs which are keyed to the axle. The outer and inner webs are held together partially by the method in which they are turned up, but principally by bolts passing through them which work in curved slots. These, then, constitute adjustable split gears, and are probably a new thing in mechanics. The gear wheels are of an especially fine grade of cast iron, and are of the same face as the pinions which mesh into them. The number of teeth

in these gears is sixty-six; they are of the involute cut, so that if the motor should be moved to or from the axles slightly, the gears will still run perfectly true, with only a little more or less closeness of meshing. The pinions on the armature shaft are set so that the one is half a tooth in advance of the other. Ordinarily, it would be a very difficult matter to get the splines on both the armature shaft and the axle and in the pinion and gears so that they would mesh smoothly when running forward and backward, and it was for the purpose of getting rid of this trouble that the adjustable split gear was designed. It is now only necessary to key the two pinions, one fixed gear, and the web of the other gear in position without any regard to their meshing. The motor is then swung into position, the hangers made to engage the axle, the caps are put on, and the motor being moved forward and backward two or three times while the bolts of the adjustable gear are slack, this gear will assume a perfectly correct position. The bolts are now tightened up and there is thus a nest of double pinions and double gears all meshing with absolute precision, no matter whether the motor runs backward or forward. The method of mounting produces a concentric motion, and by this means the driving and the driven axles are maintained absolutely parallel in two planes under all circumstances.

To allow the motor freedom to follow all the movements of the independent axles over frogs and switches, and also for taking part of the weight of the motor off the body of the axles and to throw it on to the boxes, and one end of the motor is suspended at its centre by a bolt passing through the cross girders. This bolt is adjustable, and the upper part is held by a very stiff spring in a state of compression, which spring is in turn supported by a wrought-iron saddle. The motor is then, so to speak, weighed or flexibly supported from the body of the truck. There is also a smaller spring to take up any back movement or tendency to lift of the motor. This suspension is directly in the centre of the pole piece, and the field magnets, which are grooved in the form of a circle, are independently detached from the pole pieces, one of them being put on after the motor is in place.

Because of the relation between the teeth in the pinion and the split gear, it is necessary for the armature shaft to make sixty-six revolutions before the teeth engage in the same way, and each tooth of the pinion must in turn engage every tooth in the gears. It will be seen also, since the motor is suspended at one end by the truck axle and at the other by compression springs operating in both directions, that whenever the axle is in motion there is always a spring touch, so to speak, of the pinions upon the gears. Barring friction, a single pound of pressure exerted in either direction will lift or depress the motor a slight amount.

1,500 to 2,000 pounds upon each gear. Strain has also been put upon these gears as suddenly as it is possible to close a circuit across 600 volts, and without injurious effect.

This method of mounting motors tends to produce an absolutely perfect form of gear, and has practically obviated the noise which was at first anticipated.

Designs for motors of from 200 to 300 horse power mounted on these same principles will soon be finished, and the motors constructed and put into operation.

Fig. 173.—Truck of Sprague Car.

It follows that no matter how sudden a strain, nor how great, it is impossible to strip the gears unless the resultant strain is greater than that of the tensile strength of the iron; because the moment that the motor exerts a pressure upon the gears, at the same instant do the spring supports allow the motor to rise or fall so as to give somewhat, and no matter how sudden the strain is brought upon the gears it is always a progressive one. The result in practice has been that with a weight equivalent to two tons upon each thirty-inch wheel these wheels have actually been skidded in continuous rotation upon a dry track and the strain necessary to do this amounts to from

We now come to the electrical features of the motor. The armatures shown in the illustrations have a special modified form of Siemens winding. The shafts are built up of the finest forged steel, and the body of the armature is built up with alternating layers of tissue paper and very thin iron discs, such as are used in the Edison machine, which reduces the heat loss due to Foucault currents to a minimum. The difficulties first experienced in dealing with currents of such high electromotive force and large volume have now been overcome. The bodies of the armatures are thoroughly japanned and baked, and the utmost precaution is taken in putting on the different coils of wire to insulate them both from the body of the armature and from each other by the use of a material which offers very high resistance to inductive discharges. The commutators are built of the finest copper, and no insulating material is used other than that just mentioned and fine selected mica.

One of the fundamental features of this system of electrical propulsion is to get rid of all adjustments and to reduce it to the simplest possible system of working and at the same time to maintain as high an efficiency as possible of the motors themselves. For this purpose it was necessary, because of the limited space available, to make the motors of light weight and yet capable of developing a very intense

narily extends over the first third or half of its speed.

3. Variation in the speed of the armature after full potential has been reached at its terminals.

The first characteristic is obtained by bringing the field magnets to a very high degree of saturation. Current is then admitted to the armature under perfect control, and the potential at the armature terminals gradually increased, thus increasing the current until the rotary effort is sufficient to start the train from a state of rest. When the motor is in this condition the torque, or rotary effort, is directly proportioned to the strength of the field magnet and to the current flowing through the armature.

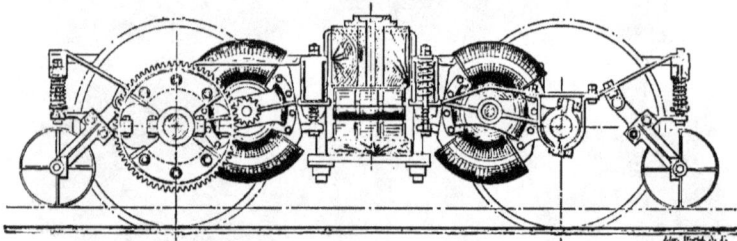

FIG. 174.—ELEVATION OF SPRAGUE CAR TRUCK.

magnetic field. The form adopted for these motors has given these qualities. The motors themselves are built entirely of the finest selected scrap iron specially forged. It was necessary further to have a wide range of speed under full potential at the armature terminals, and hence it was necessary, also, to have a wide range in the magnetic intensity of the field magnets.

A motor when acting in this manner is to be considered under three entirely different conditions:

1. When it is at rest, and it is desired to get the greatest possible torque or tractive effort. This tractive effort should be under perfect control and should necessarily be greater than that which the motor could exert for any very long continued time.

2. When exerting a continuous traction under accelerating speed. This is necessary in getting a train under way, and this effort ordi-

As soon, however, as the armature starts to rotate, a different condition exists. It is now necessary to exert a continuous traction; but the motor, on account of its accelerating speed, is generating an increasing electromotive force of its own which is counter to that of the line, and the difference between this counter electromotive force and the line electromotive force determines the current through the armature. Consequently, it is necessary, while maintaining the field magnet at the same strength, to still further raise the potential at the terminals of the armature by means of which the current is kept at the same strength. It is impossible to maintain a constant tractive effort in any other way under these conditions. The potential will soon equal the initial, and the motor will be doing its maximum work per unit of time. It is now necessary to accelerate the speed of the train, and this is done by weakening the field magnets. This principle of weakening the field magnets to increase the mechanical effect of a motor at all times when not working up to the maximum was brought out by Mr. Sprague some time ago, when he enunciated the principle, already referred to above.

In a motor with the armature and field magnet independently supplied, the work which the motor will do in a given time, its economy, and

strengthening them. The result is that, if running on a level at a certain speed and a grade is met, and it is desired to get up that grade at the same speed, it is necessary to weaken the field magnets. If the potential falls off and it is desired to keep up the same speed, it is necessary to weaken the field magnets, and, conversely, if it is desired to slow down, it is necessary to strengthen the field magnets.

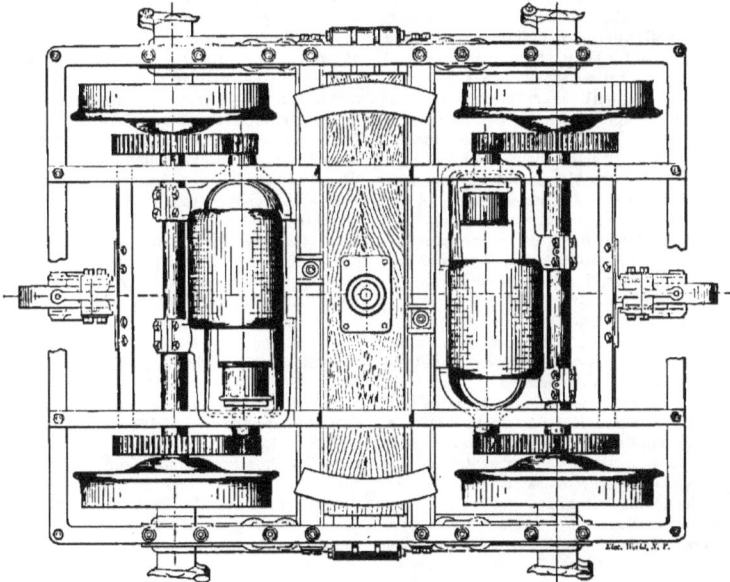

FIG. 174a.—PLAN OF SPRAGUE CAR TRUCK.

efficiency are all independent of the strength of the field magnet, provided the translating devices intermediate between the motor and whatever is the recipient of its energy are not limited as to the rate of transmission of the motor speed; and that in all cases where a motor is working on a constant potential circuit and not up to its maximum capacity, in order to increase the mechanical effect either of speed or power, or both, or to compensate for any falling off of the potential on a line, it is necessary to weaken the field magnets, instead of

A motor, when running, may be considered as a dynamo driven by a current. It generates an electromotive force dependent upon its resultant strength of field and the speed of the armature, and is independent of all other things. It follows that if the field magnet be under proper control, this counter-electromotive force is under perfect control under different speeds, and can be made greater or less in relation to the initial-electromotive force, and consequently the motor can be made to do whatever work is desired of it.

This system of handling a motor, which is an essential departure from previous methods, has been carried out to its logical conclusion in braking the train, as will be indicated later.

The winding of the field magnets of the motors is peculiar. One of the great difficulties which has invariably been met with in working with motors is the change of lead necessary to get the brushes at the point of non-sparking. It is considered necessary when dealing with large powers, if there is going to be continuous and successful running, to maintain the brushes at such a point. This change of the lead, caused by a distortion which is set up by the armature, varies with every change of load, with every change of the armature current, and with every change in the field-magnet strength. It has furthermore been the habit where any considerable power has been developed to use two sets of brushes, one for forward and the other for backward motion. Mr. Sprague has entirely obviated the necessity for doing this by an arrangement which is as simple as it is efficient. This consists in the method of constructing, winding, and connecting up the field magnets. The latter are wound with two sets of coils. One of these is a fine shunt coil which is in series with an independent regulating resistance and produces the normal poles; the other is a coarse coil in series with the armature, which tends to produce poles at right angles to the normal poles, and this circuit is included in the reversing switch, so that when the armature circuit is reversed the current in the coarse coil is also reversed. There are then four poles set up in this machine, two being normal and variable at will, the other two being abnormal, variable poles dependent upon the current flowing through the armature.

In the normal arrangement of circuits, the two sets of field coils, fine and coarse, combine to set up a resultant polar line which is distorted or rotated in the plane of rotation of the armature. Since any increase in the armature current causes the same increase in the series field coils, the tendency to distortion by these two elements will always vary to the same extent, and the resultant position will be always the same no matter what is the extent of variation of current. If the strength of the field magnet is varied independently of the armature, by changing the resistance in the shunt field circuit or by a variation of potential on the line,

while there is a tendency to change the armature distortion, there is an equal and opposite tendency to change the distortion due to the series field coils, and so this, also, has no effect. If the direction of the armature current is changed, so also is that of the current in the series field, and hence the direction of each distortion is changed; but they still oppose each other and vary equally and oppositely as before, and there is still no change in the non-sparking points. It is immaterial whether the change in direction of armature current is due to a change of terminals in changing the direction of rotation of the motor, or is caused in changing the motor into a generator by strengthening the field. Hence, the motor will run in either direction on a circuit of constant or varying potential, with a single or double set of tangential or end-contact brushes, with no change of lead, and, consequently, with no necessity for changing the position of the brushes. The position of the brushes having been once properly adjusted, it is made independent of the amount of work the machine is doing, or the speed at which it runs, or whether it is acting as a dynamo or as a motor. It is likewise independent of the strength of field and of the armature current so long as the magnetic moment of the field sufficiently exceeds that of the armature.

As will be seen from the end view, Fig. 172, there are at each end of the car three vertical switch rods, each connected by movable links with rods running through from one end of the car to the other. These rods have projecting fingers which operate the levers of three very rapidly-moving switches; the movement of these switches is independent of the rapidity of movement of the hand, which simply stores up energy until a certain point is reached, when the lever is freed and the switch thrown over automatically. These three switches are employed as follows: One for breaking the main circuit; another for reversing the armature circuit; and a third for detaching the armature partially from the line and closing it upon a local regulating apparatus. The movement of the handles on the vertical rods are similar at each end. Forward motion of one means forward movement of the car; forward movement of another means closing the main circuit; and a forward movement of the third means also a throwing off of the brake circuit. So that when a man stands at either end of the car, precisely the same movements mean the same thing as he looks up the track. In addition to these three vertical rods, there is a fourth rod which connects by a bevelled gear with a rod running through underneath the car, and provided with universal joints so as to allow of any necessary adjustment. The top of this rod carries a wheel very much like a brake wheel, and it connects with a regulator which consists of a series of resistance coils. These are so arranged that by the continuous movement of the regulator handle they are first cut out of the armature circuit, while the field is maintained at a high saturation, thereby raising the armature potential, and then cut into the field circuit in reverse manner, thereby weakening the field. This regulator governs also both steps of braking the train.

The current is taken from the centre rail by three conductors, two of which are bronze wheels working on pivoted arms under compression springs. They are provided with adjustable nuts to regulate the tension, and lock nuts to prevent the wheels dropping more than a certain limited amount when leaving the middle rail. The arrangement of contacts is such that the car will span thirty-foot spaces without breaking the circuit. The other part of the circuit comes through the wheels of the truck, so that one part of the apparatus is continually grounded. The collector and the main circuits both run to fusible cut-outs before they reach the main braking circuit, and the armatures are also independently supplied at both ends with similar cut-outs. The armatures and the field magnets are all in parallel circuit with each other.

This is the first instance in which two independent motors have been simultaneously controlled from the same regulating source, and by the methods employed it is perfectly possible to control twenty motors in the same way. When it is considered that the speeds vary from zero to 1,200 revolutions a minute, and the speeds of the two motors should be the same, it will be seen how important a step has been taken. The torque or rotary effort of these motors under slow speeds is very great, and they are able to start from rest and propel two full-sized cars up the maximum grade on the elevated railroad. The motors weigh about 1,200 pounds each.

We come now to the system of braking, which is the logical sequence of the system of controlling motors originated by Mr. Sprague. As is well known, when a motor is in operation it is generating an electromotive force. In other words, it is acting like a dynamo, and since this depends upon the strength of the field magnet and the speed, and since the field magnet strength is under positive control, it follows that this motor electromotive force can be made to equal the initial motive force and even to exceed it. When this electromotive force of the motor thus predominates, the machine will become a generator and give current to the line, and its mechanical effects are reversed so that it brakes the train instead of propelling it; and the current generated by it, and the braking power, or reversed mechanical effect, are now controllable by further increasing or re-diminishing the strength of the field, and the new dynamo can now be changed back into a motor instantly at will. The mechanical energy received by the reversed motor and delivered as electricity to the line depends upon the mass of the train and its velocity. In running on a down grade there would naturally be an acceleration of speed, but this method of braking can limit that acceleration at any desired point, or the motor can be slowed down when running on the down grade. This is done, of course, by strengthening the field magnets. Since the energy of the train is now being used to run the motors as braking dynamos, the train will be run at a certain constant speed down grade; or if the field magnets be still further strengthened, the train will slow down; this occurs also in the ordinary process of stopping. The diminution of speed, however, reduces again this motor electromotive force, and hence the field magnet has to be strengthened still further as a train slows, until the speed is reached which, with the strongest field magnet, will give a motor electromotive force equal to that of the line. This point with the motors in question is at about one-third speed, or seven miles an hour. Hence eight-ninths of the energy of a train moving twenty-one miles an hour is sent back to the line on current to relieve the generating station. In fact, the system as here set out has the advantages of a cable road, together with other advantages which the cable road does not possess; because not only do the trains running on down grades help the trains running on up grades, but those which are slowing down likewise give up their energy to the system. In fact, this system is one in which trains slowing down and running on down grades supply the current for trains running on up grades and starting, and the central station becomes a differential factor to make up for the loss of conversion and reconversion and to provide for traction and loss in the conductors.

Were the machines perfect converters of energy, this would make the power taken by a system almost independent of grades and stops, and would simply be that necessary to provide for continuous traction and for loss on conductors. Of course, this perfection of conversion can never be reached. It does, however, make a difference of forty per cent. in the power required to operate the electric railroads at the central stations, in the losses on conductors of a given size, and in the investment necessary in the central stations. As a matter of fact, there would be required at the central station only such horse power as is to-day actually developed at any one moment on the elevated railroads, which is only about two-fifths of the capacity of the motors or engines.

Hence, instead of 7,215 ampères of current being supplied from the central stations at two points, it is supplied from as many additional moving stations as there are trains being checked on a down grade and stopping. Sixty per cent. only of this current would come from the main station; that is, 4,329 ampères, or 2,165 from each.

The final step of braking is done by partially detaching the armature from the main line when its motor electromotive force is equal to that of the initial, at which moment there is no current flowing through it, and closing it upon the same local regulating apparatus which is used for regulating the speed and power, and the first step of braking. By this means the train can be brought to a full stop. All these steps of braking are under the most perfect control, but if necessary the braking can be so sudden as to cause the wheels to have a continuous skidding rotation; not such a skidding as is caused when an air brake is put on too hard, but a rotating slip which will be just enough to relieve the armature when the strain on it has come to a certain point. This is the most perfect method possible of braking, because fixed skidding is an impossibility, and the wheels will turn until the train comes to a dead stop, although where the braking power is put on too suddenly and exceeds the grip of the wheels, they will relieve themselves by slipping just enough to keep the braking at the maximum limit.

With the switch in position for the last step of braking, the car can be allowed to creep down the maximum grades at a snail's pace with a movement so slow as to be almost imperceptible.

It is the customary practice to stop at the Second avenue station of the elevated railroad, which is on a ninety-five foot grade, without the use of any shoe-brakes, although the rear truck is fitted with these and can be operated at either end of the car.

By a slight reversal of the armature effort, the car will stand at a dead rest on this grade.

The energy of the train which is expended in the last step of braking can be used in heating the car, and some interesting experiments are now being carried on at the Thirty-fourth street station.

It should be noted that at present the generating station for this experiment is situated on Twenty-fourth street, so that the current at times is carried about three-fourths of a mile. The proper electromotive force is obtained by coupling together five Edison machines in series. The wire used is No. 1 B. W. G., and is carried on the Western Union Telegraph poles. Mr. Sprague has by no means rested satisfied in developing his system of railway and carrying it to the advanced condition in which it now is, but he has been engaged in equipping the station and cars along the line of the road with Edison lamps, which are run in series from the same high constant potential circuit that supplies the car, on a system which has been developed by Mr. E. H. Johnson, the president of the company.

A résumé of the special and distinctive features of Mr. Sprague's system may not be uninteresting, and is therefore given below:

A double-track system with motors working in parallel circuit with each other on a constant potential circuit, the two tracks being supplied from the same source and from the same main conductors.

A supply at two or more points by independent batteries of automatically non-sparking machines, the points of supply being maintained at the same differences of potential.

A system of continuous main conductors intersected by switches, and sectional working conductors connected therewith through automatic safety devices.

Means for cutting out, either automatically, in case of accident, or at will, if desired, any portion of the circuit.

An automatic block signal system for day and night use.

Methods for the equalization of potential by cross connections between conductors of like polarity and on different tracks.

A very simple construction of the motor proper.

The centreing of the motor upon the axles so as to maintain parallelism between the driving shaft and the driven axle.

The method of flexibly supporting a part of the weight of the motor from the truck so as to allow perfect freedom in following the motions of the independent axles.

The method of doing away with all shock and jar and danger of stripping the gears, and the maintaining at all times of a spring touch so as to prevent any backlash and to insure quiet running.

Double driving from opposite ends of the motor shaft.

The use of fixed and adjustable split gears.

The means for getting a very intense rotary effort in starting by having an intense magnetic field and raising the armature potential gradually.

The means for maintaining a continuous and equal traction until full potential has been reached.

The method of increasing or decreasing the mechanical effects, whether of speed or power, or both, by an inverse varying of the field-magnet strength.

The method of controlling two or more independent motors simultaneously from the same source and by the same apparatus.

The use of a single resistance for both the armature and field circuits, each working independently.

The method of winding to maintain the point of least sparking at a fixed position, independent of the load, speed, or power.

The use of single sets of brushes for both forward and backward motion.

A system of braking consisting in converting the energy of the train into current, which is

delivered back to the line through the same apparatus which propels the car without any reversal of contacts, whereby a saving of at least forty per cent. would be effected in the size and capacity of the generating station, in the conductors, and in the coal and labor expended at generating stations.

The final step of braking by means of which the car is brought to rest through the same dynamic action of the motor while the field magnets are still connected with the line.

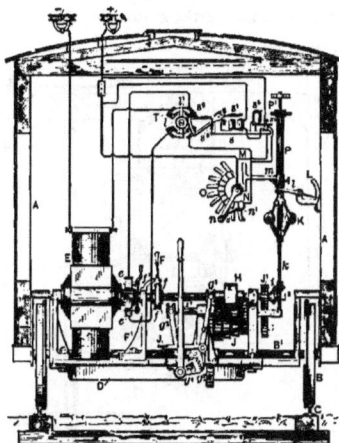

FIG. 175.—HENRY ELECTRIC MOTOR FOR RAILWAYS.

The method of lighting cars and stations from the main station.

The method of heating cars with a part of the energy of the momentum.

Another of the workers in the field of electric railroading is Mr. John C. Henry, of Kansas City, who has been busy for some time past in elaborating a system which possesses several novelties and is now going into use. Without entering into the various methods employed by Mr. Henry in distributing and taking off the current from the conductor, either overhead or underground, we will only describe the locomotive car itself and its arrangement. This is shown in Figs. 175 and 176, which represent respectively a transverse sectional view and a plan of the car.

23

The principal objects aimed at in its construction are to make each motor automatically adapt itself to every change in load and grade, and to afford safety devices by which no injurious effect could be produced through abnormal conditions of working. Mr. Henry has also adopted an arrangement by which the motor is kept at a constant uniform speed, irrespective of the speed of the car.

As will be seen by the illustrations, the motor is mounted on a frame D, which permits a direct connection to be made with the axle from the driving shaft of the motor E. For the pur-

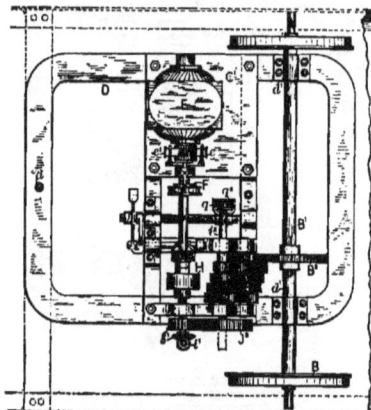

FIG. 176.—HENRY ELECTRIC MOTOR FOR RAILWAYS.

pose of automatically controlling the supply of the current to the motors, the driving shaft G of the motor is connected by the bevel gear i with the upright shaft k, carrying the governor K. A sleeve P, on the shaft k, is attached to the governor at one end, and from the sleeve there extends an arm m, carrying the rack M, which gears with the pinion N. Encircling the latter is a series of resistance coils O in electrical connection with the main conductors through brush n^1 and lever n and branch conductors. When from any cause the speed of the motor shaft varies from the prescribed limit, the rack N is drawn up or down by the governor K, which moves the contact brush n^1 over the commutator of the resistance coils, increasing or decreasing the resistance to the current.

For the purpose of automatically reversing the poles of the motor, the two opposite poles of the switch T are connected to wires running

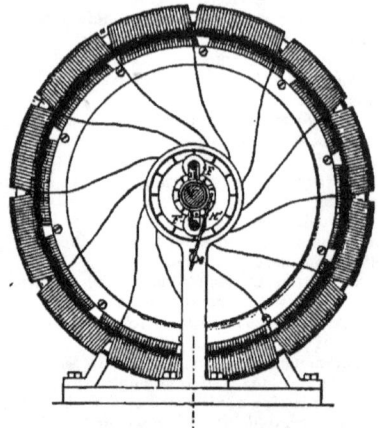

FIG. 177.—SIDE VIEW, HIGHAM MOTOR.

to the motor field and to the commutator brush e. A segment rack lever s^3 gears with the pinion of the switch, and a spring s^2, attached

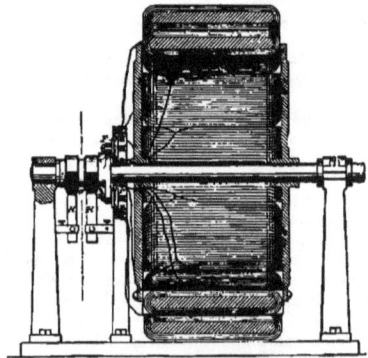

FIG. 178.—SECTION, HIGHAM MOTOR.

to the lever, keeps the rack in one position under tension. The opposite end of the lever s^3 is provided with the armature s^1, and in the field of the electro-magnets s. Two short circuits extend from branch conductors in electrical connection with the magnets s., A circuit closer s^4 is also provided, the key of which is held in a horizontal position by means of a spring s^5. The electro-magnét s is also connected to the circuit closer. With this arrangement, should the shaft of the motor attain a rate of speed above that necessary to propel the car, the governor is thrown out, thereby drawing down the sleeve P, and the projection l^n depresses the key of the circuit closer s^6, the

terlocking device is held by pins q^5. The lever g extends from the shaft G to the lower part of the supporting base to the speed gearing, to which it is pivoted. One part of the lever is extended at right angles as a foot lever, and a short portion g^2 extends downward and is tapered at right angles, so as to engage with the gear-sector g^4, which is pivoted to the frame D. A hand lever extends vertically from the gear-sector g^4, to which it is attached, and a lever g^8 is attached to the same point, and also in rigid connection with the hand lever, and is connected

FIG. 179.—PERSPECTIVE, VAN DEPOELE STREET CAR MOTOR.

magnets s become excited and draw down armature s^1 on lever s^2. This rotates the pole changer T, which changes the polarity of the motor and allows it to generate instead of drawing upon the current. When the poles of the motor are reversed and the motor is acting as a generator, the current is shunted from the resistance coils through a leak circuit, the circuit being made through the circuit closer s^6. The speed of the motor is indicated by means of the needle l on the indicator L.

The driving shaft G of the motor is provided with an intermediate friction-clutch H, placed in connection between the motor E and gear J^3. One portion of the clutch H is provided with a neck g^1, in which the end of lever g of the in-

with the end of the reciprocating plunger i^3, which is reciprocated in the direction of the speed gearing and for engaging the gear. Thus when it becomes necessary to change the speed of the car or change the relation of the gear without checking the speed of the motor-shaft, the foot lever is operated to throw the friction-clutch H apart. This releases the end g^2 of lever g from the gear-sector g^4 and the hand lever is then operated to throw the plunger in or out of connection with the various gears. The gear J^1 on the shaft meshes with gear J^2, and the gear J^2 is mounted upon the same shaft i^2 as the speed gearing, which in turn communicates power to the gear B^2, attached to the axle B^1.

While all these arrangements are by their nature designed to be more or less automatic, Mr. Henry has also introduced devices for reducing the rate of speed when required by means of the ribbon brake F. It is further evident that the contact brush of the resistance coils may also be moved by hand for starting, checking, or reversing the motor.

In the early part of this work, mention was made of the Elias motor in which two electro-magnetic rings, one within the other, acted as a motor by mutual attraction and repulsion. The same relative arrangement of armature

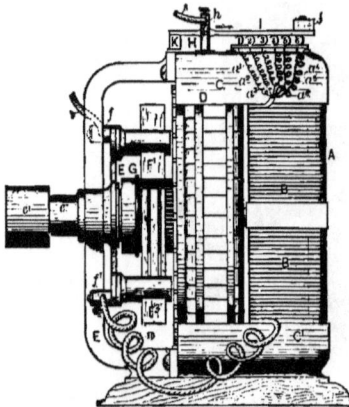

Fig. 180.—Van Depoele Motor.

and field has been adopted by Messrs. E. T. Higham and Daniel Higham, of Philadelphia. They have, however, introduced modifications which are designed to improve the efficiency of the arrangement.

The motor is shown in side view and in section in Figs. 177 and 178, and will be seen to consist of two ring magnets of the Gramme type. The inner magnet revolves while the outer remains stationary, but both are provided with commutators. The current coming in, say at the brush H, passes through the contact wheel F to the commutator-plates with which they happen to be in contact, and thence through the corresponding conductors to the

coils of both the electro-magnetic rings, there splitting and passing in opposite directions through opposite halves of each ring of coils and out through the contact wheel F' and brush H'. Thus the travelling contacts rotate the polar points of both electro-magnetic rings in the same direction as that in which the rotary electro-magnet moves mechanically and, as a result, it is said, the power developed by the motor is increased.

We have already in Chapter VII. drawn attention to and described the electric railway work accomplished by Mr. Van Depoele in his lines at Appleton, Wis., Montgomery, Ala., and in other places, but without special reference to the type of motor employed for that purpose. Hence a description of the latter will now be of interest.

The motor which is illustrated in the accompanying engraving, Fig. 179, has an armature of the well-known Gramme ring form, and the shaft rests in bearings, one of which is a bracket bolted to the lower pole piece, while the other is the neutral point of the field magnets. Having special regard to the attainment of compactness, the field magnets are given the form shown. It will be seen that the field coils are wound on the two sides of a cast-iron upright upon the ends of which are bolted the pole pieces which project at right angles and encircle the armature.

As the direction of travel of the car must be under control, two pairs of brushes are provided by which the direction of rotation of the motor can be changed at will. Each pair of brushes is attached to a brush-holder provided with a lever, by the shifting of which either pair can be brought in contact with the commutator. The end of the armature shaft carries a gear wheel, which meshes into another attached to the car axles.

In order to provide for the regulation of the motor so that it may run at different speeds, and without the use of external resistances, Mr. Van Depoele adds to the ordinary field-magnet coils additional ones, which are successively connected to each other in series and are also in series with the main field coils. This is shown diagrammatically in Fig. 180, which represents the method adopted for the automatic regulation of the motor. It will be seen that the coils a^1, a^2, etc., are brought out to spring terminals over which is placed a contact bar I. One end of this bar carries an adjustable weight J, which tends to press down on the terminal, 1, 2, 3, etc. The other end of the contact bar is provided with an iron armature K, in close proximity to the surface of the pole piece C, so that when attracted by the magnetic condition of the latter the other end is drawn away from the contact springs, thus cutting out the resistances.

This arrangement, shown in Fig. 180, is evidently intended for motors in which it is desired to keep the current and speed constant. But in the motor shown in Fig. 179, which is applied to the street cars where these conditions do not prevail, a system of hand regulation has been adopted. The auxiliary field coils are connected to a commutator which is manipulated by hand, and by means of which any speed from rest to maximum can be obtained.

We have in Chapter VI. made mention of the early work of Mr. Stephen D. Field, in the domain of electric railroading. His most recent work now deserves mention here as it is marked with the usual originality of the inventor, and is on the eve of practical demonstration.

Being, like others, impressed with the special applicability of electric motors to the propulsion of the cars on the elevated railroads in this city, and encouraged in this project, we believe, by his uncle, Mr. Cyrus W. Field, Mr. Field has for some time past devoted his special attention to the problem involved, and has so far matured his plans that

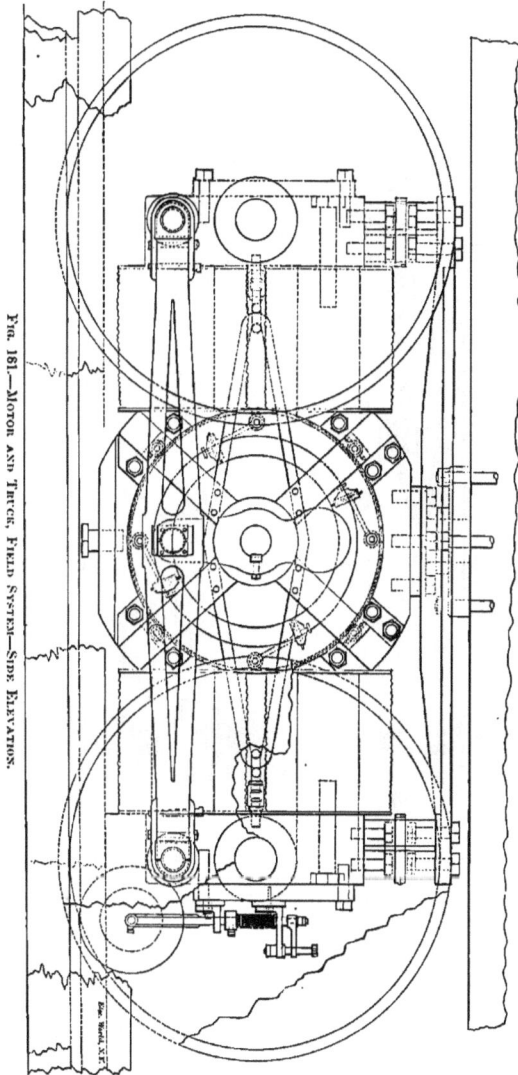

FIG. 181.—MOTOR AND TRUCK, FIELD SYSTEM—SIDE ELEVATION.

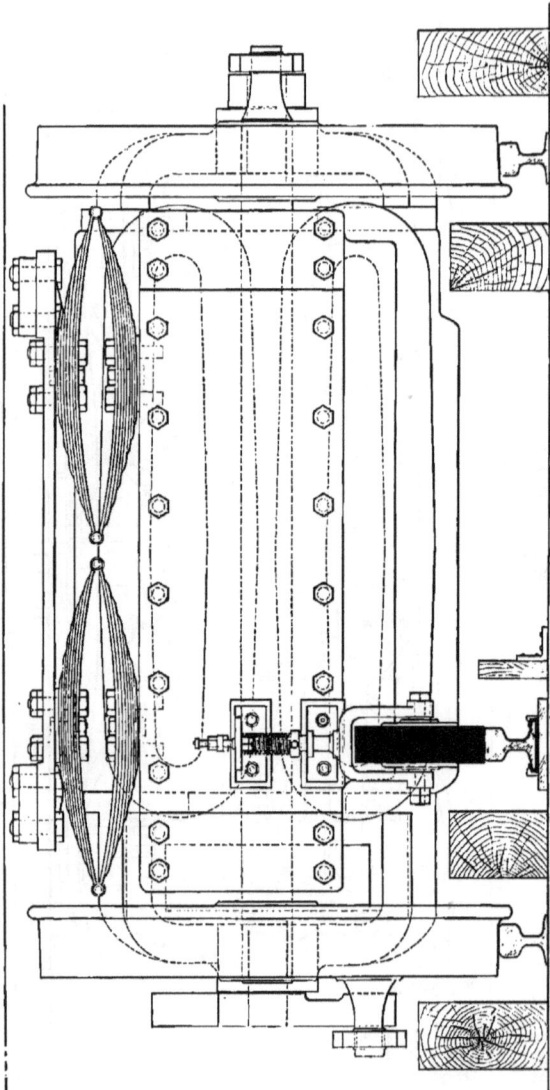

FIG. 182.—MOTOR AND TRUCK, FIELD SYSTEM—END ELEVATION.

actual work of construction is now progressing, looking to a practical test.

Taking in review the mechanical details first, it will be seen that between the wheels of the car truck a single motor is situated, the armature shaft of which is connected directly to the wheels by means of a crank and side connecting-rod similar to that employed on steam locomotives. This is clearly shown in Figs. 181 and 182, which represent, respectively, a side and an end elevation of the truck as it is being constructed. The cranks, as shown, are, for obvious reasons, keyed to the armature shaft at an angle of ninety degrees.

The manner of suspension of the motor is clearly shown in Fig. 181. The upper and lower field magnets, which form consequent poles, are held together by the usual iron connecting pieces or yokes, and through each of these passes an axle of the truck, so that the entire weight of the motor is equally distributed on both axles. The bearing, however, is not a rigid one. Although, as stated above, the axles pass through the yokes of the field magnets, it will be observed, Figs. 183 and 184, that the latter are made up of two pieces, or perhaps, to put it more correctly, that a cap is bolted to the real connecting piece at each end. Fig. 183 shows the bearing in end elevation and part of the adjoining wheel; while Fig. 184 is a sectional view which shows the usual spring interposed between the weight and the bearing. The cap maintains the spring and bearing in position and allows the motor free vertical motion without strain, due to inequalities in the road-bed.

The armature turns in bearings formed by the junction of four brass arms on each side of the armature, and these arms are in addition bolted to braces on each side, which converge and are joined to the connecting piece of the vice will be understood from the illustrations, Figs. 181 and 182, which show the contact wheel held by brackets bolted to the yoke of the field magnet.

The wheel itself is built up of alternate layers of discs 6 inches and 9 inches in diameter, of thin spring brass, so that each large disc is flexible, and in bearing upon the rail can be given a bending motion. It will be noted, at the same time, that the forked rod supporting the wheel passes through two brackets, above the upper one of which is a lever attached to the rod. This lever can be swung through an arc of 180 degrees and can be clamped in any position;

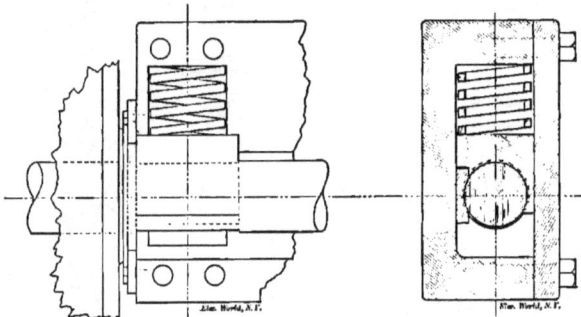

Figs. 183 and 184.—Bearing of Motor on Car Axle.

field magnets by means of bolts and turnbuckles. In this way all horizontal motion of the motor relatively to the truck is prevented, while at the same time its vertical motion is not restricted.

The body of the car rests on springs, which are bolted to the tops of the yokes, the kingbolt fitting into a bearing bolted to the centre of the upper pole piece.

There remains still another mechanical detail to be described, and that is the manner in which the current is taken from the central insulated rail. It is well known that dirt and rust not infrequently cause defective contacts and introduce resistance into the motor circuit. To guard against this, and in order to insure good contact under all conditions, Mr. Field has designed what may be called a combined contact wheel and brush. The nature of the de-

the position of the contact wheel relative to the rail corresponding to that of the lever.

Now, it will be readily understood that when the lever is in the middle position on the arc, the wheel stands as shown in Fig. 182, and only a rolling contact is maintained between wheel and rail. But if the lever should be turned slightly to either side, so that the discs are no longer parallel to the rail, a slight rubbing or scraping motion would be added to that of the rolling. By turning the lever still more and increasing the angle, the rubbing component, as it were, can be increased to any desired extent until, when the lever is at an angle of ninety degrees from its original position, the wheel stands at right angles to the rail and, obviously, rubbing alone can take place. The wheel being built up in the manner described above, acts as a resilient brush, taking off the current and keeping the rail clean. It is evident that this brushing action need only be resorted to when necessary, the wheel otherwise taking up the current by the rolling contact, a spiral matically at the non-sparking points. The reader will have noticed in Fig. 181 that four brushes are shown bearing on the commutator, but for the sake of clearness only two are

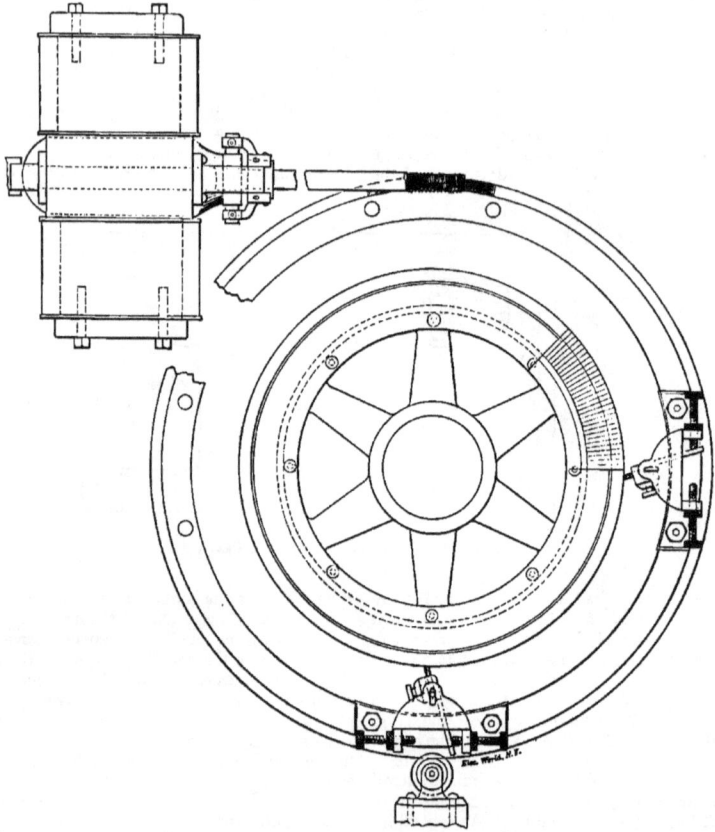

FIG. 185.—REGULATING MOTOR AND ADJUSTABLE BRUSHES.

spring being provided which presses the wheel upon the rail.

We come now to the electrical details, taking up first the manner in which the motor is regulated and the brushes are maintained auto- shown in Fig. 185, one each of the horizontal and vertical pairs. The office of the auxiliary brushes will appear presently. The brushes are all mounted upon a ring, on the outer periphery of which screw gear teeth are cut, and into which meshes a screw which forms the end of the armature shaft of a small motor.

The horizontal pair are the main brushes, while the vertical pair are what may be called the regulating brushes. The field of the regulating motor is connected in shunt to the armature of the large main motor, while the armature of the regulating motor is connected to the regulating brushes.

From what has just been said, it will be evident that when the normal amount of current passes through the motor, the regulating brushes bear upon the commutator at points of equal potential, and hence no current passes through the regulating motor. Now if while in this position any change of load or speed occurs, the diameter of commutation would be changed, and the regulating brushes not yet having changed their position, would bear upon points between which there now exists a difference of potential. This evidently would cause a current to pass through the regulating motor, which would be started revolving in a direction corresponding to the change of conditions. The turning of the little motor gearing with the ring causes all the brushes to be shifted simultaneously until the regulating brushes reach again points of equal potential, when evidently the little motor stops for want of current. The main brushes will at the same moment have arrived at the proper diameter of commutation. In this way the motor accommodates itself automatically to changes of load or speed.

There are several details in connection with this regulating device which are also worthy of notice. By referring to Fig. 186, it will be seen that while the lower main brush bears against the inner end of the commutator, the small regulating brush bears at the outer end. It will further be noted that only every fifth commutator bar is continuous, the four intermediate ones being divided near the outer end. One of these intermediate bars is shown in section in Fig. 186. The outer end is entirely insulated, and hence receives no current whatever from the motor; the insulated pieces thus serving merely as a continuous bearing for the regulating brushes. From this it will be seen that while, as usual, the main brushes are in continuous electrical contact with the motor, the regulating brushes only make contact at every fifth commutator bar. In this way the 24

regulating motor is caused to act under short impulses of current. The effect of this is that while the regulating motor is started promptly, it comes to rest very quickly when the brushes reach the neutral point where they should remain, and thus they are prevented from travel-

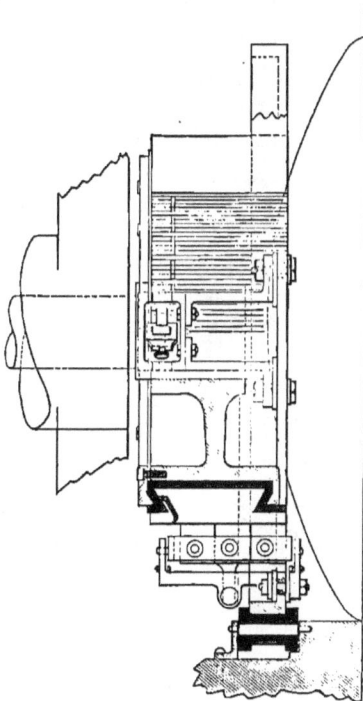

FIG. 186.—DETAILS OF COMMUTATOR.

ling beyond that point by the momentary impulses which otherwise would immediately send reverse currents into the motor.

The brush-holders being rigidly attached to the ring, some provision must be made for guarding them against injury, as they bear almost vertically against the commutator. This is accomplished in a manner shown in Figs. 185, 186, and 187, which give different views of the brush-holders. The brushes, as will be seen, are held in a clamp provided with pivots which slide in slots in the holder, the brushes being pressed toward the commutator by two springs. Now, when the direction of rotation of the armature is reversed, the brushes are pushed inwardly a short distance and then carried over until their angle of bearing is reversed, the motion being limited by the stop screws shown ; then the springs again press the brushes against the commutator as before.

In addition to the automatic method of regulation above described, resistances are provided and a reversing lever, so that the strength of current and the direction of rotation of the motor can be regulated at will. The lever is so arranged that it cannot be reversed as long as there is any current in the motor.

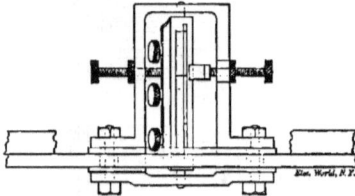

FIG. 187.—DETAILS OF BRUSH-HOLDER.

Another feature is the electrical brake, which may be applied at will and to any degree of pressure. The current operating can be regulated by a resistance switch. The switch is so arranged that at the last section, when the brake is to be taken off, a reverse current is sent into the brake coils by means of an induction coil or condenser placed in the circuit ; this momentary reverse current instantly demagnetizes the brakes, so that they fall away freely from the wheels.

These points comprise the essential details of the Field motor arrangements. The work of construction is now actively going on, and within a short while a practical trial will be made on the elevated railroads of this city.

Mr. Field has, however, looked farther abroad than the city in the application of electric railways, and is elaborating plans for an electric locomotive designed for rapid suburban transit. Our illustration, Fig. 188, shows the general design of a locomotive and baggage car com-

bined. It is proposed to employ six-foot drivers coupled direct to the armature shaft. The machine is to be a four-pole Gramme, with an armature of four feet in diameter, and a speed of from thirty to forty miles an hour will be attainable.

For several months past an electric railway has been in operation on Ridge avenue, Philadelphia, running for a distance of about two miles, and having one terminus at the Laurel Hill Cemetery. This line was constructed by the Union Electric Company, under the direction of Mr. W. M. Schlesinger, and has been in daily successful operation.

In order to avoid overhead conductors and also the use of the rails as conductors, the mains have been placed in a conduit having a slot at the top for a lever which leads the current from the conductors to the motor on the car, Fig. 189. In designing the conduit, it was found inadvisable to cut through the cross-ties of the railroad, as these form in most cases the foundation of the track. Practical experience has proved that a small conduit having good sewer connections answers all purposes. The main conduit is therefore only 9 inches deep by 5½ inches wide. It is built in sections of from fifteen feet to twenty feet, of heavy channel iron resting on the cross-ties ; substantial cast-iron wedges, resting also on the cross-ties, hold the two sides of the conduit at the proper distance from each other at the bottom and leave an opening ¾ inch in width between the lower flanges. To the sides of the channel iron, at proper distances, small angle irons are riveted to keep the slot of the conduit at the proper size. Braces are attached to the angle irons which pass through either cross-tie or stringer and are provided with nuts by means of which they can be tightened or slackened at will. An auxiliary conduit is put down below the cross-ties, built either of wood or cement. At convenient places this connects with large manholes at the sides of the track, and these again communicate with the sewers. Water and small particles of dirt fall through the opening made by the wedges between the channel-iron into this lower trough and pass from them to the manholes and sewers. But to prevent any accumulation in the upper conduit, this opening is at proper intervals increased to five inches by cutting away a part of the lower flanges of the channel iron.

The greatest difficulty experienced with underground conductors is to protect the insulation of the conductors from water and dirt. To accomplish this in the present conduit, an angle iron is riveted to the top flange of the channel iron in such a manner that one of its flanges, pointing downward parallel to the main side of the channel iron, forms one side of the slot. In the inverted trough formed in this manner the conductors are fastened, so that the contact side, i. e., the side on which the pieces rub, is the lower side. The conductors are much narrower than the trough, so that contact with its sides is impossible. Dirt or water coming in through the slot will, there-

the latter is furthermore to protect the copper from wear. The connection between the conductors of two following sections is made in boxes outside the conduit. At proper intervals, the top of the conduit is made removable, giving access to the inside. These traps are also put at every place where a connection is made with a manhole. As in opening these traps, the top plates are often handled roughly and thrown in the dirt, they are not provided with conductors, but in place thereof wood is fastened to them. The current is carried round them through insulated wires. This also prevents interruption of traffic in case of the opening of a manhole.

FIG. 188.—THE "BERKSHIRE" ELECTRIC CAR.

fore, also fall to the bottom of the conduit without interfering with the insulation.

The conductors are made shorter than the sections, and the trough is closed at the ends of the sections by means of a block of wood or other insulation, the lower side of which is in the same horizontal plane with the lower side of the conductors. These latter end within about ¼ to ½ inch of these blocks. In this manner the insulation is protected from any dirt or water coming in between the sections.

All sections are made exactly the same size, so that if one is damaged it can easily and quickly be replaced by another.

The conductor itself is a copper bar, to the lower surface of which a small angle iron is fastened. The contact pieces rub along the iron and are prevented from leaving it by the downward flange of the angle. The object of

All connections between the conductors of following sections are easily accessible, so that in case of damage to one section this can easily and without interfering with the conduit be cut out of the circuit and the current taken round it by means of insulated wire. As the sections are only twenty feet long at the utmost, the momentum of the car will easily carry it over the gap. As all motors on the cars are in multiple arc, no complicated make and break appliances are required in the conduits, and as the conductors on either side form one continuous line, testing for insulation and continuity can easily be done from the station. To convey the current from the conductors in the conduit to the motor on the car, each of the latter is provided with specially constructed frames, so arranged as to make the contact pieces perfectly independent of the oscillations of the car or any variation in the distance between the body of the car and conductors, caused either by varying loads on the former or uneven construction of road-bed. In designing these frames great care has been taken to combine simplicity with strength, the vital parts being well protected by strong cast-iron or phosphor-bronze frames. On the road now running in Philadelphia it has happened several times that large paving stones were placed intentionally at night on the conduit, but they were invariably thrown ward; and by moving the lever more or less from the central position, the speed is increased or decreased. If desired, levers can be placed on both platforms, the one not in use being secured by means of a lock.

To allow the motor to start up easily and rapidly, the field magnets are in a separate circuit which is not opened when the car stands still. The motor brushes are tangential, one pair only being required. They are connected

FIG. 189.—THE SCHLESINGER ELECTRIC CAR.

to one side by the frames without doing the slightest damage. Steel springs are used as contact pieces, a very steady and good contact being thus obtained, as shown by the ammeter.

The motors are attached to the cars in such a manner as not to interfere with the seating capacity. They are placed beneath the body of the car between the axles, and specially constructed chains transmit the power from the armature to the wheels. The car is operated by means of a single lever on the front platform. When the lever stands in the middle position, the current to the armature is interrupted, and the motor naturally stands still or gives no power. On moving the lever to the right the car runs forward; to the left, back-

with the lever in such a manner that the same motion by which the current through the armature is reversed also sets the brushes. Aside from the hand brakes, each car is provided with electric brakes of the simplest construction. The interior of the cars is lighted by incandescent lamps, deriving their current from the same source that propels the motor; and electric gongs complete the outfit.

Among the most prominent of the electric motors in general, practical use is the Edgerton, designed by Mr. N. H. Edgerton, of Philadelphia, and shown in perspective in the accompanying engraving, Fig. 190.

The pole-pieces, Fig. 191, are arranged each with three radial cores, on which the exciting coils are wound, and by which the fields are supported on the interior of a cylindrical iron shell which forms the framework of the motor, as well as the yoke-piece of the field magnets.

FIG. 190.—THE EDGERTON MOTOR.

The shell and pole-pieces form a concentrically cylindrical structure in the interior of which the armature revolves on a central shaft supported at either end by bearings situated centrally in the end caps or lids. These end caps may close the cylinder entirely or not, but usually one end is closed completely while the other is left open, as shown, for easy access to the brushes and commutator.

The armature shown in section in Fig. 191 is polar, and consists of three helices, wound upon as many radial cores, set at equal distances upon a central prism of the same number of sides. Through the central axis of this prism the shaft is placed longitudinally, and, as before stated, supported in bearings in the end caps of the motor. The outer or peripheral extremity of each of these cores is segmental in shape, coinciding in curve with the inner concave surfaces of the pole pieces between which it revolves. The helices are wound parallel with the axis of the armature as in the Siemens shuttle armature, and each is complete in itself. Similar ends of each helical wire are connected with the commutator segments, of which there is one for each helix; and the other similar ends are carried out to a common union, insulated from and carried upon the shaft.

It has been the aim of the inventor to design his motor on such mechanical lines as would

insure cheapness and simplicity of construction with least cost of maintenance. To this end the cylindrical form was adopted, as it furnishes bearings of the greatest solidity and protects completely the operative and vital parts of the motor, thus allowing of its use, without injury, in the most exposed situations. The division of the field-magnet coils into three helices for each field was adopted as the most likely way to prevent undue heating in the motor circuit, in addition to which the shortness of the cores abutting immediately upon the large surface of the outer shell furnishes, by conduction and radiation, a ready means for the dissipation of all such heat.

The polar armature was chosen by Mr. Edgerton on account of its ease of construction, and because the peripheral segments of the spools on which the helices are wound make it impossible for the motor, at its highest speed, to displace any of the wires by "tangential inertia"; further, because in the rise of temperature in

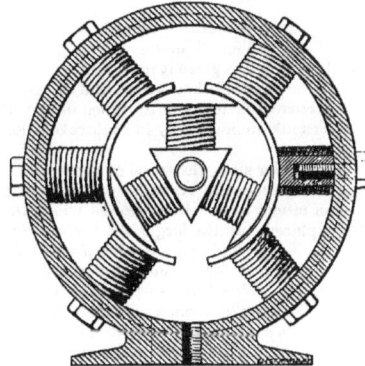

FIG. 191.—SECTIONAL VIEW OF EDGERTON MOTOR.

the motor due to flow of current, while in operation the coefficient of expansion is the same both in armature and field, which, of course, allows of the rotation of the armature in closer proximity to the pole pieces; and, lastly and principally, because, according to Mr.

Edgerton, the inductive action of the field is received first by the iron core and transferred through that to the wire, thus reducing the resistance of the armature circuit, due to counter-electromotive force.

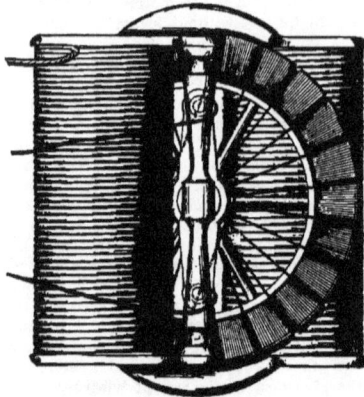

Fig. 192.—The Fisher Motor.

The armature is connected in series between the fields in the small motors, although it is perfectly feasible to place it in a shunt. When the machine is coupled in series and in operation, the current is active at all times in two of the helices and momentarily in each revolution in all three.

In the smaller sizes, the speed of the motor is regulated by means of resistances included in the main circuit in the shunt around which the motor is placed. In the larger sizes, viz., from one horse power and over, a centrifugal governor is arranged for maintaining a uniform rate of speed. With this size also the armature and commutators are changed from three to five segments; while, in those still larger, provision is made for one brush only on the commutator, while the other brush is transferred to the insulated ring of the bobbin union. As all the armature bobbins are coupled in multiple arc with this ring, it results, as a matter of course, that, with the commutator slits cut diagonally, the sparking at the commutator is reduced to a minimum.

It has been the rule, as evidenced so often in the preceding pages, in the construction of electric generators or motors, to so adjust the armature with respect to the field magnets that, in revolving, the bobbins would pass transversely through the field of force adjacent to one of the poles and then transversely through the field of force adjacent to the other pole. Recently, however, Mr. Frank E. Fisher, of the Detroit Electrical Works, has devised and patented a modification of this by locating the plane of revolution of the armature parallel to and between the two planes, each of which contains one of the field magnets. The motor is now being made by the Detroit Motor Company. The opposite ends of these field magnets are then united by a pole piece extending from one across to the other in such manner that instead of revolving through the field adjacent to the poles respectively the armature is caused to revolve within the plane containing the poles of the machine, so that the poles are opposite the periphery of the armature and diametrically opposite each other.

Our illustrations, Figs. 192 and 193, show the new design in elevation and in plan, and are so clear that no further description is deemed

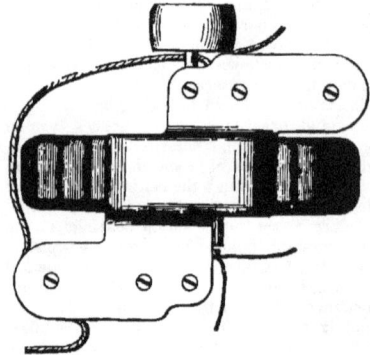

Fig. 193.—The Fisher Motor.

necessary. According to Mr. Fisher, a machine, whether motor or dynamo, constructed in this manner operates with very much less resistance, and consequently delivers a greater effective force with the same impelling current. He attributes this increased efficiency to the fact that the construction is such that the armature revolves diametrically between and in the plane containing the poles, instead of being obliged to cut through the plane transversely. Mr. Fisher, who has of late done considerable studying on the motor question, is also of the opinion that a beneficial effect in lessening the resistance to the revolution of the armature is obtained by the location of the armature with respect to the poles, so that its bobbins shall have a motion first transversely from end to end of one magnet in a direction across or through the planes of its successive convolutions of wire, and then in like manner from end to end of the other field magnet.

A patent issued recently to Mr. Elias E. Ries, of Baltimore, for an improvement in electrical railways, is of timely interest, as bearing upon the development and extension of electric street car lines.

In populated cities, as is well known, it is necessary that the conductors employed to convey the electric current to the motor cars should be carried in an underground conduit, extending along the line of the railway. As these conductors are necessarily naked or partially exposed, in order to permit of contact being made therewith by the current-collecting devices on the motor cars, one of the chief difficulties to overcome is that of maintaining proper drainage facilities and preventing water from coming in contact with the conductors at low-lying portions of the roadway, subject to such an overflow as would occur in case of unusually heavy rains, or from the accumulation of water in the conduit arising from foreign matter in, or back flow through, the drainage outlets, etc.

Mr. Ries overcomes this objection in a simple and effective manner, so that portions of the conduit may be entirely flooded with water and without in the least interfering with the flow of current to the motors on other portions of the line. This result is accomplished by automatically cutting out, under the influence of the rising water in the lowest portion of the submerged conduit section, that portion of the conduit conductors belonging to the submerged section, and shunting the current through insulated loop conductors or cables that bridge the section cut out and connect the main conductors at both sides thereof. The conductors in the submerged conduit section will remain cut out of circuit as long as the water in the conduit is of sufficient height to come in contact therewith; consequently, no escape of current from them can take place. Means—such as secondary batteries—are provided for automatically propelling the motor cars across the low-lying section when the conductors are cut out, so that it will be seen that this device renders underground electric-railway conduits perfectly practicable, even in the most unfavorable localities, and goes to settle once for all the vexed question of insulation, but permits the successful use of a shallow conduit under conditions where, without this device, a much deeper one might prove entirely incapable of protecting the conductors carried by it.

This patent also describes a number of important modifications, and forms part of a system of electrical railways now in process of development.

CHAPTER XIII.

LATEST AMERICAN MOTORS AND MOTOR SYSTEMS—Continued.

At the beginning of Chapter XII. it was remarked that great activity prevailed in the electric motor industry, and that the descriptions therein were to be taken as supplementary to the earlier portions of the present work dealing with the same branch of the subject, namely, the development of new American motors and motor systems. The activity referred to continues, and may even be said to have increased. The field of application for motors in miscellaneous stationary work has widened immensely, and the use of electric motors on street railways is becoming so general that the employment of horse or cable appears likely to cease almost entirely at an early date. Under these circumstances, certain additions are required in this, the second edition of the present work, to render the record of invention and exploitation measurably complete to date.

Taking up the subject of street railways, the first and one of the most important of the new systems that we come to is that of the Bentley-Knight Electric Railway Company. The patents owned by this company, those of Edward M. Bentley and Walter H. Knight, have especial reference to their use upon the *city* street railway, where the conductors must of necessity be supported in, and protected by, sub-surface conduits; and to the construction of an electric car, the mechanism of which is invisible and inaudible to both passenger and passer-by, in which the passenger-carrying capacity is not decreased by the presence of any part of the machinery above the car floor, and the government of which is effected by an arrangement so simple as to permit of its management by the ordinary street-car driver.

From the time that its first experiments in the transmission of electrical power were made at Cleveland in 1883–4, under the auspices of the Brush Electric Company, as noted in earlier chapters of this book, the Bentley-Knight Company's system has steadily grown in efficiency. During the past year its engineers have had the assistance of the Thomson-Houston Electric Company, of Boston and Lynn, Mass., in the perfection of a durable and reliable railway motor, and of the Rhode Island Locomotive Works, and Messrs. Nicholson and Waterman, of Providence, in the solution of mechanical problems.

The Bentley-Knight Electric Railway Company employs the constant potential system, the pressure remaining constant throughout the line, while the power is varied by the current. Dynamos giving a constant electromotive force on all parts of the line, of 500 volts, are used. They are compound wound and provided with Professor Thomson's new winding, in which the main circuit field coils closely surround the armature, and oppose the tendency to a change in the line of commutation under varying loads. The machines have, therefore, a constant lead, and require but casual attention when in operation. The efficiency of the motor *per se* is 90 per cent. The current strength employed is about 7.5 ampères. The motor will stand 30 ampères indefinitely, and 60 ampères for half an hour. Speed is controlled by a coarse resistance in the main circuit composed of iron plates standing on edge. The motor is nearly self-regulating within the limits of its work, and the resistance comes but little into play. This method is preferred to that of changing the strength of the field magnet independently, since the latter necessitates also a change in the lead. The position of the brushes is never changed either for varying load or for reversal. A chain from the resistance-lever leads to the ordinary brake-spindle, and is wound thereon oppositely to the brake-chain so that the whole control is centered in one spindle.

In the Bentley-Knight system of street car equipment, the motor and all its attendant mechanism and regulating apparatus is mounted upon the truck, is wholly independent of the car body, and can be put under any existing car without cutting or alterations, and without lifting the car above its normal height. The motor is placed under the car floor, outside of and overhanging one axle, to which it is geared, being counter-balanced by a spring connection extending under the boxes of the opposite axle. Thus the whole weight of the motor comes upon the driven axle, and a small part of the car weight is also transferred from the free axle by the leverage which the motor exerts, giving ample tractive adhesion at all times. This arrangement also ensures a rigid connection between the motor-shaft and the axle, which is essential for the proper working of the gearing, while the motor has a spring support and a yielding impact on the road at starting, making the wear on working parts and on the track very light. It moreover permits the use of a counter-shaft, and a ratio of gearing of 12 to 1. The efficiency of the electrical transmission is proportionate in a measure to the ratio of gearing permissible. Tooth-gearing is used throughout, and all journals are held in rigid castings. The brakes are hung from the truck, not from the car body, and there is, therefore, no jarring felt by the passengers when brakes are applied.

FIG. 194.—STANDARD TRUCK, BENTLEY-KNIGHT SYSTEM.

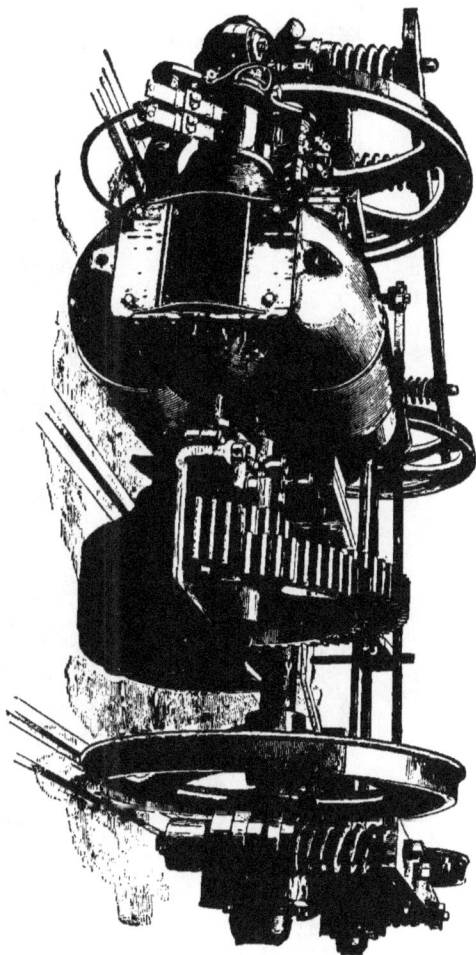

FIG. 195.—CAR MOUNTED ON STANDARD TRUCK, BENTLEY-KNIGHT SYSTEM.

The motors built for this work by the Thomson-Houston Company have cylindrical armatures 10 inches in diameter, and a speed of 1,000 or more revolutions per minute is perfectly feasible. The high ratio of gearing also brings less strain on the bearings of the armature, a matter of importance where efficiency and reliability are required.

The car is stopped by shutting off the current from the motor and applying hand-brakes. The only manipulation necessary is by the ordinary brake spindle, as described above. Turned in one direction, it releases the brakes and lets on the current in succession ; and, turned in the opposite direction, it throws off the current and applies the brakes. The spindle works within two turns. A separate lever on the dash-board is used for reversing. It only comes into play when a car reaches the end of its route.

The standard truck shown in Fig. 194 is one of those built for the North and East River Railway, of New York City, the tracks for which are now laid through Fulton street. It is hoped that the legal obstructions which have prevented the laying of the conduit will be shortly removed, and that this road will be in full operation in the near future. The motor mounted on the truck shown in the illustration will give from fifteen to twenty-five horse-power economically. The very heavy curves, grades and traffic of Fulton street necessitate ample provision of power.

The illustration of the car, Fig. 195, shows clearly the space taken up by the motor in actual use. On its first trial trip this car ran 13½ hours on a total consumption

Fig. 196.—Double Motor Truck, Bentley-Knight System.

FIG. 197.—CAR MOUNTED ON DOUBLE MOTOR TRUCK, BENTLEY-KNIGHT SYSTEM.

FIG. 198.—MOTOR CAR AND TOW, CONDUIT CROSSING, BENTLEY-KNIGHT SYSTEM.

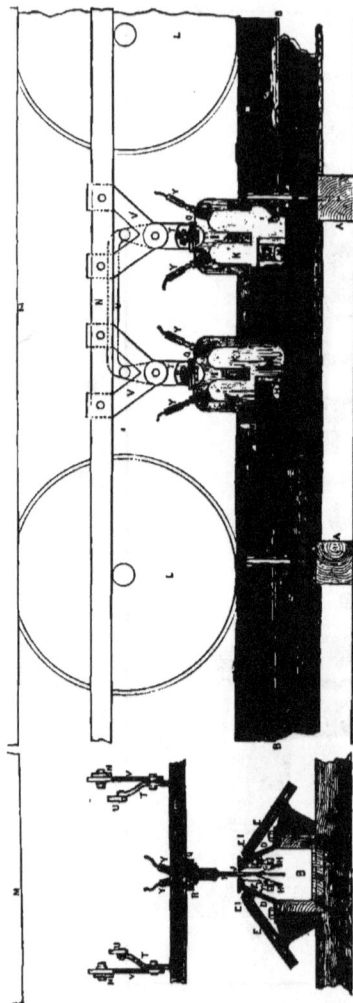

FIG. 196.—CONDUIT, BENTLEY-KNIGHT SYSTEM.

A, Ties.
B, Conduit.
C, Yoke.
D, Slot Steels.

E, Braces.
E', Ears.
E'', Leg Screws.
F, Slot Steel Lining.

G, Conductor.
H, Contact-shoes.
I, Plow Frame.
J, Plow Wearing-plates.
K, Plow Insulation.

L, Car Wheel.
M, Car Bottom.
N, Plow Supporting-frame.
O, Plow-slide.
P, Plow-guide.

Q, Upper Bearing.
R, Lower Bearing.
T, Plow-lever.
U, Connecting-rod.

V, Plow-supports.
W, Conduit Bottom.
X, Conduit Sides.
Y, Flexible Connections.

of 1,680 pounds of coal and carried over one thousand passengers, running on a bad track with heavy curves and grades. It has drawn two full-size loaded passenger cars up a 7 per cent. grade and has been timed up to a speed of 28 miles per hour.

Fig. 196 shows the very powerful motor trucks which have been furnished by the Bentley-Knight Company for the use of the Observatory Hill Passenger Railway Company, of Allegheny City, Pa. Fig. 197 shows a car mounted upon this type of truck. The grades and curves of this road are extremely heavy, the heaviest grade being $9\frac{8}{10}$ per cent. On this maximum grade a speed of six miles per hour has been regularly made by cars loaded with over fifty passengers, while on the approximately level part of the line a regular speed of 15 to 18 miles per hour has been attained. Neither snow nor rain has prevented the continuous and successful operation of this line. The cars are lighted by incandescent lamps fed from the motor circuit, and each car is fitted with the necessary contact devices to operate with both sub-surface and overhead conductors, as the road for one-fourth of its length is equipped with a conduit, the remaining three-fourths being supplied with an overhead conductor system. Fig. 198 shows a motor car and tow at a conduit crossing.

The conduit in which the conductors are carried forms a most important and interesting part of the system. In construction, the iron yokes are first set up and lined, being placed from three to four feet apart. The continuous gutter is then formed, the electrical connections between the lengths of conductor are made, and the slot-irons set on the yokes, their braces dropped into the exterior lugs of the yoke and the

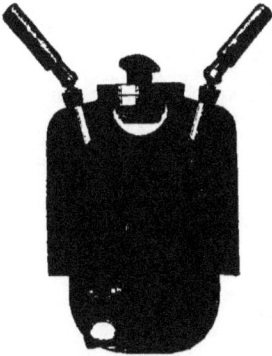

FIG. 200.—CONTACT PLOW.

slot-irons and yokes firmly bolted together, leaving a surface opening of five-eighths of an inch. The two main conductors are held in their places by heavy lag screws; they are connected by expansion joints and are of sufficient size to carry the current with but small loss of energy. Neither the traffic rails nor the conduit structure form any part of the electrical circuit, so that there is a double provision against any contact or ground. To provide for switching, a movable tongue is pivoted at the point of branching, so as to rest on the top of the con-

duit and to be readily set to close either of the branch slots and direct the contact plow into the other. A corresponding conductor tongue within the conduit is moved at the same time. The strengthening web of the heaviest yoke extends only fifteen inches below the level of the pavement.

The method of making electrical connection between the motor and the conductors in the conduit —an important point — is clearly shown in Figs. 199 and 200. For this purpose a contact-plow is employed, which consists of a flat frame hung from the car by transverse guides, on which it is free to slide the whole width of the car, and

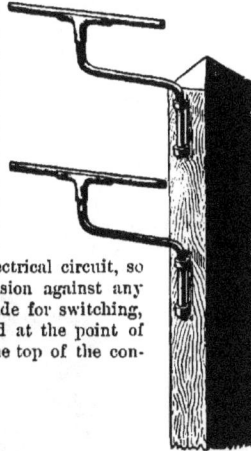

FIG. 201.—CONTACT TROLLEY, ELEVATED CONDUCTORS.

FIG. 202.
ELEVATED CONDUCTORS SUPPORTED BY POSTS AND BRACKETS AT CURB.

extending thence down through the slot of the conduit. It is provided with a swivel-joint, so as to adjust itself to all inequalities of road or conduit. This frame carries two flat insulated conductor-cores, to the lower ends of which are attached by a spring hinge small contact-shoes of chilled cast-iron that slide along in contact with the two main conductors. At the upper ends are attached flexible connections leading to the motor. This plow can be inserted or withdrawn through the slot at will, the spring hinge allowing the contact-

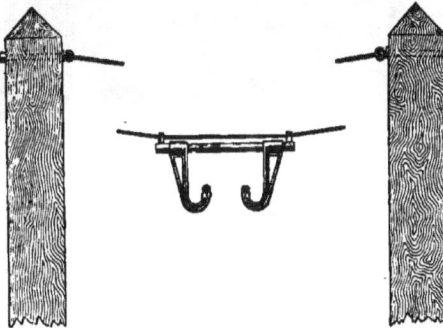

FIG. 203.—ELEVATED CONDUCTORS SUPPORTED OVER CENTRE OF ROADWAY.

shoes to straighten out into line with the conductor-cores when the plow is pulled upward and the shoes strike the insulating lining with which the slot-irons are provided. By no accident, therefore, can anything be left behind in the conduit to obstruct succeeding cars, while the plows may be pulled out at will. The plow-guides are hung on transverse axes, and are held in a vertical position by a spring-catch that gives way when the plow meets an irremovable obstruction ; and hence the plow is automatically thrown completely out of the

as to prevent leakage of the electricity from one rod to the other, or to the conduit, have proved entirely unfounded. From practical experience with a section of conduit line now in operation at Allegheny City, Pa., it is asserted that the leakage is inappreciable. Careful measurements taken after and during rain storms which lasted for days, showed no loss whatever.

Fig. 201 shows one type of trolley for use with elevated conductors. Figs. 202 and 203 show different methods of supporting elevated

FIG. 204.—FIELD'S ELECTRIC LOCOMOTIVE—PERSPECTIVE.

conduit without injury, being also immediately replaceable. The contact-shoes will stand weeks of wear, and cost very little. The frame of the plow has wearing-guards of hardened steel wherever it can touch the edge of the conduit-slot, and these are also readily renewable. Two plows are used on each contact for the sake of absolute reliability, and to prevent flashing at the contact.

The doubts which have been expressed in some quarters as to the feasibility of properly insulating the copper rods in the conduit, so

conductors now in use by the Bentley-Knight Company.

It will be remembered that in the preceding chapter a description was given of the electric motor designed by Mr. Stephen D. Field to operate upon the elevated railroads of New York City. Some experiments with it have now been made. The locomotive as it stood upon the track of the Thirty-fourth street branch is shown in the accompanying engraving, Fig. 204. The motor is mounted upon the rear truck, and the distinguishing feature is its mode of connection with the drivers. The arrangement, as will be seen, is exactly similar to that employed in the ordinary steam locomotive, and consists in the direct connection of the motor-shaft with the drivers by means of a crank and side-bar. The great advantage of this arrangement in the electric locomotive over the steam locomotive is apparent when we consider that in the latter the maximum effort is exerted on the drivers when the cranks stand vertically either above or below the center, and when on the centers no effort whatever is exerted. In the electric locomotive, however, the armature exerts a uniform and continuous effort upon the side-bar which is transmitted directly to the drivers, no matter what the position of the cranks may be. It follows from this that the starting up is much quicker than in the case of the steam locomotive, where the power of only one cylinder is available at a time.

The motor, which is series wound, is regulated by means of a liquid rheostat placed in the cab of the locomotive, shown in Fig. 205. This rheostat consists of a trough divided into two compartments filled with acidulated water. A metal plate on either side of these troughs acts as a terminal for the circuit, which is led in by the two cables shown. The speed of the motor is regulated by inserting or withdrawing from the troughs two slabs of slate, which are suspended over the troughs and can be raised or lowered by means of the long lever traveling over the sector shown at the right in the cab. By means of this liquid rheostat the resistance can be graduated from practically nothing, i. e., when the slabs are fully drawn up, to an infinite resistance when completely lowered into the troughs. On the standard

which guides the slabs there will be seen a spring-clip, and on the right-hand slab a plug. This is so arranged that when the slabs are full up, the plug presses between the spring-clips and cuts out the rheostat entirely. The reversing-switch for reversing the direction of the motor, is shown in the lower right-hand corner of the cab.

In designing the locomotive, Mr. Field constructed special brush-shifting apparatus for

FIG. 205.—FIELD'S ELECTRIC LOCOMOTIVE—INTERIOR OF CAB.

preventing sparking at the commutator with change of speed and load. This consisted of a small motor, which shifted the brushes in accordance with the action of a relay in circuit with the terminals of two auxiliary brushes placed at the neutral points on the commutator. Actual practice, however, has shown that this refinement of brush-regulation was unnecessary, the brush lead, under the influence of the peculiar speed-regulation employed, having been found to remain fixed and at an angle of 45 degrees; this, no doubt, being due to the large mass of iron employed in the construction of the field and armature.

The following table gives the weight and dimensions of the locomotive:

Weight of motor,	9 tons
Weight of armature,	1 ton
Weight of wire on armature,	600 lbs. No. 7
Weight of wire on field magnets,	1,600 lbs. No. 4
Total weight of motor, car and forward truck,	13 tons
Diameter of drivers,	3 feet
Diameter of armature,	2 feet
Length of armature,	42 in.
Wheel base,	5 feet

The generating plant was situated at a distance of half a mile from the track, and consisted of a single dynamo, built by Mr. Rudolph Eickemeyer, of Yonkers, in whose shops, also, the locomotive was built. This generator is of the iron-clad type, and showed itself fully capable of handling the load placed upon it.

The tests made, which extended over several weeks, have so thoroughly convinced Mr. Field of the practicability of the new ideas embodied in this motor that he has been preparing to demonstrate with apparatus on a large scale the practicability of electricity as a motive power for the elevated railways of New York City.

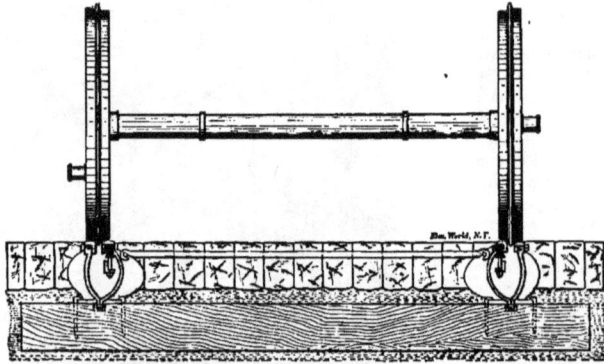

FIG. 206.—FIELD'S ELECTRIC STREET RAILWAY—DOUBLE CONDUITS AND TRACKS COMBINED.

The track on which the motor was operated has one of the steepest grades in the city, on which account it was peculiarly well adapted to show up any weakness in the system employed. One passenger car forms a load for a 13-ton steam locomotive regularly employed.

The motor easily drew one of the regular coaches up this grade at a speed of about eight miles per hour with a current expenditure of 35 ampères under an E. M. F. of 800 volts. The loss in conversion was found to be very small.

Various potentials were at times employed, 1,100 volts being used at one time with the same freedom from sparking as with the lower potential; the only change noticed being an increased speed of the motor.

Among the other novelties embodied in the motor was the "pick up" wheel of Mr. Field, already described, which operated admirably, so that no sparking whatever could be observed.

Finding that an urgent demand existed for efficient electric street railways, Mr. Field has also turned his attention in that direction. He has started out with the idea of reducing the details to the utmost simplicity, so that high potential currents can be carried with safety, and that the position of the motor on the line shall make no difference in the potential at its terminals.

Taking up the mechanical design first, it will be seen from Fig. 206, which is taken from the actual working drawings, calculated on the basis of a ten-mile road with 100 cars on each track, that two slotted conduits are employed for each line of rails, the wheel flange running in the slot. The wheels shown are 30 inches in diameter and the conduits themselves are only 8 inches high. They are built up in lengths from two sections bolted together at the bottom, and let into the wooden cross-ties. Heavy ribs are cast on the sides of the sections, which are calculated to withstand a vertical pressure at any point of 16,000 pounds to the square inch. In addition, tie-rods connecting the upper parts of the conduits lend additional stiffness to the structure and prevent any spreading or closing of the conduit slots.

It will be noted that the wheels have different treads on each side of the flange, the inner being of smaller diameter than the outer tread. On a straight track the outer, larger, tread of each wheel bears on the track. But when rounding curves, the wheel bears on the smaller tread on the inner rail, so that it has a slower motion than the outer wheel, and thus the friction usually encountered is avoided. The angle-rails, which are bolted to the tops of the conduits, are raised only one-fourth of an inch above the level of the pavement, and, being rounded, present no obstruction to ordinary traffic. These constitute the principal mechanical details of the road-bed. Special provision has also been made for drainage of the conduits.

The electrical methods employed by Mr. Field in this system are again a decided departure from past practice, and the means employed in carrying out the system are unique in conception. In Fig. 206, it will be noticed that each rail conduit has supported within it a conductor carried on insulators. The connection of these conductors with the source of power, the dynamos, is shown in Fig. 207. Here it will be seen, a dynamo (taking one each for simplicity) is placed at each end of the line. At one end (the left in the illustration) the positive pole of the machine is connected to the conductor in one slot, while the negative pole of the machine is connected to

ground ; or, what is the same thing, to the iron of the conduit. At the other end of the line, these connections are just reversed, the negative pole of the dynamo being connected to the conductor, and the positive to earth, or the conduit. Now supposing each dynamo to give 250 volts potential, it follows that the difference of potential *between the conductors*, and hence at the terminals of the motor, will be 500 volts. The circuit is made from the conductor in one conduit, through the motor, to the conductor in the opposite conduit, being completed through the two generators, conduits and axles of the cars. It will be seen that as the motor recedes from one generator, it approaches the other, so that wherever the motor may be, it will be actuated by the same

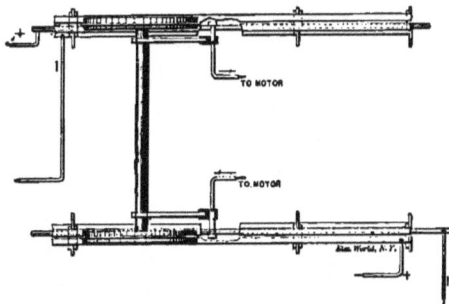

FIG. 207.—FIELD'S ELECTRIC STREET RAILWAY SYSTEM.

E. M. F., regardless of the resistance of the conductors.

The switching at either end of the line is accomplished by detaching the motor from the inside conductor, and completing its circuit with one generator only from the outside conductor to the conduit direct, thus getting the E. M. F. of only one generator. At the same time the idle contact brush passes through the path of the inside conductor, which is removed for that purpose. By this method of switching two objects are obtained. In the first place, the motor, by working on the lower E. M. F. (250 volts, instead of 500), passes the switch at a diminished speed, as it should.

But a still more important result is obtained by the use of the two conductors connected in the manner shown, and that is, that all interruption to the electrical integrity of the conductors, and hence, also, the traffic of the line, is avoided. Thus, each car will be provided with a current manipulator, by which the motor can be put in connection with either one or both of the conductors, as it is evident that each one forms a complete circuit by itself, having an E. M. F. of 250 volts; but, when combined, they give a difference of potential of 500 volts. In this way, all track switching devices are done away with.

The thorough insulation and stability of the conductors is provided for by the manner of their suspension and attachment. This is

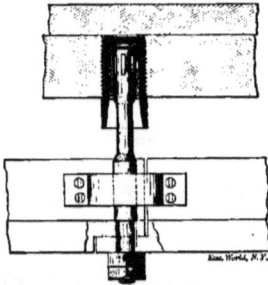

Fig. 208.

clearly shown in Figs. 208, 209 and 210, which represent longitudinal and transverse sections and a plan view respectively, of the arrangement. The conductor is secured to a steel rod embedded in a composite insulator. The inner and outer shells of the insulator consist of hard rubber, and between them there is a layer of vulcanized elastic rubber. The whole is vulcanized together so as form one piece. By the addition of the softer rubber, the conductor is given a certain flexibility of motion, so that it can follow the pressure of the contact brush without undue strain on the insulator supports or pins. At the joints of the section of copper conductors, a flexible bridge-joint is provided so as to allow for the expansion and contraction of the conductors.

Mr. Field proposes, also, to lay pipes in the conduit through which hot brine will be circulated in winter, so that all snow falling into the conduit will be melted. This device, of course, need only be put in operation during extremely cold weather. It will be noted that

Fig. 209.

the conductors are placed at one side of the slot, so that any dirt or snow falling into the conduit passes clear of the former. No extra appliances for cleaning the conduit are deemed necessary, as, being only 8 inches deep, it can be easily cleared of any refuse with a shovel let into the slot, and by means of which the accumulation can be removed to the drains which are provided at short intervals.

The street car designed for use on this system is shown in side elevation in the accompanying illustration, Fig. 211. It is mounted at its front end on the usual pair of wheels and axle, but at the rear it rests upon a 4-wheel bogie truck, so that the car can turn very short curves. Upon the same truck the motor is mounted; it is geared directly to the wheels by connecting rods attached to the cranks on either end of the armature shaft. The lever on

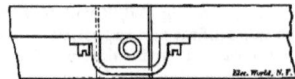

Fig. 210.

the front platform is connected by two rods with the shifting devices on the motor, and a single movement is sufficient to start, reverse and stop the motor, and to connect it with either conductor in the conduits. At the front end of the car two plows keep the slots clear of obstructions, and behind them the contact arms project into the conduits.

In the month of February, 1885, a company was organized in Denver, Col., under the name of the Denver Electric and Cable Railway Co., for the purpose of building and operating an electric railway in the streets of Denver. It adopted a plan proposed by Mr. Sidney H. Short, then Professor of Physics in the University of Denver, which embraced a series electric railway; and, as an experiment, a short piece of track, between three and four hundred feet in length, was laid in a circle in the University grounds.

A small car, named "Joseph Henry," was built for the track and carried many hundreds of people. The success of this little road encouraged the company to make more extended experiments in this direction.

this working conductor were connected by means of wire with the dynamo. When a car was placed on the track and had its brushes or contact springs, one on each of the parallel conductors of any section, the electrical switches at each end of that section would at once disconnect one of these conductors at one end and the other at the other end, leaving the only path for the current along one conductor, through the motor to the other conductor and to the line again. These sections could be made any convenient length, from that of an ordinary car to a block or a mile. A car could be on every section and the same current would pass through all of them.

Mr. Short maintained that greater economy in electric railways was to be had by the use

FIG. 211.—THE FIELD ELECTRIC STREET CAR.

This first road was built on the following plan : Two bare conductors were laid side by side along the track in a small conduit placed between the rails. The conductors were small and supported on insulators attached to the cross-ties, and they were cut into sections in order to test the working of the system. Switches operated by the current connected the two conductors in multiple arc, making them practically one conductor, having the conductivity and sectional area of the two. These electrical switches at the same time connected the ends of the sections of this double conductor, making one continuous conductor along the entire length of the track. The ends of

of a constant current of small quantity, by running the cars in series like arc lamps or telegraph instruments, and varying the electromotive force with the power required to operate the line of road. This main principle has not been varied from since the beginning, but many modifications of the details of construction have been made from time to time.

The two conductors used in the first arrangement were intended to carry the current in the same direction at the same time along the track, and, as stated above, they became, electrically, one conductor. At the junction of the two sections there were four contacts to be operated by an electrical circuit-closer, which

was necessarily too complex and delicate a
piece of mechanism to keep in good working
order, especially when placed on the street,
where it was liable to injury. Mr. Short, therefore, set about to devise some means of combining the two wires, in order to do away with
one pair of contacts at the junction of the sections, and thus to lessen by one-half the chance
for leakage from the insulating supports. He
found only one thing to do, *i. e.*, to shorten the
sections of the conductor to the length of one
car or train of cars. This increased the num-

means for keeping the spring or circuit-closer,
shown in Fig. 212, open as long as the car was
over it. Mr. Nesmith proposed that a long,
slender bar of insulating material be stretched
between the two current-collectors or brushes,
this bar to slide between contacts at the ends of
the sections of the conductor and keep them
apart so long as the the car is passing, as shown
in Fig. 213. In July, 1885, this second plan was
tried on about three hundred feet of track with
the same car used in the first experiment. A
brush at each end of the car made contact with

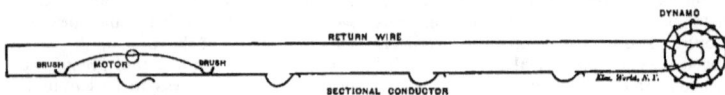

FIG. 212.

ber of circuit-closers greatly, but at the same
time it increased the flexibility of the road, as
cars could be run as close together as they
could be placed upon the track. On this plan
a contact brush must be placed under each end
of the car, or train of cars, to make electrical
connection with the conductor, so that there
will always be a circuit-closer between these
brushes, as shown in Fig. 212. The circuit-
closer between these two brushes must, how-

two sections of the conductor and a leather
strap stretched between the two brushes kept
the springs apart as the car passed over the
circuit-closer. This experiment was a highly
successful one, and the desired end seemed to
have been attained.

The Denver Electric and Cable Railway Company in consequence ordered the material to lay
one-half mile of track upon this plan, and also
two motors and one dynamo of the Short type.

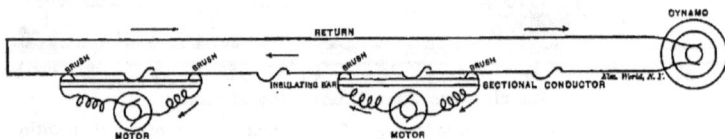

FIG. 213.

ever, be kept open as long as it remains between these brushes, and must close the line as
soon as it passes from between them.

Electrical means for accomplishing this were
devised, but were open to the same objection as
the one referred to above, and with the large
increase in number this objection became still
greater. It thus became evident that a mechanical circuit-breaker must be used. Mr.
Short presented this plan to Mr. John W.
Nesmith, a practical mechanic of long experience, and asked him to devise some mechanical

The electric conductor used in this trial was a
slotted copper tube of about 1 inch inside diameter. Each tube was about 16 feet long and imbedded in asphaltum and tar. The whole was
incased in an iron shell, and a slot was cut
through at the top. These centre rails were
spiked to the ties. Between the ends of these
sectional conductors were mechanical contact
closers. The car was provided with a contact
brush at each end, which fitted into the tubular
conductor like a gun-wiper brush. Between
these brushes was stretched a long flexible rub-

ber cylinder which almost filled the tube. As the car moved along, the forward brush was pushed through the tube and contacts at the ends; the rubber rod followed and kept the contacts apart until the rear brush had passed, and the current was forced to pass from one brush through the motor and back to the other brush.

rent-collecting brushes, and mechanical circuit-closing springs were placed at intervals in the conduit, as before. By this arrangement it was necessary to provide careful insulation only for the exposed ends of the sections of the conductor. This was a great stride towards success with the conduit system, but the

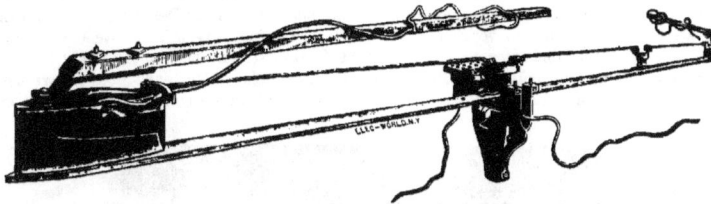

Fig. 214.

In October 1885, this half mile of track was completed, and for the first time a car driven by electricity made its appearance in Fifteenth street, in Denver. The experience of a few days showed this plan to be impracticable on

change in the arrangement of conductors made a new device for connecting the current-collectors necessary. This was accomplished by the construction of a collecting bar shown in perspective in Fig. 214, as it appears when be-

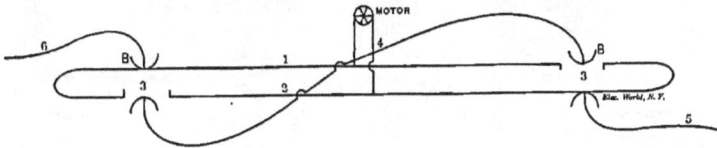

Fig. 215.

account of the difficulty of keeping the conductor insulated. A very little sand and dust in the tube would stop the brushes. The most serious difficulty to overcome, however, was the leakage due to imperfect insulation.

tween a circuit-closer on the line. Fig. 215 shows the arrangement of the bar diagrammatically, and Fig. 216 is a diagram of the circuit connections. As will be seen from Fig. 215, three adjoining sections of the under-

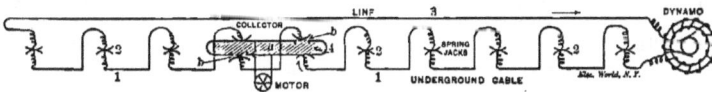

Fig. 216.

To overcome this difficulty, an ordinary underground cable was substituted for the bare conductor. This, by Mr. Nesmith's suggestion, was laid in the ground outside of the conduit, leaving the conduit for the passage of the cur-

ground cable are connected with the two circuit-closers, BB. But when the current-collector is removed from the circuit-closers, the latter are in contact, and the line is continuous, as shown in Fig. 216.

The current-collector is made of a slender bar of insulating material, somewhat longer than the section of the conductor between circuit-closers. The metal strips, 1 and 2, are fastened around the entire edge of the bar, except at 3 3, where there are short insulated breaks, as shown. Wires pass from these metal strips, 1 and 2, to the motor. The current passes from section 5 to metal strip 2, through the motor to metal strip 1, and thence to section 6. The middle section of the conductor 4 is cut out of circuit when the current-collector is in the position shown in the figure. If the bar be

and mud. The tracks were flooded and the severest possible tests were made to prove its merits and it never failed. Upon this a new company, the United States Electric Company, was organized in August, 1885, for the purpose of developing the system. In June and July of 1886 this new company put down under contract 3,100 feet of an electric railway, upon the last plan, for the Denver Electric and Cable Railway Company. This track was laid on Fifteenth street, between the Court House and the Post Office. Ordinary cross ties were used, upon which were strung 20-lb. T-rails. Notches

FIG. 217.—CAR ON THE DENVER, COL., ROAD.

moved in either direction it will be seen that no flashing can occur, nor can the line be short-circuited. This current-collector is 2¼ by 1½ inches in section, and about the length of a car. It is supported in the conduit by attachments passing down through the slot. In November, 1885, the company put down on the same street 500 feet of the new line of insulated wire for one of these current-collectors, and it worked to perfect satisfaction. During the winter of 1885 and 1886 this short line was worked in all kinds of weather, in snow, rain

were cut in the centre of the ties to form the conduit, and planks were spiked to the ties so as to cover the notches and to make the slot. The edges of the slot were protected by strap-iron.

Under one of the planks was laid a Waring lead-incased cable of No. 3 B. & S. gauge. Circuit-closers were placed every 25 feet and the cables attached to them. The two coaches built in Denver were placed upon the track. Each car had a motor of the Short type, made in Denver, and was supplied with both reversing gear and speed gear. Each motor weighed about 1,500 pounds. Tests were made of the efficiency of these motors, when they were developing from five to eight horse-power with a current of 25 ampères, and they showed a return of 80 per cent. Our illustration, Fig. 217, shows a longitudinal elevation of one of the first two cars on a section of the new road-bed and conduit.

On these cars the motors were placed on frames which were hung from the axles of the car trucks, and carefully insulated from them. On the armature shaft of each motor was placed a rawhide pinion. This pinion drove an iron gear on a counter-shaft which had its bearings attached to the same frame. The counter-shaft carried another pinion, which meshed with the large gear of the driving axle of the car. This method of gearing was direct and positive,

three and a half miles were soon fully equipped. A new conduit of cast-iron is used on this line, and is shown in longitudinal perspective in Fig. 217 and in transverse section in Fig. 218. The slot is five-eighths of an inch wide, and the conduit is enlarged to 9 × 12 inches and is of substantial construction. The new truck and gearing, with Brush motor, is shown in Fig. 219. It will be seen that the ordinary street car wheels and axles are provided with an iron frame supported upon spring-pedestals, upon which the motor is fastened. On the shaft of the motor is a pinion which drives a gear wheel on a short counter-shaft ; the latter again carries another pinion which meshes with a large toothed gear wheel keyed to the axle of the car wheels. The greatest ratio between the pinion and the gear is 1 to 5, and hence very little noise is made. The car body and truck

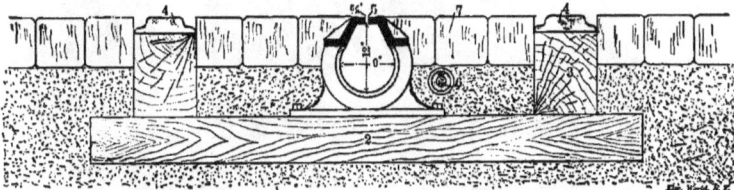

FIG. 218.—CROSS SECTION OF CONDUIT AND TRACK.

made little noise, and admitted of a simple and efficient reduction of speed from the motor armature shaft to that of the car axle.

The small engine and dynamo first placed in the plant were used to furnish the power.

On the 31st of July this road was started, and the car moved from one end of the track to the other without a mishap. This road was run constantly for three months, two men being required to operate it, one at the engine and one on the car.

In the meantime the Denver Electric and Cable Railway Company consolidated with another local company, and the new association took the name of the Denver Tramway Company. This company contracted with the Denver Tramway Construction Company for the equipment of its lines with the improved system. The work was immediately begun, and

are each complete in themselves and are put together simply, so that cars that are now being pulled by horses can, without much expense, be adapted to the electrical system.

The conduit shown in Figs. 217 and 218 is so arranged to admit of its being put down on existing horse-car lines. It is necessary to remove the pavement only in the centre of the track, to dig a trench down to the cross-ties and to put down the cast-iron conduit, which is made complete in sections of 8 feet. Before the pavement is relaid, the lead-incased cable is laid along at one side, as shown, all insulated and complete. The change from horses to electricity is made in this way, without interfering with the regular running of the road.

The current which feeds the motor is used to light the cars with incandescent lamps, and to operate an alarm-gong to warn persons of the approach of the cars. The latter can be moved backward and forward and governed from either platform. According to Mr. Sidney H. Short, to whom we are indebted for many of the foregoing details, indicator cards were taken from the engine while three cars were in service. These showed an average of 32 h. p. delivered to the driving belt. A constant current of 40 ampères was passing over five miles of No. 3 B. & S. gauge conductor. The resistance of the circuit was 7 ohms; hence the circuit alone used 15 h. p., which left 17 h. p. for three cars, the friction of belting, shafting, dynamo, etc. This gives an average of 5.66 h. p. for each car. One of the three cars, it may be

of an inch wide, in which the contact wheel travels, bearing directly upon the conductor within.

Mr. Fisher employs two methods of attaching the motor to the car. In one he suspends the motor under the car, and in the other method, which is illustrated in Fig. 220, the motor is placed on the front end of the car, which is inclosed by a cab, the motor man acting also as a conductor, having charge of the fare-box.

The length of the road is 3½ miles, and, being a suburban one, comparatively high speeds can be attained. Thus, it is said that the average speed attained is 18 miles per hour. At pres-

FIG. 219.—MOTOR AND TRUCK, DENVER ELECTRIC RAILWAY.

added, was climbing a grade of 350 feet to the mile at the time of the tests, and it is safe to assume that it was using one-half the power delivered to the cars. Hence each car can be run at the expense of 5 h. p. delivered, after the power necessary to overcome the line resistance has been deducted.

A road has recently been put in operation at Detroit, Mich., by Mr. Frank E. Fisher, one of whose stationary motors is illustrated on page 194. As will be noticed from Fig. 220, current is supplied by a third rail, supported upon insulators placed in a conduit 8 inches square. The conduit has a slot opening three-fourths

ent there are running on this road four cars equipped with motors of 10 h. p., giving them sufficient capacity to haul an extra car with full load. The generating dynamo is of 25,000 watts capacity, and delivers the current at a potential of 500 volts to the cars. The road has an extremely neat and compact station at the Detroit City end of the line, convenient to the railroad from which fuel supplies are obtained.

While it is acknowledged that, other things being equal, a conduit for the conductors is better adapted to heavy city traffic than an overhead system of conductors, there are still

FIG. 220.—MOTOR CAR ON THE FISHER ELECTRIC RAILWAY.

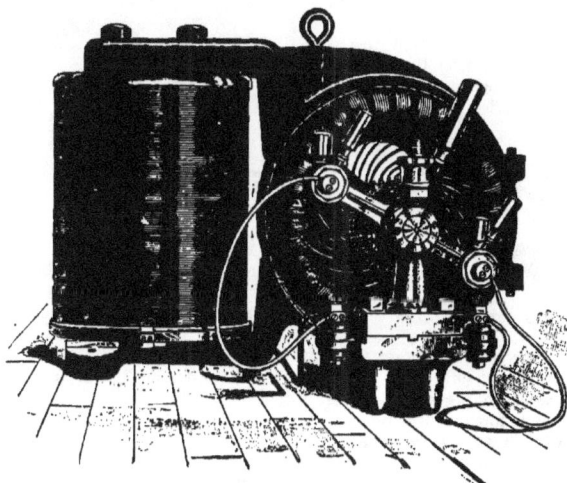

FIG. 221—FISHER ELECTRIC MOTOR.

some who object to the slot running along the street, and indeed more than one attempt has been made in the past to avoid the use of the slot and to establish connection with underground conductors by other means.

In attempting to solve this problem, Mr. W. E. Irish, of Cleveland, Ohio, hit upon the idea that if a conductor could be inclosed in an

FIG. 222.—IRISH'S ELECTRIC RAILWAY SYSTEM.

elastic conduit, a car passing above it might, by pressure, make contact with the conductor within and thus establish a connection; and this connection, made directly under the moving car, would be immediately broken when the car had passed on, this action being due to the elasticity of the conduit.

The manner in which this has been carried out is shown in the accompanying illustrations. Here Figs. 222 and 223 represent the arrangement adopted by Mr. Irish, in section and

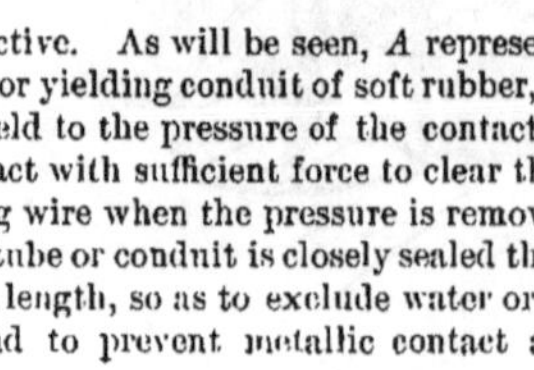

FIG. 223.

perspective. As will be seen, A represents an elastic or yielding conduit of soft rubber, which will yield to the pressure of the contact wheel and react with sufficient force to clear the conducting wire when the pressure is removed.

The tube or conduit is closely sealed throughout its length, so as to exclude water or moisture and to prevent metallic contact at any

point except through the proper connections. A channel for carrying the tube is formed in a line of timbers or blocks of stone B, the channel being shown as having flat parallel sides and an open top. The timbers or blocks carrying the tube are laid along the rail-post track between the rails, flush with the surface of the roadway, two lines being used, one to carry the outgoing conductor and the other the return conductor.

The tubes A carry the line wires or conductors C at the bottom of the oblong track therein. These conductors are uncovered and uninsulated except as to the rubber tube A, which

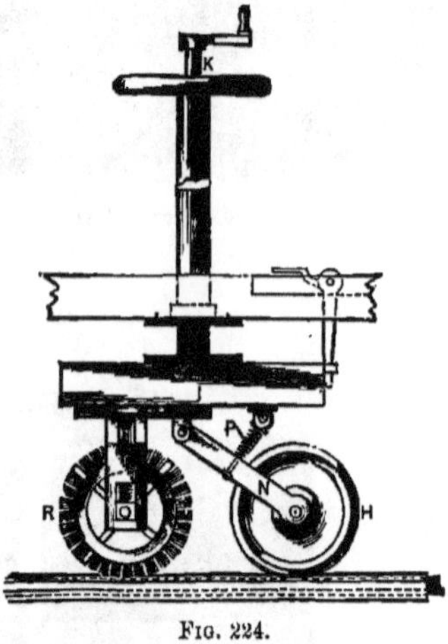

FIG. 224.

forms a covering and insulation, so that contact may be made within the tube at any point in their length. Attached to the tube along its upper surface are short rail pieces D, having small flanges at their sides, which rest on shoulders on the tube, and when in their normal position are flush with the roadway and top of the timbers.

Inside of the tube A, and corresponding to the rail pieces D in length, are contact pieces E, flanged laterally at the top, and having a central portion which rests in the tube above the line wire or conductor C, and normally out of contact therewith. The rail and contact pieces are insulated from each other by the rubber tube except where they are connected by screws, by which they are firmly united. The short rails and inside pieces are arranged in pairs, and the pairs are insulated from each other by having a sufficient space between them at the ends. This will allow one section or pair to be depressed without interfering materially with the next on either side, the rubber to which they are secured being sufficiently flexible for this purpose.

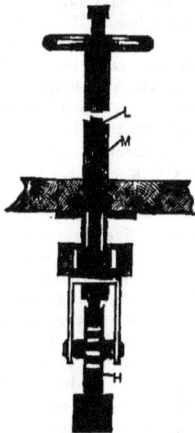

FIG. 225.

In Figs. 224 and 225 H represents the contact wheels, similar in construction, one of which is attached to either side of the centre of the car and electrically connected with the motor upon the same. The wheels are each pivoted on a frame or turn-table I, which is arranged to be reversed upon the car according to the direction the car is to travel, and to be raised or lowered, according as the wheel is to be thrown into or out of contact with the track. Figs. 224 and 225 show the mechanism by which the several movements are effected. Thus the position of the table I and its attachments is reversed by means of a rod K rigidly fixed thereto and having a crank lever at the

top, and it is raised and lowered by means of the tube L, which is provided with a hand wheel and works inside the sleeve standard which supports the parts of the platform on the car.

The contact-wheel H is hung in a pivoted frame N on a shaft on which it has some lateral play to adapt it to travel on a track that may be somewhat out of alignment, and the wheel has a concave tread which enables it more certainly to keep the track. P is a tension spring, bearing upon the wheel frame N with such pressure that, as the wheel passes over the sectional track, it will depress the sections successively, so as to bring the inside pieces E in contact with the main wire within the tube. The wheels H act alike in this particular. The moment a wheel depresses a section sufficiently, the circuit is closed through that section; the screws connecting the short rails and the inside contact pieces being the medium for the passage of the current through the rubber tube; and as one section is not cleared before connection is established through the next succeeding section by the wheel passing from one section to another, the flow of the current is made continuous while the contact wheel is down. When the sections have been passed by the contact wheel, the elasticity of the rubber carries them successively back to their original position.

The conductors in the bottom of the rubber tube are coupled together in sections in suitable lengths by means of a variable expansion joint which will allow of expansion and contraction under varying temperature without, to any appreciable degree, altering or affecting the electrical resistance. One satisfactory way of doing this is by placing a sleeve over the ends of the conductor where they meet, inside of which is a spiral spring which bears against the ends of the conductor. This spring is compressed by the expansion of the conductor and elongated when the conductor is contracted, but always makes effective contact with the conductor and with the sleeve. The sleeves have about the same resistance as the conductor, but the contact between the sleeves and conductor, or line wire, is more or less imperfect, and this imperfect contact is made up for by the springs.

In front of the contact wheel is a combined brush and scraper R, supported upon the same frame with the contact wheel, and designed to run in front of it and clear the exposed section surface rail of dirt, snow and other obstructions, so that perfect contact may be made with the rail or track. Connected with the conduit described above, and beneath the same, is a second conduit, Fig. 222, of cylindrical form with a V-shaped opening longitudinally in its top and having lateral flanges by which it is bolted to the timbers or blocks carrying the first-named conduit. The conduit is designed to carry telegraph, telephone, electric light, and other wires, and has a V-shaped covering, underlaid by a sheet of packing of soft rubber, which extends laterally over the flanges and upper conduit timbers. By unscrewing the top, the wires can be reached

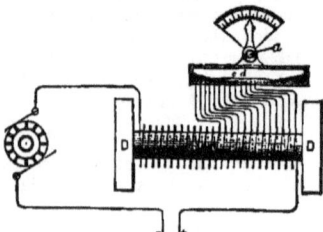

FIG. 226.

at any point, and the tube can again be perfectly sealed by screwing the top into position.

One of the methods of automatic regulation devised by Mr. Irish is shown in Fig. 226. This consists of a tube having a concave longitudinal section, and a pivot on which it may be adjusted to give it any desired inclination. As shown, the tube is resting in a horizontal plane, and this throws the mercury contained in the tube into the centre. The series of wires $c\,d$ extend into the tube and make connection with the corresponding coils on the field magnets of a motor D. When the mercury lies as shown in Fig. 226, all the coils on the motor whose wires, $c\,d$, are covered by the mercury are cut out, and only those are energized whose wires are not so covered and connected. The

charging of the field magnets and the amount of electrical energy therein is therefore under convenient and easy control through the tube. Obviously, if the tube be tilted to the left it would convey the mercury forward and thus make connection with one set of contact points after another, until at last, if carried far enough, the entire field would be cut out and the motor would stop. On the other hand, the entire energy of the current may be thrown into the motor by making the extreme adjustment to the right, and relieving the contact points of the connecting fluid.

The motor employed by Mr. Irish is very flat and of novel design. It is supported under the car body between the wheels, and is so arranged that all or any part may be examined, oiled, or replaced from the inside as well as the outside of the car in a few minutes. The motor shaft is furnished with two armature coils and two driving wheels, the latter being on each end of the shaft. The maximum power is obtained when both the fields and the armature coils are active, although the car can be run with one field and both armature coils, and also with both fields and one armature coil, or with one armature and one field active.

The electric railroad in the Lykens Valley, Pa., collieries was designed by Mr. W. M. Schlesinger for the purpose of hauling the coal mined in the upper part of the workings out of the mine. It is the first and only electric mine railroad built in America, and exceeds, as regards power and length, any of those built in Europe. The following table shows the relations between the different roads:

Where running.	System.	Length.	Speed.	Weight of locomotive.	Largest weight pulled.
		Feet.	Miles.	Lbs.	Tons.
Zankeroda	Siemens and Halske...	2,028	6	3,520	13½
Paulus and Hohenzollern	Siemens and Halske....	2,460	..	4,200	...
Lykens Valley	Schlesinger .	6,300	6	15,000	100

The plant in Lykens was put in operation on the 26th of July, 1887, and has been working without trouble ever since, hauling with ease trains of from 10 to 21 cars, weighing from 50 to 100 tons, according to whether loaded with coal or rock. As soon as the gangway is completed and the works open up, the motor is to haul from 200 to 250 cars a day, or bring out from 900 to 1,125 tons of coal. The trains are hauled out at a speed of 6 miles an hour (the maximum speed allowed in the mines), making the round trip, with necessary stoppages at the termini for shifting, in from 25 to 30 minutes, whereas it took the mules three times as long. On several occasions the motor has already hauled out trains at a speed of 10 miles an hour, and once even as fast as 15 miles an hour.

Outside of its greater economy, the electric mine locomotive has other great advantages over the steam locomotive and over mules; its greatest feature is the entire absence of smoke, steam and sulphur, and the nonconsumption of oxygen. How important, especially the first-named feature is, only those who have been in gangways in which steam motors are running can fully understand; and it is due to this, also, that the electric motor can be put in places where the steam-engine could not be used; for the latter, on account of the unhealthy gases, etc., it gives out, is limited to places near the fans, so that these gases are at once removed from the mine, as it is dangerous to allow the air, after once having passed through the steam locomotive, to go to places where men are at work. With electric motors this precaution is unnecessary.

As regards the non-consumption of oxygen by the electric motor, it saves in the drift at Lykens 20,000 cubic feet of air per minute, which would be required for a steam locomotive. But outside of this, another great advantage of the electric locomotive lies in the fact of that it has but few working parts, and that these can be got at with the greatest ease. The plant in Lykens was built by the Union Electric Co., of Philadelphia. The drift in which the motor runs is one of the upper drifts of the Lykens Valley collieries, the entrance to it being on the side of a mountain. Like all gangways it is higher at its further or inside end, but the average grade is below that usually employed, being only 3 feet in 100, 5 feet in 100 being the usual grade. In most places the road is perfectly level; there are three or four

down grades, two of which are very heavy, and two reverse grades, one being of 1½ per cent., up which the loaded cars have to be hauled. The road outside the mouth of the drift is about 300 feet long, and enters the mountain on a curve having a 105-foot radius. For 1,500 feet the road is practically straight; then comes an S-curve, with a 90-foot radius, and a series of smaller curves up to 2,400 feet from the entrance; the next half mile is nearly straight, but at the end of it is a 90-degree curve, having a 30-foot radius; and 200 feet further is another 90-degree curve, with less than a 25-foot radius.

The siding down which the empty cars are taken at present, so as to be out of the way of the loaded train (or trip), is between these two curves. In a short time the gangway leading to the regular turnout, which is about 1,000 feet further in, will be completed, and the above-mentioned siding will be abandoned. In coming in with the empty cars the motor finds the loaded ones generally standing distributed on the curves and between them. The motor has to collect them and then to push them up a 1-per cent. grade and round the 25-foot curve, until the siding for the empty cars is cleared. This is the heaviest work the motor has to perform, especially when the cars are loaded with rock. The start from the inside is always made while the train is standing on one or the other, or, if a long train, on both these curves. The gangway is so low in some places that it is impossible to walk upright.

The cars in use weigh, empty, 3,300 lbs., and have a capacity of 94 cubic feet each. The wheels are 18 inches in diameter, and run loosely on the axles, like ordinary wagon wheels; the axles are rigidly attached to the car body. The result is that very soon the hole in the wheel through which the axle passes is worn larger, and unless carefully and frequently oiled the cars run very heavy, especially as the wheels are often worn oval or flat, due to the insertion of sprags to prevent the cars from running away down hill, and to stop them when mule power is used. The heaviest work performed by the motors, so far, was at one time to push a train of 12 cars, loaded with rock and weighing about 150,000 pounds, up grade round the 25-foot curve; another time a train of twenty cars was started on this curve, only three cars and the motor standing on or beyond it, the other 17 cars standing at right angles to the direction in which the motor was pulling.

A 25-pound rail is used as the conductor, which is fastened to props inclined a little from the vertical, as they were easier to put up in that position. About half way between the top and bottom of the props, and mostly at a height of 22 inches above the track and 15 inches to the side, blocks are fastened, and to these are screwed the rails, which are insulated by rubber. In places where men or mules have

and thus throw them down, without in the least injuring them. In not one single instance was any one of these miners prevented from resuming work at once, nor did they feel any after-effects from the shock. The joints in the rails are all carefully made by placing brass plates under the fish-plates and bolting these down tight, so as to press the brass into all the uneven parts of the iron rails. The track rails are used as the return conductor, and these are connected together in the same way as the conductor. At a distance of every 100 yards the two track rails are connected by heavy copper wire, so that repairs can be made

FIG. 227.—SCHLESINGER MOTOR FOR LOCOMOTIVE WORK.

to cross the conductor, it is raised to a height of 5 feet 9 inches, and recedes to 24 inches from the track, the ascents and descents being made gradual. Outside the drift the conductor is 5 feet 9 inches high all along. The props are placed 10 inches apart on the lower side of the gangway, so as to be out of the way of the miners, although the E. M. F. used (350 to 400 volts) is not dangerous. As an instance, several miners have already, accidentally or purposely, touched the conductor and received shocks from the same, which had no other effect than for the instant to numb their legs

to the road while the motor is inside without interrupting the current.

The engine-house is situated at the mouth of the drift on the side of the mountain. The steam-engine used is an old one, which has in been use in the mines for more than 25 years. It is an old style, long stroke, and was formerly used for hoisting purposes. Since it was built it has undergone a good many changes, so that at present the steam-ports are entirely out of proportion to the size of the cylinder. This engine will most likely be soon replaced by another. The steam is conveyed to it from boilers 1,000 feet away, and a great saving in power might have been effected had the engine-room been placed nearer the boilers ; but as a steam-pipe passed near the present location of the engine-room, it was thought cheapest to place the latter near the entrance of the drift.

The engine is run at a speed of 60 revolutions, and from a 9-foot band wheel the power is transmitted by a 10-inch belt to a counter-shaft, from which the power is transmitted to the generator. This generator is a 50 h. p. series wound dynamo of the Manchester type. The armature is of the drum type, 12 inches in diameter, and makes 700 revolutions per min-

is reset. The generator is at present in charge of a boy 18 years old, who also acts as conductor to the train. The dynamo has been built with special regard to efficiency and simplicity, its main feature being that the commutator can be replaced by an ordinary mechanic without testing instruments, within one hour, and without the possibility of a mistake being made.

The motor represented in two views, in Figs. 227 and 228, is capable of giving 35 h. p., and is designed especially for locomotives having to haul heavy loads. It is 27¼ inches broad, 43½ inches long and 22 inches high, and weighs,

FIG. 228.—SCHLESINGER MOTOR FOR LOCOMOTIVE WORK.

ute. The commutator is placed outside the bearings and has 32 segments. The cores of the field magnets are forged out of picked scrap, the pole pieces being of cast-iron. The whole machine weighs 1¼ tons, and takes up a space of about 3¼×5¼ feet, and is 2¼ feet high. A safety cut-out is placed on the dynamo, which puts a large resistance into the circuit whenever the current exceeds a fixed limit, thus reducing it and preventing the destruction of the generator, as well as motor, armature. A large gong in the engine-room rings whenever this safety device acts, and until it

with counter-shaft and gear, 1,500 pounds. The armature is of the Siemens drum type, 9¼ inches in diameter, and has a core 10¼ inches long.

The wires are wound within troughs fastened to the core of the armature, and are thus prevented from being displaced either to one side or the other by the pull exerted by the magnets on them. This pull, with a motor, varies from one side to the other as the direction the motor runs in is altered, and tends not only to loosen the wires, but also, by causing them to move, gradually to destroy the insulation.

The armature makes about 1,000 revolutions, the countershaft about 400. The bearings of the latter are part of the same frame in which the two armature bearings are cast. This keeps the two shafts parallel under all conditions, and insures an easy running of the gears, even with the most trying alteration in the load. After six months' running, the teeth of the pinion showed a wear of $\frac{1}{24}$ of an inch; *i. e.*, about $\frac{1}{48}$ of an inch on each side of the tooth.

The 32-part commutator is placed outside the bearings, and is constructed in such a manner that it can be taken off and replaced by a new one within an hour and a half, it being unnecessary to disturb any of the other parts of

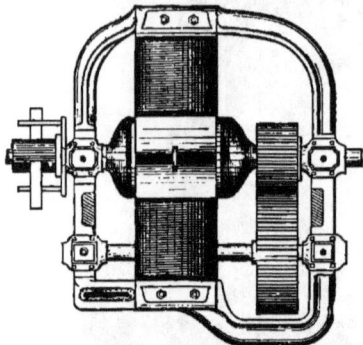

FIG. 229.—SCHLESINGER RAILWAY MOTOR.—PLAN.

the machine, or to make any electrical connection otherwise than to solder the wires into grooves in the commutator bars. As all these wires are so arranged that they can be placed in no other groove than the one to which they belong, it requires no electrical knowledge or training to accomplish this. The position of the commutator outside the bearing has the further advantage of allowing at all times an easy inspection of it and the brushes, and facilitates the proper setting or adjusting of the latter.

The motor, of which Fig. 229 is a plan, is securely fastened by eight bolts to a strong wooden truck having 30-inch wheels and a

wheel base of 40×40 inches. The truck is built for a weight of 15 tons. The motor is entirely boxed in, one of the countershaft bearings and the spur wheel passing through one end of the box, while at the other end is a door through which the commutator can be reached and the brushes set. The top of the box is also removable. On either side are large compartments, well braced, for the placing of ballast. This consisted at first of scrap iron, as shown in the engraving, Fig. 230, but now iron plates, 2 inches thick, specially cast for the purpose, have been put on. The seat for the driver is at one end and placed sideways, so that he can run the locomotive either way without having to change his seat. With his left hand he operates a powerful hand-brake, and with the right the regulating lever. At first the regulation was effected by two levers, one to reverse the motor and one to start, regulate the speed and stop the motor; but now both these actions are performed by one lever. By moving it out of its central position away from the driver, the motor is run forward; by moving it toward the driver, it is reversed. The speed is regulated with the same lever by means of resistance coils placed in an open box underneath the foot-rest for the driver.

The necessity for simple means of operating the car is obvious, when it is considered that the motor runs in a dark tunnel with a large number of sharp turns, and the walls or sides of which are liable to give way at any moment, and the circumstances being such that a breakdown of this nature is generally not recognizable until the motor is within but a few feet of the place. An electric lamp was attached in front of the driver, to act as a current indicator, but although it simply hung in the wires, the jar of the car was so great that several lamps were broken in a very short time. A plain ammeter, consisting of a solenoid with a long pointer, now takes its place. The motor, with ballast, weighs about 7½ to 8 tons.

The current collector, as shown in Fig. 230, consists of a frame having two vertical and two horizontal wheels. In this frame is attached a rod which passes through a tube and the collector is pressed against the conductor rail by means of a strong spiral spring. The tube is fastened to a movable arm 4 feet long, attached to a locomotive. This arm allows the collector to adapt itself to all vertical inequalities of the rail, while the tube, spiral spring and rod take up all horizontal variations of the rail. This collector is connected directly to the switch-box, and the return current passes partly through the chains and partly through small shoes rubbing against the car wheel, then

pecially designed for the purpose. The links are made of phosphor bronze and the pins of steel, the latter having a wearing surface of two square inches. From the second countershaft to the car axles, steel chains with thimbles are used. The introduction of so many countershafts, of course, reduces the efficiency of the motor, but it was unavoidable at the time; it is intended, however, to alter this soon.

FIG. 230.—LOCOMOTIVE ON THE SCHLESINGER ELECTRIC ROAD, LYKENS VALLEY MINE.

through the latter, and from them to the track rails. In addition to this, to prevent sand and dirt on the track interfering with the running, metal brushes are added, which either run continually on the rails or can be pressed against them by hand.

The chain transmitting the power from the first to the second countershaft has been es-

The engraving, Fig. 231, is made from a photograph recently taken, and represents the motor hauling twenty loaded cars and one passenger car containing six persons.

One of these motors has been running for the last nine months in the Lykens Valley Coal Company's mine at Lykens, Pa., hauling daily about 500 tons gross weight over a road

6,300 feet long. This load is gradually being increased, and it is expected to haul about 1,000 to 1,300 tons gross weight daily. Some of the work this motor has accomplished was to haul a train weighing about 150 tons, consisting of 31 cars and 380 feet long, round two curves, the one having a radius of 20 feet, and the other 30 feet, and the distance between them being 180 feet. This same train had afterwards to be started while standing partly on a level and partly (100 feet) on an up grade of 15 inches in 100 feet. The average number of cars hauled by this motor in one trip varies between 10 and 20, these cars being partly loaded with coal and partly with rock or slate. A second motor of the same type will in a short time be in operation in another part of the mine, where it will have to haul daily 300 loaded cars, representing a gross weight of about 1,500 tons, and 300 empty cars weighing about 450 tons, so that the total load of this motor will be nearly 2,000 tons.

An appendix to the first edition of this work gave a description of the Julien accumulator and traction system, which were being introduced here at the time of its publication by M. Ed. Julien, engineer and electrician, of Brussels. Such has been the success of the demonstration that at the present time ten Julien cars are building, and a charging station is being equipped for the Fourth avenue road of this city. The Julien car, which has been in continuous service for some months, is shown in Fig. 232. The new cars will be of different type and will embody many improvements in construction, the work being carried out by Mr. C. O. Mailloux. Various other trials with storage cars have been made in New York and

FIG. 231.—VIEW OF TRAIN AT THE MOUTH OF THE MINE.

other cities, by far the most important of them being those made by Mr. Anthony Reckenzaun, at Philadelphia, for the Electric Car Company of America.

On the car there employed are two Reckenzaun motors, supported by two small trucks, similar to those on Mr. Reckenzaun's street cars in Europe. Each motor weighs about 500 lbs., and the pair, when working to their fullest capacity, are capable of giving 30 horse-power collectively. Such power will scarcely ever be needed, but it can be called into requisition should circumstances demand it. The car has been tested on the experimental line, and it

street car in the world. It is one of the most carefully designed in every detail.

The frames of the four-wheeled trucks are made of wrought iron ; they have a very light appearance, yet they are of ample strength to support the maximum load with safety. The wheels are only 26 inches in diameter, which, when revolving at 103 revolutions per minute gives a speed of 8 miles an hour when the motors run at 824 revolutions, the armature speed being reduced 8 to 1 by means of Mr. Reckenzaun's worm gearing.

The speed of the car is regulated by a switch which causes the motors to work in series,

FIG. 232.—THE JULIEN ELECTRIC STREET CAR.

was ascertained that the current, when mounting the grade of 5¼ per cent. (264 feet to the mile), was only 80 ampères, and on the level 25, and this current was supplied by 74 storage batteries ; but there are actually 120 cells on board, stowed away under the seats on long boards, which run on rollers to facilitate the speedy removal and replacement of the whole battery. The seats are of the usual height and width, but they are 22 feet long, accommodating about 34 people, so that with the available standing room and platforms a hundred passengers can be crowded into the car. It is said that this is the handsomest and largest electric

singly or in parallel circuit, so that all the cells are always used when the car is in motion, whereby they are discharged uniformly, a result which will be appreciated by those who have had experience with storage batteries. The cells were manufactured by the Electrical Accumulator Company, of New York. Large as this car is, it goes round the sharpest curves with remarkable ease, and altogether works well, the motion of the whole apparatus being absolutely silent.

The use of eight wheels removes the objections raised by street railway men with regard to the additional weight of storage batteries, so that by distributing the load over 8 points on the rails there is no need of providing stronger roads. Pivoted trucks, with short wheel bases, facilitate the movement round curves very much ; there is no jerking as in ordinary four-wheeled cars, and, therefore, the flanges of both the wheels and the guard-rails are preserved. In the large car under notice the wheel base is only three feet eight inches.

Turning now to the newer stationary motors brought out in America, the first that claims attention is that designed by Professor Elihu Thomson for the Thomson-Houston Electric Company, and whose use in street railway work has already been touched upon. For some time past the work of developing a line of electric motors suited for the transmission and distribution of power from central stations and

FIG. 233.—THOMSON FIFTEEN H. P. ELECTRIC MOTOR.

The use of four driving wheels actuated by the positive and smooth action of the worm gear offers important advantages in mounting steep grades, and the employment of two distinct motors (one on each truck) decreases the chances of an absolute stoppage to a minimum, because in case of accident to one machine the other is sufficiently powerful to bring the car home.

for other purposes, particularly electric railroading, has been carried on by Professor Thomson. The object held in view was to construct machines of the highest type mechanically and electrically. Beginning with motors of one and one and a half horse-power, larger sizes, up to full fifteen horse-power capacity, have been rapidly brought out, and still larger sizes are in process of construction. The sizes now built are one and one and a half horse-power, three, five, seven and a half, ten and fifteen horse-power. The motors are built for constant potential circuits of 110, 220, 400 and 600 volts, as needed. The proportioning is such that, supplied with a constant potential, they are practically self-regulating as regards speed, though the load be varied from nothing up to full power, or the reverse. At the same time the brushes on the commutator run without spark, and are not shifted in position during extreme changes of load on the motor. In other words, the non-sparking points of the commutator remain at one position without

arrangement, and the field magnet coils are in shunt to the armature.

The armature core is so well laminated and the resistance of the armature conductor is so low that loss by Foucault currents or local currents in the iron, and by internal resistance, is very slight as compared with the output of the machine. The consequence is that the motor keeps practically cold during the running, and is capable of delivering power considerably in excess of its rated capacity.

Another prominent new motor is that designed by Mr. W. Baxter, Jr., for the Baxter Electric Manufacturing and Motor Company, of

FIG. 234.—SMALL BAXTER MOTOR FOR ARC CIRCUITS.

change, notwithstanding the greatest variations of load. The proportioning is such, it is claimed, as to secure the highest efficiency of conversion of electrical energy into mechanical energy. Tests have shown that over 90 per cent. commercial efficiency can be attained at full load.

As will be noted in the illustration, Fig. 233, the poles of the field magnets—the bodies or cores of which are round in section—project upward and inclose the armature, the section of the core of which is nearly square. The winding of the armature is a modified Siemens

Baltimore, of which early types are here illustrated. Fig. 234 shows a motor of the smallest size, built for arc-light circuits, and intended principally for running sewing-machines, job printing-presses, pumps, fans, etc. When intended for pumps, fans, or any other purpose where a constant speed and power are required, it is made without the ring surrounding the commutator and brushes shown in the illustration. This ring is only used when the nature of the work is such that the motor has to be started and stopped very often, or when it is necessary to vary the speed at will. In this case the brush-lever is made movable around its axis, and the electrical connections between the field-coil terminals and brushes are maintained by springs attached to the brush-holders, and so arranged that their free ends press against metallic linings on the interior surface of the ring.

A spring attached to the upper end of the brush-lever holds it around against a stop. When in this position the brushes are raised off the commutator and no damage can be done by turning the armature backward. At the same time, the contact-springs that press

reaches the maximum. Upon the wooden base, and underneath the armature, is a switch by which the whole motor is cut in or out of circuit. These small machines are complete in themselves, and require no auxiliary attachment for regulation. The commutators are made of cast-steel, and are calculated to last for years. The journals are self-lubricating, and will run for weeks without attention.

This type of motor is made in three sizes, namely: $\frac{1}{16}$, $\frac{1}{12}$ and $\frac{1}{6}$ horse-power. The same patterns are used in all, the difference in capac-

FIG. 235.—LARGE BAXTER MOTOR FOR ARC CIRCUITS.

against the interior of the ring are in such a position as to cut out the armature. The motor is set in motion by rotating the brush-lever in a direction opposed to the tension of the spring. The first movement throws the brushes on to the commutator, so that the circuit may be closed before the armature is cut in. In this position, the velocity will be very slow; but as the lever is rotated through an angle of 90 degrees, it gradually increases until it

ity being effected by winding more or less wire on the field. Their efficiency ranges from about 70 per cent. for the $\frac{1}{6}$, to 65 per cent. for the $\frac{1}{16}$ horse-power. This efficiency might be considered very low for large motors, but for motors of this size shows good design.

Figs. 235 and 236 show the general appearance of the larger sizes of the Baxter motors. Fig. 235 is a constant current, and Fig. 236 a constant potential, motor. There is no difference in design between the two, but the method of regulation differs in each.

The constant potential machine is wound so as to run at a constant speed, but as the same principle cannot be applied with a constant current without too great a loss in efficiency, a mechanical governing arrangement is used in the latter type. This is illustrated in Fig. 235. The governor proper is carried on the outer end of the shaft, and is located within the shield shown in front of the motor. The de-

highest efficiency is between one-half and two-thirds the full capacity, and that at one-third and full load is about the same. On this account this type of motor is well adapted to a system of distribution of electrical energy from a central station.

The resistances of a 10-horse-power motor for a 10-ampère current are as follows: Armature, .75 ohm; field, 3.75 ohms; total internal resistance, about 4.5 ohms; difference of potential without load, 45 volts; total difference of

FIG. 236.—LARGE BAXTER MOTOR FOR INCANDESCENT CIRCUITS.

vice by which the action of the governor is made to regulate the speed is located on top of the pole-plate. The principle of regulation consists in changing the magnetic intensity of the field by a variation of the ampère turns in the magnet coils.

With this system of regulation, down to a certain limit the efficiency rises; below that it begins to decrease until it becomes the same as at maximum load. Practice shows. that the

potential of motor when developing 10 horse-power, about 792 volts; loss by internal resistance, .0567, or a little more than 5½ per cent. There are two layers of wire on the armature, the number of turns being 320; as this sets up a counter E. M. F. of about 750 volts, it is at the rate of 2.28 volts per turn. Reducing this to work, it means an output of more than 1,000 foot-pounds per turn, or nearly 300 foot-pounds per foot of wire.

The weight of wire on the armature is about 12 pounds; weight of iron, about 75 pounds; weight of wire on field, about 160 pounds; iron in field, 1,100 pounds. These figures show that a very small amount of wire is used, but that otherwise the machines are very massive. On this account the reaction of armature on the field is practically nothing, and therefore the brushes require no lead; hence the load may be varied at will without causing sparking, as the diameter of commutation remains unchanged.

The Baxter constant potential motor is almost identical in appearance with the constant current machine just described. The regulation is accomplished by the method of winding; hence the governor and its attachments are removed; the switch is also replaced by simple binding-posts, as these motors are provided with an independent automatic cut-out and hand-switch combined. Constant speed is obtained by a simple shunt winding.

From an electric standpoint Mr. Baxter considers the constant current method of distribution to be the better. This superiority, he states, is not due to any defects in the constant potential motor, but is owing to the fact that there are certain advantages in the constant current system, in virtue of which it is possible to obtain results that are beyond the reach of the constant potential system, no matter on what principle the motor may be constructed.

The motors manufactured by the C. & C. Electric Motor Company were designed to meet the demand for a small and simply constructed machine, in which the correct principles of dynamo construction were not all violated. It was sought to make those elements which are known to be the essential features of a thoroughly good machine—high circumferential speed of armature, low internal resistance and strong magnetic field, etc.—the first consideration, and then to design shapes and invent methods of manufacture which would enable the electrical requirements to be carried out most cheaply. The result of careful consideration of these things has led to the construction of a motor weighing about twelve pounds, and of the proportions shown in the engraving, Fig. 237, and which is made on the American plan

of automatic machine work and interchangeable parts, to a degree comparable with the manufacture of watches and other well-established American manufacturing industries.

Each pole-piece is equal in cross-section to the core inside the field-coil, and both are made of one piece of the softest domestic iron (Burden's Best), with the fibre running lengthwise, or in the direction of the lines of force. These cores and pole-pieces are made from round bar-iron of the size of the core, and are struck or drop-forged between dies having exactly the shape of the finished magnet, even including

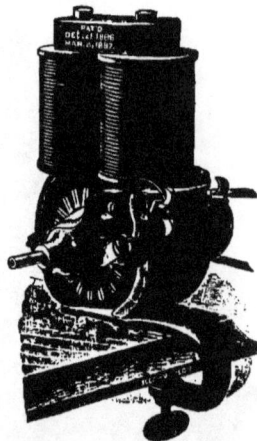

FIG. 237.—THE C. & C. MOTOR.

the small lugs on the outside to complete the horizontal support for the field-coil washers and insure a true winding of the wire. By the process of drop-forging the inner surface of the pole-piece is bent around a rounded part of the die which exactly represents the space required for the armature, thus insuring a true and smooth circular space for the latter. The outside of the forging is likewise left by the die smooth and finished, and the irregularities or "fins" are all brought to the sides of the pole-piece, where they may be trimmed off by the same trimming operation which is necessary to make a seat for the bearing-plates.

Since the circular space for the armature is the important thing to preserve, all of the further operations of cutting and fitting the forgings and of assembling the motor are all carried on with reference to this circle. The forgings are milled off bright at the upper end to receive the yoke, the forging being clamped in a vise which consists in part of a round iron block, which fits into the concave side of the pole-piece and represents the armature space, and which is set at a fixed distance from the

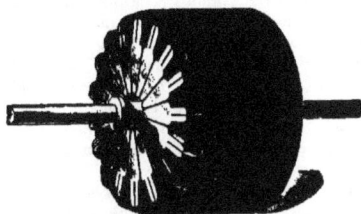

Fig. 238.—The C. & C. Armature.

milling tool. The result of this operation is that when any two trimmed forgings are screwed to a straight yoke in the usual way, the concave portions of the pair will agree in forming the required circle for the armature space. The forging is then clamped in another vise, consisting in part of a similar round block, and the two edges of the pole-piece are trimmed off parallel, so as to bring all of the pole-pieces to the same width, a straddle cut being taken by two milling cutters mounted on the same arbor. The forging, which is thereby finished to gauge in all respects, is next clamped against another round block inside of a box drilling-jig, and all of the screw-holes are drilled at one operation while the forging is held against the round block ; so that when the bearing-plates and other parts are screwed on, they will be true to the armature space.

The field-coils are then wound between washers driven on the forgings, and the field is completed by screwing to the top a yoke having a cross-section equal to the cores, which gives uniform magnetic conductivity throughout the frame. Cast or stamped brass bearing-

plates are then screwed to either side of the pole-pieces, and in these bearings revolves the spindle or shaft, carrying a Gramme ring armature, shown in Fig. 238.

The iron of the armature-core consists of semi-circles punched from thin sheet-iron, on one side of which tissue paper has been pasted before punching. These semi-circles are laid together with the ends of alternate rings projecting at either edge of the built-up half-cylinders so that the edges of the two half-cylinders so formed will mate together or interlock. The half-cylinders are then mated together at one edge, and a rivet is passed through, uniting them like the parts of a hinge. Upon the split-ring so formed is slipped a flat helix of wire of a length sufficient to form the entire winding of the armature and consisting of only one layer, so that the operation of slipping it on is very simple ; if there happens to be any defect in the insulation it can only short-circuit or render useless a single convolution instead of an entire section. In order to get the required E. M. F. the wire used is flat, as shown in Fig. 239, and wound on edge, by which means it is possible to obtain the necessary number of convolutions with the necessary conductivity with only a single layer.

By this plan all complication due to unsymmetrical winding and unequal position of dif-

Fig. 239.—The C. & C. Armature Winding.

ferent convolutions in the magnetic field are avoided, as well as the various and complicated inductive effects of one section against its neighbors when built up close to each other in a good many layers in the usual way. As the winding is a simple or progressive helix, none of the convolutions overlap each other, and consequently there can be no serious short-circuiting ; and the bad inductive effects of the current in one section running in the opposite direction to the current in the other sections as the coils pass the point of commutation are reduced to a minimum. In addition, this winding has the advantage of producing a mechanically balanced armature, as it is obviously impossible to get the wire wound on deeper or in larger quantity on one side of the armature than on the other. The flat or beam-shape of the wire also serves to stiffen the winding, and prevent its flying out by centrifugal force, thus rendering a circumferential outside lashing unnecessary. As the number of convolutions of wire on the armature is of great importance to the efficiency as well as the power of the motor, great attention has been paid to this point, and a ribbon is now used for one of the windings (type *E*), which is eleven times as wide as it is thick, and in some of the types of motor is made trapezoidal in cross-section with the narrow edge out, the inclination of the sides of the wire being sufficient to allow the sides of consecutive convolutions to lie flat against each other on the inside of the ring. By doing this, the amount of copper and consequent conductivity is slightly increased and the insulation is better protected, since it does not bear against the insulation of the next wire in a single line, as with a round wire, but in a flat surface. In fact, after a winding is in place, the entire accessible insulation can be scraped off both inside and outside the rim without injury.

This winding, which is shown in part in Fig. 239, is wound by an entirely automatic machine upon a flat mandrel equal in length to the circumference of the armature, so that when the mandrel is wound full with one continuous piece of wire there are just enough turns to fill one armature. The winding is divided into seventeen sections by seventeen equally distant convolutions, which project about a quarter of an inch farther out than the others, and furnish means of making excellent soldered connections with the commutator. In forming the winding the end of the wire from the reel is fastened to a suitable catch at one end of the mandrel, which is then started revolving, the wire being held upright or on edge against the mandrel by a suitable guide-arm, which is arranged to rise and fall so as to follow the surface of the mandrel as it revolves. A large ratchet-wheel, divided into a number of teeth equal to the number of convolutions which the armature can carry, is impelled one tooth at a time at each revolution of the mandrel, and the completion of the revolution of this index-wheel automatically stops the winding-machine at the moment when the exact number of convolutions are wound, and calls the attention of the operator to the fact that the winding is completed. The projecting convolutions or loops of the continuous winding which mark the termination of each section are formed in the following way : a ratchet-wheel and cam driven by the revolving mandrel control the movement of a rounded finger of steel, which slides along the edge of the mandrel, and is automatically brought directly under the oncoming wire at the instant when a particular convolution is about to be wound. The result is that this convolution is wound over the steel finger and made correspondingly higher than the rest. The finger is removed instantly by the cam at the moment before the next convolution begins to be formed. The whole operation of winding a complete and symmetrical armature of three hundred and forty turns of wire eleven thousandths of an inch thick, and one hundred and ten thousandths of an inch wide, and transferring it from the mandrel to a wooden spit, occupies eight and one-quarter minutes.

The commutator which consists of sector-shaped pieces of copper with projecting tails to embrace the loops above referred to, is secured by taper-rivets to a fibre-washer, and the latter is attached to a wooden block, through the centre of which the spindle is driven. This block is forced into the centre of the wound armature ring, so as to bring the commutator even with the end of the armature and cause the high convolutions or loops of the winding to project between the tails of their respective commutator strips. These tails are then pinched together and soldered to the high loops, the insulation having previously been scraped off the wire at these points.

The best proportion for the different dimensions of the field-magnets of this motor were determined by building a magnet of approximately correct design, and then gradually increasing the diameter of the iron and decreasing the amount of copper wound on it and decreasing the length of the iron and trying experimentally the effect of enlarging the mass slightly at different points. The results of each change were noted carefully, and but a single change was made at one time, so that its effects should not be confused. By this process the present model was arrived at, and upon comparing it with some of the latest and

tention, with the recent ideas of dynamo construction ; while the machines with which it is compared are some of the largest types. The electrical efficiency of the machine is about 70 per cent., and the net commercial efficiency has been pronounced by prominent electric-light engineers who are using it to be about 55 per cent.

The machine described has a capacity of 2,000 ampère-turns each on the field and armature ; that is, the magnets reach the best point of sat-

SCHEDULE OF TYPES OF C. & C. MOTORS.

Size.	Power in h. p.	Height in inch's	Weight lbs.	WINDING.					
				A	E	F	G	L	N
				20 amp., for battery, or U. S. circuit.	10 amp., or Brush, etc., circuit.	6½ amp. Thomson-Houston circuit.	Field in mult. arc for signaling bells, electro-plating, etc.	100 volts incandescent circuit.	100 volts incan. circuit for constant running.
No. 1	¼	7¾	13	Field, No. 10 wire.	Field, No. 12 wire.	Field, No. 14.	Field, No. 19 in mult. arc.	Field, No. 23.	Field, No. 23 extra pull.
				Armature, flat wire.	Armature, flat wire.	Armature, flat wire.	Armature, flat wire.	Armature, No. 27 wire. Takes ½ amp.	Armature, No. 27 wire. Takes ½ amp.
No. 3	½	10½	50	Field, No. 10.	Field, No. 12.				
				Armature, No. 16 wire.	Armature, No. 19 wire.				
No. 5	1	12½	77	Field, No. 10.	Field, No. 12.				
				Armature, No. 16 wire.	Armature, No. 18 wire.				
No. 8	2	15½	127	Field, No. 10.					
				Armature, No. 15.					

most efficient dynamos, embodying the latest and best ideas, especially Dr. Hopkinson's latest machine, it has been found that this little machine is proportioned almost exactly like the large Hopkinson dynamo. The field in both cases is produced by from 500 to 600 ampère-feet of winding for each square inch of field-magnet cross-section. The comparison is interesting, because this is probably the smallest extensively manufactured motor that has been made in strict accordance, as was the in-

uration with that current. As a great many of the motors are used on constant current circuits, and the various commercial circuits differ in current strength, it was necessary to make a separate class or type of winding for each current by varying the number of turns so as to produce 2,000 ampère-turns on either circuit. The size of wire used was also varied inversely, so as to produce the same sized coil and thereby maintain the same efficiency of copper in each type.

The preceding schedule will be of interest to many, as showing the arrangement of sizes and windings.

The windings for different circuits have been designated by letters, while the different sizes of motors are known by numbers.

FIG. 240.—C. & C. SHUNT-WOUND MOTOR.

Thus, motor type 1 E No. 500 is the 500th ¼-horse power motor that has been wound for 10-ampère current; while motor type 5 A No. 10 is the tenth 1-horse power motor that has been wound for 20 ampères.

The C. & C. motor, illustrated in Fig. 240, is shunt wound, so that it runs at a constant speed, whether running free or heavily loaded, and it is belted to a light shaft which runs lengthwise under the sewing tables. The operation and speed of each sewing machine is controlled by an individual treadle, which throws the belt of that machine into, or out of, connection with the main shaft by a friction clutch. This arrangement gives to each machine the full advantage of the whole power of the main shaft to start up quickly, so that a great deal of working time is saved by the operator being able to have her machine started instantly at full speed. Though the load occa-

sionally put upon the motor, when all the sewing machines happen to be in operation at once, is considerably in excess of the rated capacity of the motor, and though again at times none of the machines are in operation and the motor is consequently running on "no load," its speed never varies more than 100 or 200 revolutions from its normal speed of 1,800.

Fig. 241 is also a ⅓-horse power motor wound for constant speed on the 110-volt incandescent circuits. This motor has a resistance of about 16 ohms, runs at about 2,300 revolutions with no load and at about 1,800 when fully loaded, at which time it takes about 1¼ ampères.

The iron work and frame are the same as in all the ⅓-horse power motors made by the C. & C. Company; but the winding is new as applied to very small motors, and is identical in plan with the winding used on the largest and best dynamos, i. e., the field is fed by independent connection to the line wires, and is thus kept at constant strength.

FIG. 241.—NEW C. & C. SHUNT-WOUND MOTOR.

The armature is wound with about 3,000 turns of wire, so that the counter E. M. F. equals the direct when running at 2,300 revolutions. A very slight slowing down reduces this counter E. M. F. enough to allow the full strength of current to enter the armature, and therefore the motor puts out its full power with slight decrease of speed. The switch for stopping and starting is attached to the motor, so that no connections to a separate regulator-box are required.

FIG. 242.
C. & C. MOTOR WITH WHEELER REGULATOR.

The motor is fitted with a number of improvements, including a new cylindrical commutator, as perfect and thoroughly well constructed as those used on the largest machines. The shaft can be unscrewed and taken out to have a longer or shorter one put in its place. In setting up, for example, to drive a pump in a private house, the motor is screwed to a bracket near the pump and belted to it. The wires are connected to the binding posts at either side of the switch knob, and the apparatus is ready to start. The action of the knob is simply to send the current through the field when the brass sector reaches the first spring clip, and through both armature and field when the sector touches both springs.

In Fig. 242 is shown a motor provided with the Wheeler regulator. This regulator consists of a multipolar connected armature and field, the field being wound in sections, which are proportioned so that they will divide the main line current with the armature, always taking that percentage which will give the

most efficient output for the machine afforded by the condition of running, as determined by the speed switch. This switch is connected so as to be simply the means of short-circuiting more or less of the coils of the field. In this way the current flowing through the armature is also controlled, without having actual access to the armature. The paths for the current of the arc light circuit between binding post and binding post of the motor are through the brushes and armature, through the coils of the field, and through the short circuiting switch and the coils of the field which are not short-circuited. There are thus three paths for the current, and there is absolutely no danger of injury from the accidental opening of any one of these paths.

The small $\frac{1}{8}$-horse power motor, shown in Fig. 243, is interesting as being also made with a complete regulator. A further advance has

FIG. 243.

been gained in this smallest machine by fitting it with a cylindrical commutator in all respects like those used on larger sizes. The armature core is supported by brass spiders into which the shaft is screwed, overcoming the difficulty of an unbalanced armature, and producing a perfectly interchangeable machine in motors to be used continuously on power lines. The bearings also are made heavier.

Quite recently Dr. Orazio Lugo, of New York, has constructed a motor which, though

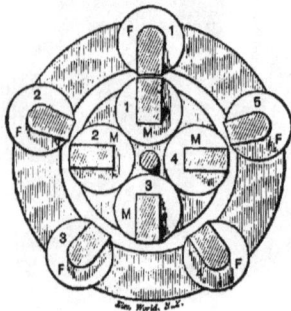

FIG. 244.—THE LUGO MOTOR.

resembling some older types in appearance, nevertheless possesses some decidedly novel features.

The new motor of Dr. Lugo is designed to utilize the maximum effects found in placing moving solenoids in proximity to fixed solenoids in such relation that, as the armature

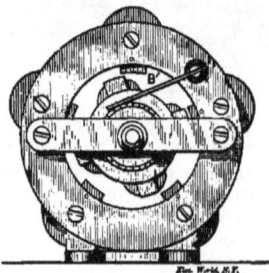

FIG. 244A.—THE LUGO MOTOR.

coils move, a maximum number of lines of magnetic forces is cut in the passage of each armature bobbin past each field magnet bobbin, and these effects are made successive, so that there is a continued application of such suc-

cessive effects. And, in addition to utilizing these effects, due to a specified relation of the solenoids themselves, the combined effects of these solenoids and their cores also are employed.

By creating successive magnetic circuits through successive armature and field bobbins in pairs, until the series has been gone through once in each complete revolution of the armature, each armature bobbin being in circuit at least once in a complete revolution with each field magnet bobbin, and all of the bobbins, both of the armature and field magnets, being allowed to rest magnetically by being cut out of circuit during a fraction of each revolution of the armature, the evil effects of Foucault currents are said to be largely avoided.

The motor, which is of very simple construction, is shown in perspective in Fig. 244 and in

FIG. 244B.—THE LUGO MOTOR.

cross-section in Fig. 244A. As will be seen, it consists of a series of bobbins arranged in a circle to the number of five, and within which there revolve four similar bobbins attached to the shaft of the motor. All these bobbins are provided with cores and short pole pieces, which revolve in close proximity to each other.

The revolving armature coils are connected to a commutator fixed to the shaft, upon which bears a stationary brush B', shown in the end view, Fig. 244B. The connections of the individual armature bobbins to the commutator are seen in Fig. 244c, which shows also the cross-connection between the commutator strips. The other side of the armature shaft carries a revolving brush, which bears against a fixed commutator to which the field bobbins are connected. This is shown in the end view, Fig. 244D, and the commutator in detail in Fig. 244E. Finally, Fig. 244F shows the circuits diagrammatically, with the bobbins laid side by side,

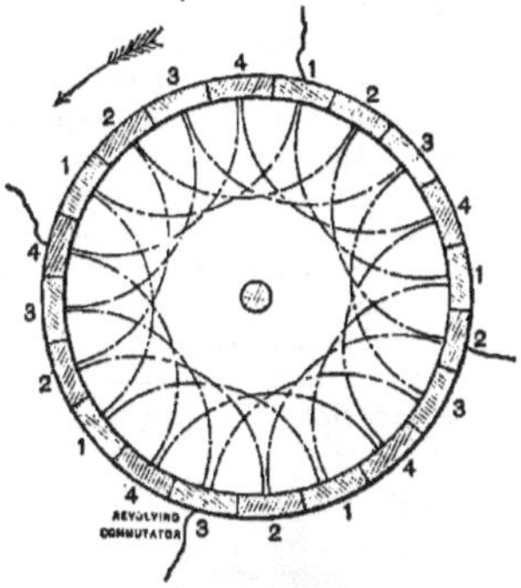

FIG. 244C.—THE LUGO MOTOR.

and showing the manner of their connection one with the other.

It will now be readily understood that if a current be passed through the armature and field bobbins M^2, F^2, the pole pieces are in such a position as to cause M^2 to be attracted

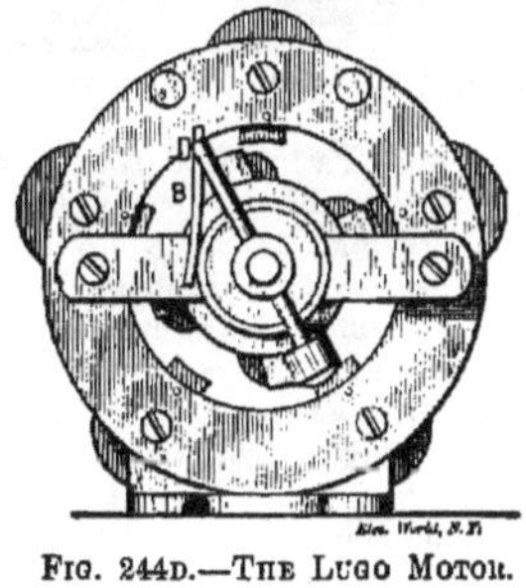

FIG. 244D.—THE LUGO MOTOR.

to F^2 and cause rotation in the direction of the hands of a watch. This attraction continues until the pole pieces M^2, F^2, have arrived exactly opposite each other, or in the position in which M^1, F^1 is shown. At this instant, however, the current is shifted to the bobbins M^3, F^3, which will then have arrived at the relative

position formerly existing between M^2 and F^2, and the magnets act in succession as follows : M^4, F^4; M^1, F^3; M^2, F^1; M^3, F^2; M^4, F^3; M^1, F^4; M^2, F^5; M^3, F^1; M^4, F^2; M^1, F^3; M^2, F^4; M^3, F^5; M^4, F^1; M^1, F^2; M^2, F^3; M^3, F^4; M^4, F^5, and back to the starting point M^1, F^1.

Thus, in every revolution of the armature shaft each armature coil is placed in circuit with each field coil; and, there being four of one and five of the other, it follows that there are twenty successive pulls at as many radial positions of the shaft, each pair of coils being in circuit successively only during one-twentieth of a revolution of the shaft. It will be noticed that the stationary commutator is divided into four groups of segments, five in

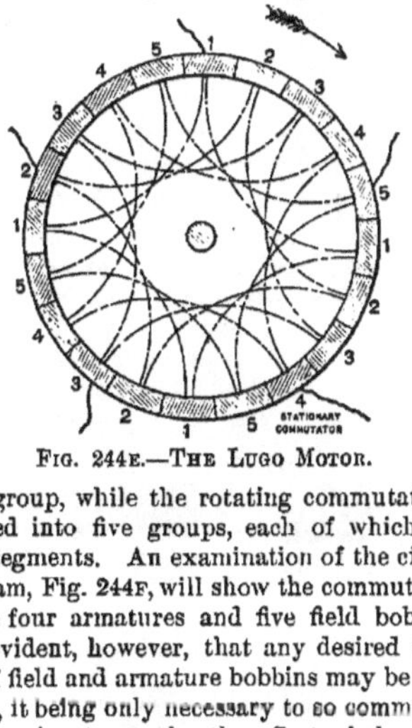

FIG. 244E.—THE LUGO MOTOR.

each group, while the rotating commutator is divided into five groups, each of which has four segments. An examination of the circuit diagram, Fig. 244F, will show the commutation of the four armatures and five field bobbins. It is evident, however, that any desired number of field and armature bobbins may be combined, it being only necessary to so commutate the circuits as to make the effect of the magnetic pull on the bobbins successive in its action.

With a motor constructed in the manner just described, Dr. Lugo claims to utilize the power due to the attraction of the pole pieces of the armature and field magnet bobbins, as well as that due to the parallelism of the windings of the bobbins as they approach each other during the rotation of the armature, thus obtaining an increased effect from those magnetic lines of force which are ordinarily radiated to create reverse inductive effects detrimental to the efficiency of the machine.

The arrangement is also claimed to avoid the evil effects due to self-induction, for the reason

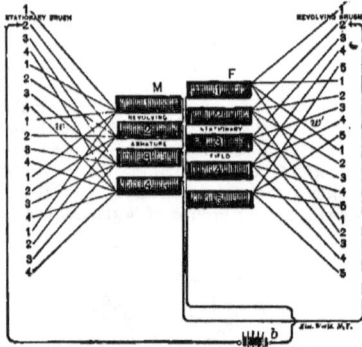

FIG. 244F.—THE LUGO MOTOR.

that only a minimum amount of effective wire is in circuit at any time, and that, only, at the time when it is needed to give the best effects. By creating the effective field and armature circuits at stated intervals and only when needed, and permitting all of the field or armature bobbins to rest or be magnetically or

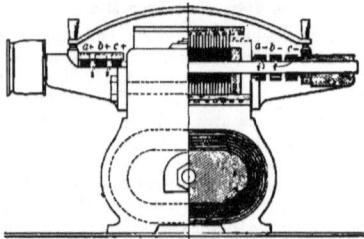

FIG. 245.—PATTEN'S ELECTRIC MOTOR.

electrically discharged at different portions of the armature rotation, heating and waste of energy is said to be prevented. The Foucault currents are also reduced to a minimum by reducing the effective field and armature cir-

cuits with their magnetic cores to a minimum, while obtaining their maximum effect. The sparking at the brushes is likewise avoided by reason of the fact that there is no magnetic lead in the field or armature, inasmuch as the field travels around just in advance of the armature as each bobbin is cut in.

Among other novelties which the machine is claimed to possess by the inventor is, that both field and armature circuits are conjointly cut out at different points in the rotation of the armature, thereby utilizing only that portion of the combined circuit which is needed to propel the motor or create a current in the machine when used as a dynamo. The efficiency of this motor is said to be very high, and the machine is very light and simple in construction.

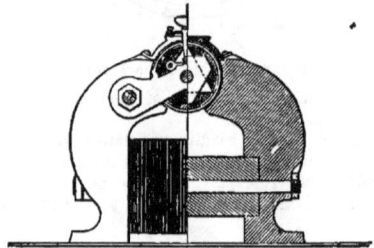

FIG. 246.—PATTEN'S ELECTRIC MOTOR.

Lieut. F. Jarvis Patten, U. S. A., has recently constructed a motor of the novel design shown in Figs. 245, 246 and 247. Its novel features consist mainly in a new system of armature winding, commutators and connections, as well as some mechanical details worthy of notice. In the accompanying engravings, Figs. 245 and 246, are side and end elevations of the motor, half of each being shown in section.

A noticeable feature of the machine is an exceedingly short magnetic circuit and a slightly unsymmetrical field, the purpose of which will be explained later. The motor has a comparatively large proportion of iron, a single magnet core and coil forming the yoke of a very short and stout electro-magnet, thus bringing the centre of gravity of the entire machine to the lowest possible point. To the core there is bolted by a single transverse bolt the two curved pole-pieces of cast-iron, which are so shaped as to reduce the magnetic lines of force to their shortest length, while the armature fits so closely that there is but $\frac{1}{12}$ of an inch of air space in the magnetic circuit.

Secured to the spindle is a three-coil armature of peculiar construction. The coils are wound longitudinally around three poles, radiating at angles of 120 degrees from the spindle. The armature may therefore be regarded as a

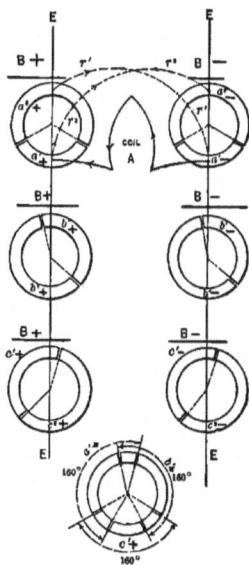

FIG. 247.—PATTEN'S ELECTRIC MOTOR.

drum armature, with three coils and a minimum amount of inactive wire; the ends being shortened in the proportion that the length of a chord of 120 degrees, or less, is to that of a diameter of the same circle. These three coils of the armature are all continually in circuit in parallel arc, so that the armature consists of a three-coil multiple-arc winding, thus reducing materially the armature resistance, as by the peculiar arrangement of the field described none of the coils of the armature are at any time cut out of circuit.

This system of construction admits further of a non-sparking commutator for all positions of the brushes, and is the result of the peculiar commutator connections rendered necessary. These are shown in Fig. 247.

As will be seen, there are practically two commutators, one revolving under the positive and the other under the negative brush, and the two brushes are placed upon the same side of the spindle and at opposite ends of the armature. This arrangement, it will be noticed, leaves the entire commutator exposed, so that it can easily be cleaned while the machine is running. In Fig. 247 the vertical lines E E, represent a vertical diameter of the armature spindle, and each segment of the commutator is shown in its proper relative position thereto. An examination of the diagram of connections will show that each of the three coils is constantly in circuit, and is entirely independent of the other two. The action of a single coil during one revolution will therefore describe that of all. Thus, in Fig. 247, the coil A is shown with its terminals connected to two partial commutators, one revolving under the positive, and the other under the negative brush. The coil is secured by one terminal to the segment $a^1 +$, and after traversing the armature has its other end secured to the segment $a^1 -$, which revolves under the negative brush and occupies exactly the same angular position on the spindle as the first segment. This last segment, however, is also connected by a free conductor $r^1 r^1$ back to the second segment $a^2 +$ of the first half commutator; and the segment $a^2 -$ of the second half is likewise connected by the free conductor $r^2 r^2$ back to the short segment $a^1 +$ of the first half.

The other coils of the armature are similarly connected to their corresponding commutator segments, the only difference being that the corresponding segments for the different coils are placed in rotation around the spindle, so that their middle lines make with each other angles of just 120 degrees.

If now the current be regarded as flowing direct from the positive to the negative brush, and the result during a single complete revolution of the coil A be traced, it will be seen that while the short segments $a^1 +$ and $a^1 -$ are under their respective brushes, the current will flow direct in the coil A from the positive to the negative brush. When, however, the short segments pass out of action the current will flow from the positive brush $B +$, through the long segment $a^2 +$, the free conductor r^1, over to the short segment $a^1 -$ (no longer under the negative brush), then back through coil A in a reverse direction to the short segment $a^1 +$ (no longer in contact with the positive brush), and thence through the free conductor $r^2 r^2$ to the long segment $a^2 -$, and out through the negative brush as before.

It is therefore plain that the current flows direct in any coil while the short segments are under the brushes, and in a reverse direction whenever the long segments are in contact with the brushes; and this change takes place alike for each of the armature coils in rotation.

The relative amount of direct and reverse current taken in a revolution by each coil will depend upon the relative amount of the spindle circumference covered by the long and short segments, and this must necessarily depend upon the amount of distortion or unsymmetrical arrangement of the field, which, for well-known reasons, must have comparatively narrow limits. This distortion of the field amounts practically to changing the diameter of commutation from a right line (diameter) to a broken line, consisting of two radii meeting in the centre of the spindle, an effect involving, it is claimed, no disadvantages under the conditions as secured and provided for, viz., that both brushes are on the same side of the spindle.

If, now, the further condition be borne in mind that the coils must of necessity radiate at equal angles of 120 degrees from the spindle, and we give to the field poles a relative position, as shown, such that the points on the armature circumference where the current must change direction are at the extremity of radii that are inclined at some angle between 140 and 160 degrees, it will result in giving to the armature a continuous unbroken circuit, and as the three coils are always in parallel arc, there can be no short-circuiting of the coils due to the brush bearing upon two different segments simultaneously.

This effect is shown by the lower diagram in Fig. 247, in which the short segments are shown as covering arcs of 160 degrees each of the spindle circumference. Any two successive ones must, therefore, overlap by an angle of 20 degrees, from which it results that there is no point during a complete revolution at which the armature circuit is broken. For, as the current changes from any one coil at any point, there are two others in contact with the same brushes that form a continuous circuit. The resultant effect amounts practically to making a set of commutator segments that cover $3 \times 160 = 480$ degrees, each single part overlapping the other by an amount equal to $\frac{1}{2}$ (160—120)—20 degrees.

Another result of this peculiar form of armature construction is the complete elimination of a dead centre, there being at all points of a single revolution a continuous rotary torque, which may be expressed as the resultant effect of three separate tangential efforts which can never be so placed as to neutralize each other.

The demand for motors on arc light circuits or for constant current has led Mr. Wm. Hochhausen, the electrician of the Excelsior Electric Company, to design a machine, which, with fixed brushes, should regulate so as to keep constant speed with a variable load, and without the interposition of external resistance. The motor, which is illustrated in the accompanying engraving, Fig. 248, has a single magnetic circuit in which the armature is included. The latter is mounted on bearings at the top of two arms which rise from the base, which also constitute the bearing of the electro-magnets, which have wrought-iron cores and cast-iron pole-pieces.

The regulation of the motor is effected by varying the intensity of the magnetic field to correspond with the load. For this purpose the two field-coils are divided into ten sections, the ends of which are brought to consecutive strips, shown at the side of and below the armature.

The governor is of the centrifugal type, and is held in an extension bearing at one end of the armature shaft. The governor acts upon an arm which extends downwardly and operates upon a contact-maker which touches the various contact strips to which the field-coils are connected. Thus, when the motor, which is series wound, runs with full load and at normal speed, all the sections of the field-coils are in action; as the load diminishes the governor expands from the momentarily increased speed and cuts out successive coils in the fields. This reduces the magnetic strength of the latter, and brings the motor back to the same speed as before. Conversely, when the load is increased, the speed is reduced for an instant, the gov-

When the motor runs without load, all the field-coils are cut out, so that the resistance is that of the armature alone, or 1 ohm. In that case the energy absorbed by the motor is 100 watts, or a little over $\frac{1}{8}$-horse power.

Mr. Hochhausen has also designed a motor of somewhat similar appearance, but necessarily different in details of construction, for incandescent circuits.

At the American Institute Electrical Exhibition last year, Mr. W. E. Hyer, of Newburgh, N. Y., exhibited a small motor designed by him, and embodying several novel features.

FIG. 248.—NEW HOCHHAUSEN MOTOR.

ernor contracts and puts additional field coils in circuit to correspond to the increased speed.

The machine illustrated is designed for 3-horse power and runs at 2,000 revolutions per minute, taking a 10-ampère current, such as is largely employed in arc lighting. Its weight is 250 pounds. The resistance of the armature is 1 ohm, and that of the full field also 1 ohm. Hence, the energy lost in the motor when running at full speed, with 10 ampères, is 200 watts, and, as the motor delivers 3-horse power or 2,238 watts, its efficiency is thus about 90 per cent.

The illustrations, Figs. 249 and 250, show a sectional and perspective view of it. As will be noticed, the armature is placed directly within the helices of the field-coils, which are wound on spools of non-magnetic material. Both field coils and armature are surrounded by an iron shell, cast in two parts, and having the bearings extending horizontally across the open ends. This construction serves to close the magnetic circuit, and so completely is this accomplished that no external magnetism can be detected. By means of this construction, it is claimed, very high efficiency is obtained, that of the small $\frac{1}{8}$ horse power motor reaching 65 per cent., with a corresponding increase in the larger sizes.

The armature of this motor is of the Gramme form, and its core is built up of rings of soft

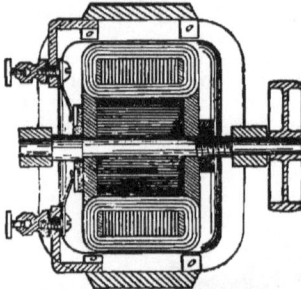

FIG. 249.—THE HYER ELECTRIC MOTOR.

sheet-iron, insulated magnetically from each other, and thus entirely avoiding eddy currents. It is wound in sections varying in number according to the size of motor, and is secured to the shaft between two brass discs. One of

FIG. 250.—THE HYER ELECTRIC MOTOR.

these fits against a shoulder in the shaft, and the other is forced against the winding by means of a nut on the shaft. The commutator is a flat one, and is placed against the end of the armature, and the brushes are secured to

lugs cast to the frame. The motor illustrated is of a rated capacity of $\frac{1}{10}$-horse power, but it may be worked up to $\frac{1}{8}$-horse power without injury. It occupies a space of 4×4 inches, is 5 inches high and weighs 6½ lbs. The motor of $\frac{1}{4}$-horse power is 7×6 inches by 7 inches high, and weighs 30 lbs. The motors, with the exception of that of $\frac{1}{10}$-horse power, are compound wound, and show good regulation.

The illustration, Fig. 251, represents an electric motor recently brought out by the Hawkeye Electric Manufacturing Company, of Oskaloosa, Ia., and designed by their electrician, Mr. Thone. As will be seen, the armature is

FIG. 251.—THE THONE ELECTRIC MOTOR.

of the disc type, and two field-coils are employed. The machine is shunt wound, and designed to be self-regulating, without necessitating the shifting of the brushes or the employment of rheostats. A switch is attached to the motor, by which it is thrown into circuit gradually, thus preventing an abnormal flow of current at starting. These machines are built in sizes ranging from $\frac{1}{6}$ to 10-horse power. Those up to 4-horse power are adapted to circuits ranging from 50 to 110 volts, and the larger sizes from 110 to 220 volts.

Figs. 252, 253 and 254 illustrate two types of motors designed by Mr. Geo. F. Card, and recently brought out by the G. F. Card Manufacturing Company, of Cincinnati, O. These machines are of the bi-polar type with Gramme ring armatures.

FIG. 252.
THE CARD CONSTANT-POTENTIAL MOTOR.

The constant-potential motor, type "B," shown in the engraving, Fig. 252, is wound for a current of 2½ ampères, the field and armature being in series. With a current of 1½ ampères, the motor attains a speed of 5,000 revolutions per minute. It measures 6×7½ inches, stands 5 inches high and weighs 9 lbs., without the base. A resistance-box of lamps, of lower voltage than the dynamo, arranged in parallel, is used on the incandescent circuit. By turning them out singly, and increasing the resistance, the speed of the motor can be reduced at will. By the use of a larger number of lamps, of the same voltage as the dynamo, a corresponding increase in the number of variations in the speed can be effected. The current can be thrown off from the motor entirely, without disturbing the lamps, and then their light, be-

ing nearly up to full candle-power, can be utilized by a reflector. It will be noticed that, in this style of motor, besides the usual field magnets, an additional branch is added, which arches from pole to pole and encircles one of the bearings of the shaft.

The constant-current motor, illustrated in Fig. 253, is series wound, and is designed for a current of five ampères and to run at a speed of 5,000 revolutions per minute. In order to be able to operate the motor on a 10-ampère circuit, a shunt resistance of carbon is employed, by means of which five different speeds can be obtained.

FIG. 253.—THE CARD CONSTANT-CURRENT MOTOR.

One of the features of the Card motors is the reversible commutator, shown at the side of the armature in Fig. 254. This is so arranged that all the sections can be removed and replaced without disturbing a wire. If worn on one face the sections can be reversed; or if worn on both faces, so as to be serviceable no longer, new ones can be inserted. The inventor, after an experience of nearly a year with a dynamo

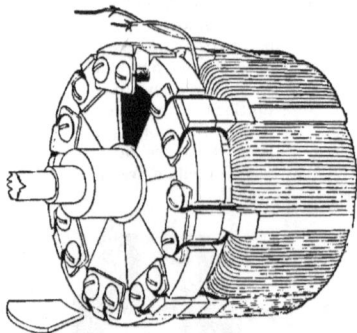

FIG. 254.—ARMATURE OF CARD MOTOR.

of considerable size, in which this form of commutator is used, claims that the uneven wear on the brushes, which might be supposed to interfere with the practical workings of the device, is an entirely negligible quantity. This arrangement of the brushes also admits of an

FIG. 255.
DIEHL COMBINED SEWING-MACHINE AND MOTOR.

observation being taken on both at the same time, and is convenient in setting the brushes.

There is to-day probably no domestic labor-saving device in more general use than the sewing-machine, and it ranks rightly as one of the prominent inventions of this century. While, however, it saves a vast amount of

manual labor, its continuous use for hours, entailing the employment of a treadle, has called for methods of driving the machines by auxiliary power; and in large factories they are frequently coupled to lines of shafting. This method of driving has not been applicable to the case of isolated machines, whether in shops or private dwellings, and hence the advent of

FIG. 256.—FIELD MAGNET OF DIEHL MOTOR.

the electric motor, which permitted each machine to be independent of any other, was welcomed, because it offered a ready means of accomplishing in a convenient manner what was heretofore impracticable. The small mo-

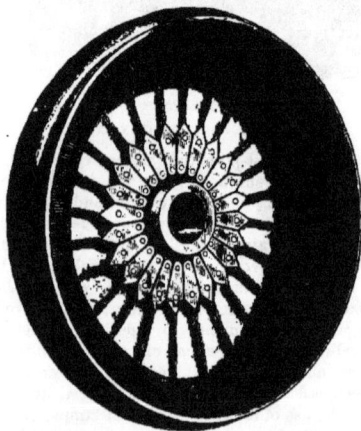

FIG. 257.—DETAIL OF ARMATURE WINDING.

tors have, as a rule, been attached to the board on which the sewing-machine is mounted, and then belted to the shaft of the latter.

It was to avoid the necessity of belting, and at the same time do away with the presence of an auxiliary machine on the board for driving, that Mr. Philip Diehl, of Elizabeth, N. J., one of whose ingenious motors has already been described, conceived the idea of combining the motor and sewing-machine into a practical unit.

The simple and elegant manner in which he has accomplished this is shown in the engraving, Fig. 255. The motor, it will be seen, is completely housed within the fly-wheel of the machine, and connected directly with the driving-shaft, so that all gearing is obviated. The details of the arrangement will be readily understood from Figs. 256 and 257, which show respectively the field magnet and armature of the motor. The magnet, which consists of a single piece, is wound with wire connected to the two terminal brushes shown. This magnet is permanently fixed to the hub through which the shaft passes. The armature shown in perspective in Fig. 257 is of the Gramme type, and is held in position within the rim of the wheel. The wires leading from the periphery connect to the commutator at the hub, and the brushes on the magnets bear against the segments.

The wires leading to the motor pass up through the hollow casting of the frame, and are connected to a switch, by which the machine can be started and stopped at will. The fly-wheel is provided with a clutch or stop motion in connection with the shaft, so that it may be connected with the latter, or turned loose, as is common in sewing-machines—the wheel being disconnected from the shaft when winding bobbins. This is accomplished by a turn of a thumb-nut at the rear end of the machine. By unscrewing this nut entirely, the armature may be slid out completely, so that it may be examined should necessity require. This also exposes the field magnets and brushes, so that they can be easily gotten at for examination and attention. The entire motor is put together in a most compact and neat form, and it adds greatly to the value of the sewing-machine as a labor-saving device.

CHAPTER XIV.

LATEST EUROPEAN MOTORS AND MOTOR SYSTEMS—CONTINUED.

NOTWITHSTANDING the variety of methods and devices employed in transmitting the current from the central station to the motor on the car, to which attention has already been called, there are still new ones to make their appearance, and some of them must be credited directly to European ingenuity. Among the most recent of these is the system of Messrs. Pollak and Binswanger, the chief point of novelty of which consists in the ingenous method adopted for transferring the current from an insulated conductor to the motor on the car.

Mr. Pollak uses the rails as the first conductor, but the second conductor is completely insulated under the road, and does not communicate directly with the exterior. The current enters the car by means of a third rail placed in the middle of the track. Fig. 258 shows this arrangement in longitudinal section, and Fig. 259 in transverse section. This rail, which should be made of soft iron, is divided into segments of about 3 to 4 metres in length, insulated from each other electrically by means of wood and fibre. Each segment is formed of two parallel bands of iron, R, separated by

FIG. 258.—POLLAK AND BINSWANGER'S ELECTRIC RAILWAY.

a non-magnetic body, such as wood, as shown in Figs. 260 and 261.

Ordinarily these segments do not communicate with the insulated conductor; each of them is fitted with two metallic boxes K, Fig. 261,

Fig. 259.—POLLAK AND BINSWANGER'S
ELECTRIC RAILWAY.

firmly fixed below, and into which penetrate the branches of the principal conductor. These branches abut on pieces of soft iron n, termi-

rail. These then attract the piece of iron m, which closes a contact, bringing the segments into communication with the principal conductor. Consequently, at this moment the brushes resting on the central rail can collect the current. When the car has passed, the segments are demagnetized, the iron contacts fall back, and the communication between the segment and the principal conductor is broken.

The boxes K, in which these contacts are made, are closed hermetically and half filled

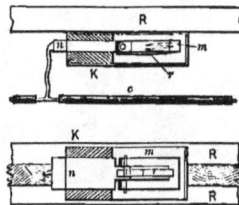

with petroleum, which prevents moisture from gathering on the different parts.

The insulation of the boxes may be rendered practically perfect, so that losses from faulty insulation can take place only between the central segment placed beneath the car and the extreme rails on the surface of the earth. The loss, being limited to so trifling a length, is insignificant. The cost of laying the line is not high in comparison with the systems having underground channels; the central conductor

Fig. 260.—POLLAK AND BINSWANGER'S ELECTRIC RAILWAY.

nated by a piece m, also of iron, jointed to the former, and connected with it besides by the spring r.

The car is fitted with a powerful magnet, N S, Figs. 258 and 259, the poles of which magnetize, by induction, the two segments of the

being well insulated, there is no occasion to make a drain or to interfere with the road. The chances of deterioration are claimed to be very slight, as no delicate part is exposed. The car always covers the segments which communicate with the underground conductor, so that there is neither danger of short-circuiting nor of giving shocks to men or horses traversing the road.

The system permits of the use of high potentials, and consequently effects a great reduction in the loss in the conductors.' In fact, with a difference of potential of 500 volts, 6 ampères suffice to furnish four electrical horse-power, which is enough for propelling a small car under ordinary conditions.

FIG. 262.—THE IMMISCH MOTOR.

The different figures accompanying the text refer to the reduced model of this tramway as it has been presented to and worked before the Paris International Society of Electricians.

Instead of bringing the current by the central rail, and letting it return by the two others, it will often be advantageous to make use only of two rails properly so-called. One of them is divided into segments, like the central rail just described while the other remains as it was. In this case the wheels placed on one side of the car have to be insulated from their axles. The central rail is especially advantageous when it is required to convert a line already existing.

In order to employ gear-wheels the inventor makes use of slow speed motors which revolve only about 500 times per minute, so that a single transmission by a cog-wheel is sufficient to transmit the rotation from the dynamo to the wheels of the car, which at their normal speed make from 100 to 120 turns per minute (16 kilometres per hour, with wheels of a diameter of .8 metre). The illustration, Fig. 258, made from a model, however, shows a worm gear employed as the means of transmission.

Considerable attention has lately been drawn in England to the Immisch motor, which embodies some novel features, especially in the

winding of the armature. The machine of 6 horse power, is shown in perspective in Fig. 262 and the manner of winding the armature is shown diagramatically in Fig. 263. In the diagram only eight coils are indicated, although 48, 96 or more may be employed. The commutator is of the bisected type, and the coils are joined to two adjacent segments of the commutator on the two rings, of which one has an angular advance equal to one-half the width of the commutator bar. The two brushes side by side upon the two rings, are connected together so that only one pair is shown in the figure.

Starting from one ring of the commutator under the brush, say, with the coil marked 1, it crosses to the other side of the armature and joins the connection leading to coil 6; but if we follow this line backward to the commutator, we arrive at a segment under the same brush from which we started. Similarly coil 5 connects with coil 2, and is short-circuited in the same way by the other brush. It will be observed that the magnetic axis of the armature itself would be situated underneath the coils which are short-circuited. The remaining connections are easy to follow; the two halves of

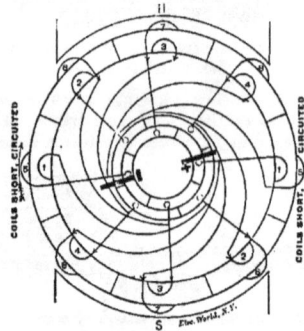

FIG. 263.—THE IMMISCH MOTOR.

the circuit can be traced through coils 6, 3 and 4 in series on the one side, and coils 8, 7 and 2 in series on the other. It is to be observed, however, that this arrangement only amounts to the same as having an armature of normal type with a brush of wide face, so that its contact with two adjacent commutator bars is considerably prolonged.

It was thought probable when the first experiments were made, that difficulties might be experienced from the heating of the coils dur-

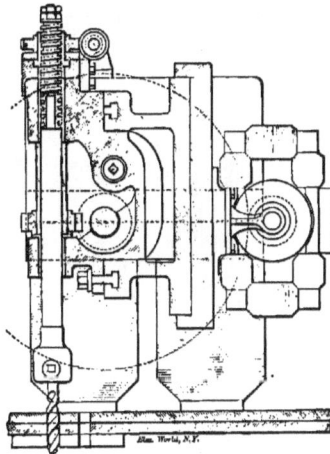

FIGS. 264 and 265.—RIVETING-MACHINE.

ing the period of short circuiting, which occurs twice in every revolution. No such effect has practically been found to occur so long as the field poles are properly proportioned, and, in the case of a motor, the brushes may even be shifted some distance to either side of the normal position, without producing either sparking or any increase of heating. But curiously enough this does not apply to a dynamo of similar construction, the position of the brushes having to be adjusted with some care. So long as this is attended to, the machine runs with perfect smoothness, but as soon as the brushes are displaced, although no sparking takes place, yet considerable and rapid heating is the result. It is claimed, however, that this is of little practical importance, for when the position of the brushes has been once determined, they can be rigidly fixed. It is said that the non-heating of the coils proves that during this part of the revolution the algebra-

ical sum of the number lines of force passing through it is constant.

The average efficiency of one of these motors run at between 1,400 and 2,200 revolutions, and delivering from .98 to 1.76 horse power, was 71 per cent. In a larger motor of from 4.5 to 5-horse power, 85 per cent. efficiency was obtained.

It is often of great convenience to be able to use certain tools in places whither it has been difficult to convey energy. Of late years the use of hydraulic machinery has, to a certain extent, enabled apparatus to be used in such situations, but it does not fully supply the requirements. Attempts have been made to use electricity for the purpose, and one of the most successful of these is to be found in the apparatus designed by Mr. Rowan, of Glasgow, intended more especially for ship work; but, of course, the apparatus is useful for many other purposes.

Fig. 264, shows a small riveter, operated by means of a helical cam, shown in side elevation

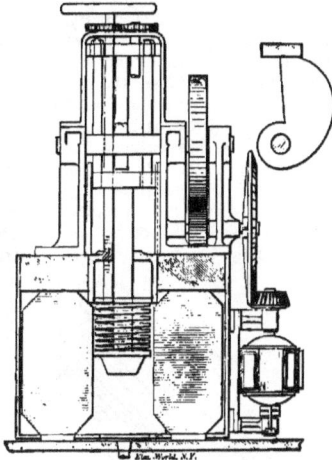

FIG. 266.—DRILLING-MACHINE.

at Fig. 265, which works in the line of the hammer-rod, between cross heads connected by four rods, to the lower of which cross-heads, the hammer-rod is attached. The machine has holding-on magnets, and a spiral or volute cam, which, through an anti-friction roller on an arm, lifts the hammer-rod against a spiral-spring, which, when released by the cam, oper-

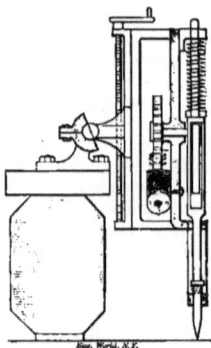

FIG. 267.—CHIPPING-MACHINE.

ates to produce the blow of the hammer. The spiral-spring is compressed between the hammer-head and a disc or plate, working in a circular guide-box, its position, and consequent-

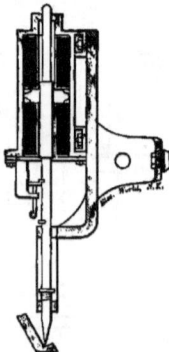

FIG. 268.—CALKING-MACHINE.

ly the amount of compression given to the spring, being regulated by two screwed spindles working through the top of the guide-box, and moved by gearing.

The illustration, Fig. 266, shows an arrangement of an electro-magnetic drilling-machine for a single drill spindle, capable of being traversed horizontally along the frame to which the motor is attached. It represents also, a cross-section of a multiple drilling-machine.

Another application of electricity in a kindred way, is shown in Figs. 267, 268 and 269, which represent calking and chipping-machines. In the first of these the tool is shown, actuated by an electric motor through gearing, lifting-cam and spring, as in the riveting-machine. The other illustrations, Fig. 268 and 269, show in section and elevation, an arrangement of solenoid coils for producing a reciprocating motion so as to deliver blows. The tool is supported by a holding-on magnet, to which it is attached by an arm and traversing screw

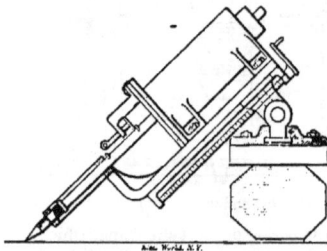

FIG. 269.—ELECTRIC CALKING-MACHINE.

and a ball and socket, or swivel-pin joint, which allow its position and angle to alternate at will.

A noteworthy instance of the successful transmission of power by means of electric motors, has been afforded by some work recently performed in Switzerland, by the Oerlikon machines, designed and built by Mr. C. E. L. Brown, of Zurich. Fig. 270 is an engraving of the Oerlikon machine, and Fig. 271 shows the circuits and the disposition of the generators and motors, two of each being employed. The generators and motors are similar in construction, but differ in some of their proportions and in the winding. The field magnets are formed of two vertical pillars of wrought-iron, which are united above and below by cast-iron pole pieces, the pillars simply fitting into borings in the latter. The lower pole piece is cast in one piece with two supports for the armature. The driving pulley (not seen in the cut) is inside the armature bearing. The dynamo frame instead of being bolted to the floor, fits in a slide-rest on a firmly secured bed-plate, and by turning a hand-wheel the tension of the belt may be varied at will while the machine is in motion. The exciting bobbins of the field magnets are wound separately on spools which slide easily over the wrought-iron cores. The whole

Figs. 272 and 273 represent the arrangement for a Gramme ring, and it will be noticed that the sections of the windings in the armature immediately below the surface may be comparatively large, thus counteracting their comparatively confined position for radiation, and making it possible that the same number of conductors can be obtained outside and inside. It is obvious that with the conductors entirely inside the armature, the latter can be run exceedingly

FIG. 270.—THE BROWN (OERLIKON) MOTOR.

construction is exceedingly simple, compact and neat.

The principle novelty, however, is to be found in the peculiar construction of the armature. The latter is a modified Pacinotti-Gramme ring with an unusually large iron section; but, departing from the customary method of placing the windings or conductors on the surface, Mr. Brown places them in special borings immediately *below* the surface of the armature.

close to the pole pieces; the air-space between armature and pole pieces is reduced to a minimum, and the conductors move through a most intense magnetic field.

The experiments were made by a committee of engineers and scientific men, with a view of ascertaining the total commercial efficiency of the transmission plant; but as the machines were in this case placed side by side, the results could only be taken as approximately correct. It is evident that, in such an arrangement of machines and resistances erected within the limits of a covered workshop, the insulation of the circuits presents no difficulty whatever, whereas in the actual installation, when many miles of overhead wires must be used, the insulation becomes a matter of some difficulty, and atmospheric influences may also have some effect upon the performance of the plant. These considerations induced the makers to

of October, 1887, with the plant as actually installed. Before quoting the results of these trials, it will be well to briefly refer to the general arrangement of the installation. At Kriegstetten, there is a water-power available, representing about forty actual horse power, and the problem was to carry as much of this power as possible to a mill in Solothurn, the distance being 4¾ miles as the crow flies; but, allowing for deviations, the length of each circuit may

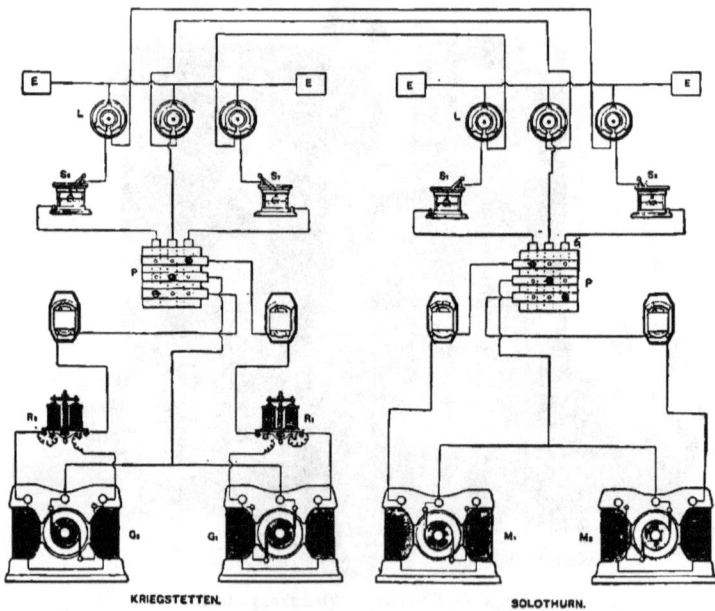

FIG. 271.—TRANSMISSION OF POWER BY OERLIKON MACHINES.

arrange for some further trials with the plant as actually installed. A committee was appointed, under the presidency of Professor Amsler, of Schaffhausen, the well-known inventor of the planimeter, and other well-known gentlemen were members, among them Professor Weber, of the Zurich Polytechnic School. This committee have lately issued their official report on the trials made on the 11th and 12th

be taken as about five miles. There are at Kriegstetten two generating dynamos, and at Solothurn two motors, coupled up on the three-wire system, as shown in the illustration, Fig. 271. Each dynamo weighs 3 tons 12 cwt., and has a Gramme armature 20 in. in diameter and 14 in. long, the normal speed being 700 revolutions per minute. Referring to the diagram of connections, G_1 and G_2 are the generators at

Kriegstetten, and M_1 and M_2 are the motors at Solothurn. R_1 and R_2 are electro-magnetic switches, which automatically come into action and short-circuit the exciting coils in case of the current rising beyond a certain limit. This provision was introduced in order to guard against the destruction of the generator in case a short-circuit should take place somewhere on the line. The current from each of the generators passes through an ammeter and then to a plug switch-board, P, to which is also connected the balancing wire joining the negative brush of G_2 with the positive brush of G_1. The balancing wire is then carried direct to the middle one of the three lightning arresters, L, and then to the middle wire of the line, while each of the outside wires is led through a liquid

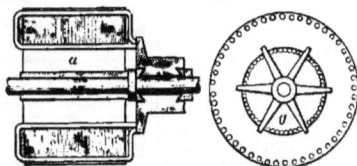

FIGS. 272 AND 273.—THE BROWN DYNAMO.

switch, S_1 S_2, then to a lightning arrester, to the line. Each lightning arrester consists of a circular metal disc, the edge of which is provided with projecting teeth, and situated in a concentric metal ring, the internal circumference of which is also provided with teeth, but not touching the teeth of the disc. All the discs are connected with a common earth wire and two earth plates, E E. The same provision against lightning is made at the motor station. The switches S_1 S_2 are of peculiar construction, and consist of a vessel containing a conducting liquid and a perforated metal ball dipping into it. When the current is to be switched off, the handle is turned so as to raise the ball out of the liquid; but the circuit is not immediately interrupted, since the liquid within the balls issues in fine streams out of the perforations, and so maintains the connection for a short time after switching off. As the liquid in the ball gets exhausted, and the streams become thinner, the resistance of the

liquid connection is gradually increased to infinity, and thus causes the current to gradually diminish to zero. The line wires are supported on Johnson & Phillips' patent fluid insulators, and the average span is about 130 ft.

Two sets of experiments were made. On the 11th of October only one generator and one motor were tested, while on the 12th of October both generators and both motors were tested. In the latter test the balancing wire was cut out of circuit as of no importance, when, as in these experiments, it was quite easy to regulate the load of each motor so as to fairly divide the work between them.

Electrical measuring instruments were fitted up at both stations in rooms sufficiently distant from the machinery so as not to be influenced by stray magnetism. The current was measured by large tangent galvanometers, and Thomson mirror galvanometers, standard cells, and potentiometers were used to measure the pressure. The object in measuring the current at both ends of the line was to ascertain whether any appreciable leak took place. In addition to these purely electrical measurements, observations were made at the generator station regarding the water level in the head and tail race of the turbine, the position of the regulator on the latter, and the speed of the dynamos and turbine. After the transmission trial on the 11th of October was completed, the armature of the dynamo was taken out and replaced by a plain spindle, provided at the end with a brake. The turbine was then started again under exactly the same conditions as were noted at the previous trial, and the power absorbed by the brake was measured. The comparison between the power thus measured and the electrical energy given out by the generator is evidently the commercial efficiency of the latter. On the following day, both generators and both motors were tested in the same condition as prevails in actual practice, with the only exception that, as already mentioned, the balancing wire was cut out of circuit. This alteration, which could obviously not increase the efficiency of the whole system, was made to simplify the measurements. The power absorbed by the generators was computed on the basis of the previous day's trial from the observed conditions under which the turbine worked, while the power developed by the motors was on both days directly ascertained by means of a friction brake fitted to a first motion shaft common to both motors. A small correction was made for the power absorbed by this shaft when running idle. The tables subjoined give the results as published by the committee.

An inspection of these figures will show that there is practically no loss of current by leakage on the line. In some cases the current measured at the motor station is slightly below that measured at the generator station; but the discrepancy is exceedingly small, and evidently due to personal or instrument errors, since in some other cases the current received by the motors appears to be even slightly larger than that sent out by the generators, which is obviously impossible. The second table also shows the influence of the air temperature upon the total resistance. The third table gives the power, and the fourth the efficiencies in percentages. It will be noticed that when one generator and one motor only were used, the commercial efficiency was slightly over sixty-eight per cent.; but when both generators and both motors were used, this efficiency rose to about seventy-five per cent., which is clearly due to the higher voltage employed.

I.—ELECTRICAL MEASUREMENTS.

Time of trial.	Electromotive force.		Terminal pressure.		Current measured at.	
	Generators.	Motors.	Generators.	Motors.	Generators.	Motors.
11th Oct.	1231.6	988.6	1177.7	1041.2	14.20	14.17
" "	1237.0	1016.8	1186.8	1066.1	13.24	13.28
12th "	1836.5	1575.4	1758.3	1656.1	11.48	11.42
" "	2120.0	1806.9	2058.0	1905.1	9.78	9.79

II.—RESISTANCES AND LOSS OF PRESSURE.

Time of trial.	Resistance of machines.		Line resistance.	Pressure lost in line.		Temperature of air centigrade.
	Generators.	Motors.		Calculated.	Measured.	
11th Oct.	3.741	3.716	0.228	130.9	136.5	+ 7.5
" "	3.741	3.710	0.228	122.3	120.7	+ 7.5
12th "	7.251	7.000	0.044	108.7	97.2	+ 3.2
" "	7.240	7.043	0.040	88.4	92.8	+ 3.2

III.—DETERMINATION OF ENERGY.

Time of trial.	Internal electrical horse-power.		Terminal electrical horse-power.		Actual horse-power.	
	Generators.	Motors.	Generators.	Motors.	Supplied to generators.	Obtained from motors.
11th Oct.	23.76	19.03	22.72	20.02	26.15	17.85
" "	22.27	18.34	21.35	19.23	24.54	16.74
12th "	28.64	24.46	27.34	25.71	30.87	23.21
" "	29.20	25.21	27.37	26.13	30.87	23.05

IV.—PERCENTAGE OF EFFICIENCIES.

Time of trial.	Electrical efficiency.		Commercial efficiency.		Total efficiency of transmission.	Remarks.
	Generators.	Motors.	Generators.	Motors.		
11th Oct.	90.7	93.7	86.8	89.1	68.3	One generator and one motor.
" "	90.6	91.3	86.9	87.1	68.2	
12th "	92.8	94.8	88.5	90.3	75.2	Both generators and both motors.
" "	91.6	91.4	88.7	88.2	74.6	

CHAPTER XV.

ALTERNATING CURRENT MOTORS.

AT the time that the authors began the present work, the electric motor had only been thought of in connection with continuous currents, which then occupied almost exclusively the attention of those interested in the distribution of electricity. Since that period, however, the distribution of electricity by means of alternating currents has reached very large proportions, especially in this country. While, at first, the alternating current was only employed for the purpose of lighting, it soon became evident that, in order to attain the full scope of its usefulness, it must also be made available for the distribution of power. In other words, it became necessary to construct alternating current motors.

As far back as the year 1868, Wilde, in experimenting upon alternate current machines, discovered that they could be coupled together in parallel without interfering with each other. He also tried the experiment of coupling two machines together and driving one as a generator, which delivered its current to the other, the field magnets of both being independently excited. On placing the stationary armature, with its coil, in a suitable position, in relation to the magnet-cylinder for producing electro-magnetic rotation, and setting the generator armature in motion, the motor armature with its coil oscillated rapidly in arcs of very small amplitude, the oscillations corresponding in number with the alternations of the current. As the amplitude of the oscillations in this experiment was limited by the inertia of the armature, and in order that the effect of one pulsation only on the armature might be observed, contact was made and broken suddenly between the connections by a kind of tapping motion, with the result that the stationary armature was suddenly jerked around nearly a

quarter of a revolution, sometimes in the direction in which it would have been driven by the belt and at other times in the opposite direction, according to the polarity of the alternating wave which happened to pass at that instant.

These experiments, which were published in the *Philosophical Magazine* for January, 1869, attracted little attention, and had almost been forgotten until recalled by a paper read by Dr. John Hopkinson before the London Institution of Civil Engineers in 1883, on "Electric Lighting," in which he deduced theoretically the results obtained experimentally by Wilde.

Although the motor effect obtained by Wilde consisted only in the oscillation of the armature, Dr. Hopkinson showed that continuous revolution could be maintained. Without entering here into a full treatment of the subject, it may be stated generally that the required conditions under which this takes place are that the lag of electromotive force of one machine behind that of the other shall be greater than a quarter period, the lag of the current being as usual either equal to or less than a quarter period behind the resultant E. M. F. Another result arrived at was that one machine can be driven as a motor by another even if its E. M. F. is greater than that of the latter.

Subsequently, Dr. Hopkinson, in conjunction with Prof. W. G. Adams, verified these conclusions experimentally on three large De Meritens alternating machines at the South Foreland light-house, during their investigation on "Light-House Illuminants."[*] In those experiments, two of the machines were connected in parallel and clutched together until they had attained their usual speed, when they were

---

[*] For a full description of these experiments the reader is referred to a paper read by Prof. W. G. Adams before the Soc. of Tel. Eng. & Elecns. Nov. 13, 1884, entitled "The Alternate Current Machine as a Motor."

unclutched and each was driven by its own belt. The E. M. F. on open circuit remained steady—the machines continuing to rotate in unison—and was the same as that of one of the machines when tested by itself. No current passed along the connecting wires. The circuit was now closed through an arc lamp; the machines continued to run as steadily as before, although a large current of 221 ampères was passing through the arc. Lastly, the lamp circuit was broken, the machines were short-circuited on one another, and the belt was thrown off one of them. It continued to run at the same steady speed, being driven as a motor by the current from the other machine. Other experiments were made, all confirming the theoretical conclusions, but their enumeration is needless and would lead us too far. We can only refer the reader to the original papers and to Professor S. P. Thompson's work on "Dynamo-Electric Machinery," which contains an excellent summary of them, as above outlined.

The experimenters also showed that the speed of a motor running thus is perfectly steady, but is accompanied by the serious disadvantages (1) that the motor can only run at one speed, which depends on the speed of the generator; (2) that it has to be brought up to this speed by some extraneous means before it can be run as a motor at all; (3) that if any of the conditions (such, for instance, as the load being too great) are unfavorable it pulls up and stops altogether. One way of bringing the motor up to the required speed was to drive it by a belt until the right pitch is attained and then to throw the belt off. Another method employed was to start the generator slowly and turn the motor by hand until it fell into step with the generator; the speed of the latter was then increased, when it was found that the motor also increased its speed, keeping pace exactly with the generator. This, of course, involves the serious disadvantage of having to stop or at least slow down the generator every time the motor has to be started.

Although the feasibility of operating alternate current motors was thus established, the subject, as stated above, called for little attention, however, until the general intro-

duction of the alternating current system of distribution. One of the first to recognize the importance and value of such a machine was Professor Elihu Thomson, who, in a classical paper read before the American Institute of Electrical Engineers in May, 1887, entitled "Novel Phenomena of Alternating Currents," drew attention prominently to the subject and described some entirely new forms.

It is well known that an alternating current passing in a coil or conductor laid parallel with, or in inductive relation to, a second coil or conductor, will induce in the second conductor, if on open circuit, alternating electromotive forces, and that if its terminals be closed or joined, alternating currents of the same rhythm, period or pitch, will circulate in the second conductor. This is the action occurring in any induction coil whose primary wire is traversed by alternating currents, and whose secondary wire is closed either upon itself directly or through a resistance.

In 1884, while preparing for the International Electrical Exhibition at Philadelphia, Professor Thompson had occasion to construct a large electro-magnet, the cores of which were about six inches in diameter and about twenty inches long. They were made of bundles of iron rod of about $\frac{1}{8}$ inch diameter. When complete the magnet was energized by the current of a dynamo giving continuous currents, and it exhibited the usual powerful magnetic effects. It was found also that a disc of sheet copper, of about $\frac{1}{16}$ inch thickness and ten inches in diameter, if dropped flat against a pole of the magnet, would settle down softly upon it, being retarded by the development of currents in the disc due to its movement in a strong magnetic field, and which currents were of opposite direction to those in the coils of the magnet. In fact, it was impossible to strike the magnet pole a sharp blow with the disc even when the attempt was made by holding one edge of the disc in the hand and bringing it down forcibly towards the magnet. In attempting to raise the disc quickly off the pole, a similar but opposite action of resistance to movement took place, showing the development of currents in the same direction to those in the coils of the magnet, and which currents, of course, would cause attraction as a result.

The experiment was, however, varied, as in Fig. 274. The disc D was held over the magnet pole, as shown, and the current in the magnet coils cut off by shunting them. There was felt an attraction of the disc or a dip toward the pole. The current was then put on by opening the shunting-switch and a repulsive action or lift of the disc was felt.

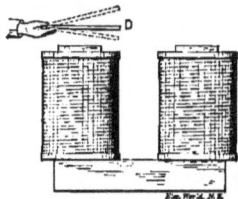

Fig. 274.

The actions just described are what would be expected in such a case, for when attraction took place, currents had been induced in the disc D in the same direction as those in the magnet coils beneath it, and when repulsion took place, the induced current in the disc was of opposite character or direction to that in the coils.

Now, let us imagine the current in the magnet coils to be not only cut off, but reversed back and forth. For the reasons just given we will find that the disc D is attracted and repelled alternately; for, whenever the currents induced in it are of the same direction with those in the inducing, or magnetic coil, attraction will ensue, and when they are opposite in direction, repulsion will be produced. Moreover, the repulsion will be produced when the current in the magnet coil is rising to a maximum in either direction, and attraction will be the result when the current of either direction is falling to zero, since in the former case opposite currents are induced in the disc D in accordance with well-known laws; and in the latter case currents of the same direction will exist in the disc D and the magnet coil. The disc might, of course, be replaced by a ring of copper or other good conductor, or by a closed coil of bare or insulated wire, or by a series of discs, rings or coils superposed, and the results would be the same.

The account just given of the effects produced by alternating currents, while true, is not the whole truth, and Professor Thomson supplements it by the following statements:

"An alternating current circuit or coil repels and attracts a closed circuit or coil placed in direct or magnetic inductive relation therewith; but the repulsive effect is in excess of the attractive effect.

"When the closed circuit or coil is so placed, and is of such low resistance metal that a comparatively large current can circulate as an induced current, so as to be subject to a large self-induction, the repulsive far exceeds the attractive effort."

Professor Thomson calls this excess of repulsive effect the "electro-inductive repulsion" of the coils or circuits.

This preponderating repulsive effect may be utilized or may show its presence by producing movement or pressure in a given direction, by producing angular deflection as of a pivoted body, or by producing continuous rotation with a properly organized structure. Among the simple devices realizing these conditions are the following:

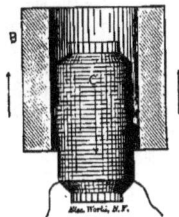

Fig. 275.

In Fig. 275, C is a coil traversed by alternating currents, B is a copper case or tube surrounding it, but not exactly over its centre. The copper tube B is fairly massive and is the seat of heavy induced currents. There is a preponderance of repulsive action tending to force the two conductors apart in an axial line. The part B may be replaced by concentric tubes slid one in the other, or by a pile of flat rings, or by a closed coil of coarse or fine wire insulated, or not. If the coil C, or primary coil, is provided with an iron core (such as a bundle of fine iron wires), the effects are greatly increased in intensity and the repulsion with a strong primary current may become quite vigorous, many pounds of thrust being producible by apparatus of quite moderate size.

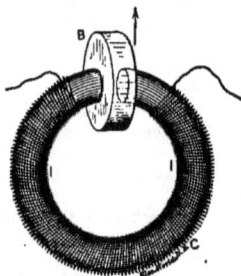

FIG. 276.

The forms and relations of the two parts C and B may be greatly modified with the general result of a preponderance of repulsive action when the alternating currents circulate.

It will be evident that the repulsive actions will not be mechanically manifested by axial movement or effort, when the electrical middles of the coils or circuits are coincident. In cylindrical coils in which the current is uniformly distributed through all the parts of the conductor section, what Professor Thomson terms the "electrical middle," or the centre of gravity of the ampère turns of the coils, will be the plane at right angles to its axis at its middle.

If the iron core takes the form of that shown by II, Fig. 276, such as a cut ring with the coil C wound thereon, the insertion of a heavy copper plate B into the slot or divided portion of the ring will be opposed by a repulsive effort when alternating currents pass in C. This was the first form of device in which Professor Thomson noticed the phenomenon of repulsive preponderance in question. The tendency is to thrust the plate B out of the slot in the ring excepting only when its centre is coincident with the magnetic axis joining the poles of the ring between which B is placed.

We will now turn our attention to the explanation of the actions exhibited, and afterwards refer to their applications. It may be stated as certainly true that were the induced currents in the closed conductor unaffected by any self-induction, the only phenomena exhibited would be alternate equal attractions and repulsions, because currents would be induced in opposite directions to that of the primary current when the latter current was changing from zero to maximum positive or negative current, and so producing repulsion; and would be induced in the same direction when changing from maximum positive or negative value to zero, so producing attraction.

This condition can be illustrated by a diagram, Fig. 277. Here the lines of zero current are the horizontal straight lines. The wavy lines represent the variations of current strength in each conductor, the current in one direction being indicated by that portion of the curve above the zero line, and in the other direction by that portion below it. The vertical dotted lines simply mark off corresponding portions of phase or succession of times.

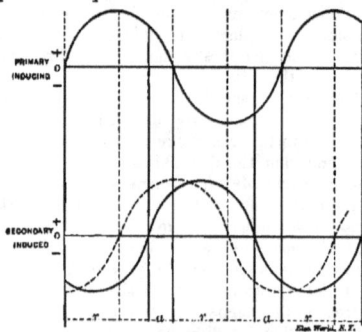

FIG. 277.

Here it will be seen that in the positive primary current descending from m, its maximum, to the zero line, the secondary current has risen from its zero to m^1, its maximum. Attraction will therefore ensue, for the currents are in the same direction in the two conductors. When the primary current increases from zero to its negative maximum n, the positive current in the secondary closed circuit will be decreasing from m^1, its positive maximum, to zero;· but, as the currents are in opposite directions, repulsions will occur. These actions of attraction and repulsion will be reproduced continually, there being a repulsion, then an attraction, then a repulsion, and again an attraction, dur-

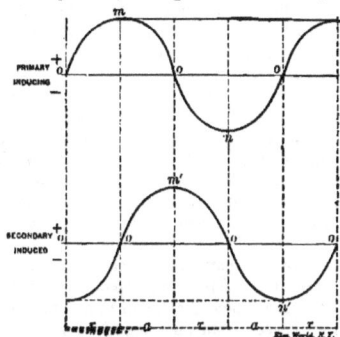

FIG. 278.

ing one complete wave of the primary current. The letters r, a, at the foot of the diagram, Fig. 277, indicate this succession.

In reality, however, the effects of self-induction in causing a lag, shift, or retardation of phase in the secondary current, will considerably modify the results and especially so when the secondary conductor is constructed so as to give to such self-induction a large value. In other words, the maxima of the primary or inducing current will no longer be found coincident with the zero points of the secondary currents. The effect will be the same as if the line representing the wave of the secondary current in Fig. 277 had been shifted forward to a greater or less extent. This is indicated in diagram, Fig. 278. It gives doubtless an exaggerated view of the action, though from the effects of repulsion which have been produced it is by no means an unrealizable condition.

It will be noticed that the period during which the currents are opposite, and during which repulsion can take place, is lengthened at the expense of the period during which the currents are in the same direction for attractive action. These differing periods are marked r, a,

etc., or the period during which *repulsion* exists is from the zero of the primary or inducing current to the succeeding zero of the secondary or induced current; and the period during which *attraction* exists is from the zero of the induced current to the zero of inducing current.

But far more important still in giving prominence to the repulsive effect than this difference of effective period, is the fact that during the period of repulsion both the inducing and induced currents have their greatest values, while during the period of attraction the currents are of small amounts comparatively. This condition may be otherwise expressed by saying that the period during which repulsion occurs includes all the maxima of current, while the period of attraction includes no maxima. There is then a *repulsion due to the summative effects of strong opposite currents* for a *lengthened period*, against an *attraction* due to the summative effects of *weak currents* of the *same direction* during a *shortened period*, the resultant effect being a greatly *preponderating* repulsion.

It is now not difficult to understand all the actions before described as obtained with the varied relations of coils, magnetic fields and closed circuits. It will be easily understood,

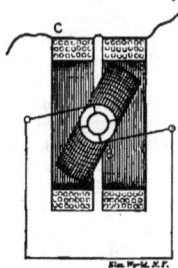

FIG. 279.

also, that an alternating magnetic field is in all respects the same as an alternating current coil in producing repulsion on the closed conductor, because the repulsions between the two conductors are the result of magnetic repulsions arising from opposing fields produced by the coils when the currents are of opposite directions in them.

This principle has been applied to the construction of an alternating current motor which can be started from a state of rest, and a number of designs of such motors are practicable.

One of the simplest is as follows: The coils *C*, Fig. 279, are traversed by an alternating

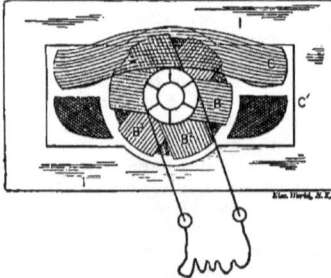

FIG. 279A.

current and are placed over a coil *B*, mounted upon a horizontal axis, transverse to the axis of the coil *C*. The terminals of the coil *B*, which is wound with insulated wire, are carried to a commutator, the brushes being connected by a wire, as indicated. The commutator is so constructed as to keep the coil *B* on short circuit from the position of coincidence with the plane of *C*, to the position where the plane of *B* is at right angles to that of *C;* and to keep the coil *B* open-circuited from the right-angled position, or thereabouts, to the position of parallel or coincident planes. The deflective repulsion exhibited by *B* will, when its circuit is completed by the commutator and brushes, as described, act to place its plane at right-angles to that of *C*, but being then open-circuited its momentum carries it to the position just past parallelism, at which moment it is again short-circuited, and so on. It is capable of very rapid rotation, but its energy is small. Professor Thomson has, however, extended the principle to the construction of more complete apparatus. One form has its revolving portion or armature composed of a number of sheet-iron discs wound as usual with three coils crossing near the shaft. The commutator is arranged to short-circuit each of these coils in succession, and twice in a

revolution, and for a period of 90 degrees of rotation each. The field coils surround the armature and there is a laminated iron field structure completing the magnetic circuit.

Figs. 279A and 279B will give an idea of the construction of the motor referred to. *C C¹* are the field coils or inducing coils which alone are put into the alternating current circuit. *I I* is a mass of laminated iron, in the interior of which the armature revolves, with its three coils *B*, *B²*, *B³*, wound on a core of sheet-iron discs. The commutator short-circuits the armature coils in succession in the proper positions to utilize the repulsive effect set up by the currents which are induced in them by the alternations in the field coils. The motor has no dead point and will start from a state of rest and give out considerable power, but with what economy is not yet known.

A curious property of the machine is that at a certain speed, depending upon the rapidity of the alternations in the coil *C*, a continuous current passes from one commutator brush to the other, and it will energize electro-magnets

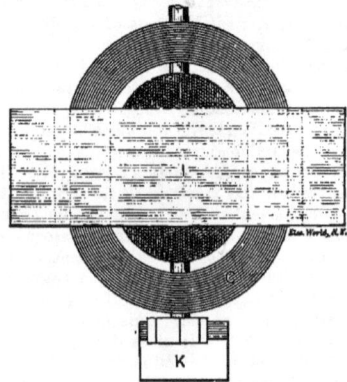

FIG. 279B.

and perform other actions of direct currents. Here we have, then, a means of inducing direct currents from alternating currents. To control the speed and keep it at that required for the purpose, we have only to properly gear the motor to another of the ordinary type for alternating currents, namely, an alternating current dynamo used as a motor.

Taking up the principle of electro-inductive repulsion enunciated by Professor Thomson, Lieutenant F. Jarvis Patten, U. S. A., has de-

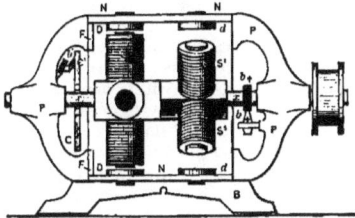

FIG. 280.

signed a motor in which the same principle is reversed, with the apparent effect of producing a more continuous rotary effort as well as an increased moment of rotation by virtue of placing the point at which the effort is applied farther from the axis of rotation and thereby giving to the force at work, however small, a much greater lever-arm. Figs. 280 and 281 are end and side elevations of the machine, and Fig. 282 is a diagram of circuits and connections. A polygonal frame $F F F$ is connected by lateral strips $N N$, and all is supported upon the base $B B$. Secured to the lateral strips $N N$ are two sets of copper discs $D D$ and $d d$. These discs constitute the armature

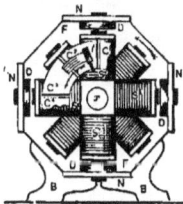

FIG. 281.

of the machine; they are arranged peripherally, as shown, fixed to the frame structure, and consist of circuits of high self-inductive capacity that remain permanently closed.

A spindle $x x$, Figs. 280 and 281, carries a revolving switch $C^1 C^2$, or sunflower, which is

secured to the spindle and turns with it. This sunflower commutator constitutes one terminal of the machine, while an insulated ring $b+$, and its contact brush secured to the support $P P$, constitutes the other. Under each set of discs, $D D$ and $d d$, and secured to the spindle x, is a set of solenoids ; those of one set being placed at an angle of 45 degrees to those of the other set, from which it results that these two sets of solenoids will each come alternately into action with respect to its own set of copper discs. Each set of solenoids has its four coils connected in series and the two sets are arranged in two independent circuits. Both have one terminal secured to the contact $b+$, and the other terminal to the alternate segments of the revolving sunflower. This system

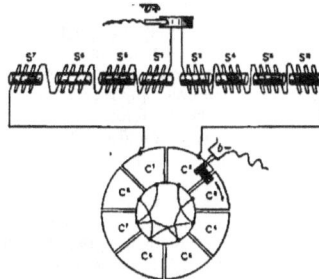

FIG. 282.

of connections is shown in Fig. 282, in which the eight-part commutator is shown with alternate segments connected to each other in separate series. The coils S^1, S^3, S^5, S^7, form one set of revolving solenoids connected in series from the rubbing contact $b+$, which forms one terminal of the machine, to the commutator segment C^1, and thence to all the odd-numbered segments. In like manner the other set of solenoids S^2, S^4, S^6, S^8, are similarly connected in series to all the even-numbered segments, C^2, C^4, C^6, C^8. From this arrangement it results that, as the spindle revolves, carrying these solenoids as a sort of fly-wheel, the alternating current will be sent in rapid succession first through one set of solenoids, and then through the other, and by suitably placing this commutator upon the spindle it can be made to send the current through the different sets of solenoids during the periods of maximum effort of repulsion between each set and its corresponding set of copper discs.

The moment of rotation is evidently a maximum under this system of construction, and by increasing the number of sets of solenoids that follow each other in action, the effort of the rotary torque may be made nearly continuous and constant.

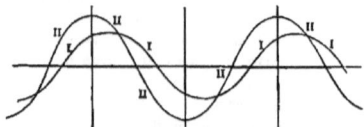

FIG. 283.

It is evident that this motor is not designed for heavy work. Its efficiency is not high, and its form renders such an application quite impossible, but it can find a place in the smaller industries, where light motors fill a definite requirement, and render an alternating current circuit a paying one during the working hours of the day.

In discussing the actions which take place in an alternating current motor, Dr. Louis Duncan, in a paper read before the American Institute of Electrical Engineers in February, 1888, considers the case of an alternating current motor built like an alternating dynamo, the field excited by a continuous current. If $A B C$, Fig. 283, are the poles of the field magnets, supposed to be excited by a continuous current, then the curve $I I$ will represent the counter or motor E. M. F. If the motor is running slowly there will be two or three reversals of current in the time it takes a coil to go from A to B, Fig. 284, and the product $C E$ will be nearly zero, the positive and negative parts being almost equal. Whether it is + or — is somewhat a matter of chance, and if it is — from A to B it is very likely to be + from B to C. Thus the armature is pushed first in one direction, then in the opposite, and there is no definite tendency for it to rotate as a motor. This brings us to the first difficulty, viz., a simple alternating current motor will not start itself. Let us suppose, however, that it has been started by some

means and has reached such a speed that in the time of a reversal a coil shall have moved over the distance between two similar poles. We will have the state of things in Fig. 283, and the armature will continue to rotate. Now the position of the curve $I I$ is fixed; we will consider the position of I as determined by the position of the armature coils when the current in them is zero; for instance, if the speed increases a little the curve will advance as shown by the dotted line. The total work transformed is the product of the two curves; from 1 to 2 it is —; from 2 to 3 it is +; from 3 to 4 —; from 4 to 5 +. The result is, — [1 to 2 + 3 to 4]; + [2 to 3 + 4 to 5]; part of the time then the machine is working as a dynamo, part of the time as a motor; the difference of the values represented by the brackets gives the mechanical work that is really available. Now while the available work is the difference of these areas, the difference becoming small as the load is decreased, the heating is the *sum* $C^2 R$ for all the values of the current, and is independent of the position of its curve.

Looking at the figure again, we see that from 1 to 2 the armature is pushed forward by the current in the line; from 2 to 3 it is pulled back, since it is acting as a dynamo feeding into the line, and can only get the energy to produce the current by decreasing its speed and drawing from its energy of motion. The armature oscillates then, and it is evident that the amount of its oscillation depends on the

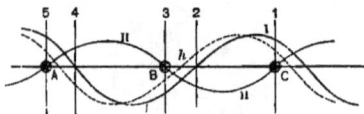

FIG. 284.

kind of work it is doing. If it is driving heavy wheels or machinery having considerable inertia, it will only have to slow down slightly when it becomes a dynamo. If it is lifting weights, the amount of oscillation will be considerable.

It is evident that there is a certain position of the curve I that will make the available work a maximum. If the motor is doing all of its possible work the curve will take up this position; as the load is decreased the speed will increase for an instant until the curve has shifted forward into such a position that the sum of the products—$E\,C$—is equal to the work done. In fact, we can plat a curve representing values of the distance from the point the current curve crosses the axis to the same point when the work is zero corresponding to different loads.

Now the value of this lag cannot be greater than $O\,p$, Fig. 285, for although we might ex-

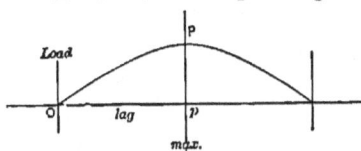

FIG. 285.

tend our curve on the other side, yet a slight increase in our load will cause the armature to fall back, decreasing the available work, and suddenly stop. It is, in fact, in unstable equilibrium.

Suppose we have found the value of the lag that will give us a maximum value of the work, and have calculated this value by the ordinary mathematical methods employed, the real work we can obtain is less than this, for the current curve oscillates on both sides of the maximum, supposing we *could* work at the maximum; that is, it is only for a very short time in the best position; so that if we take the sum of the work for a period, it will be less than the calculated maximum. In reality we must work slightly in advance of the maximum, for if we were too near it the curve would fall behind P, and will then be in unstable equilibrium and stop. The practical maximum will then vary according to the amount of the oscillation; that is, according to the nature of the work being done.

Dr. Duncan, in discussing the possible forms of alternating current motors, suggested the following: (1) An ordinary series motor; (2) an alternating dynamo reversed, the field being excited by a continuous current; (3) an alternating dynamo reversed, the field being excited by the alternating current first cor-

rected by a commutator on the shaft; (4) the arrangement suggested by Professor Thomson, which has been described above.

In the first type—a series motor—there is no difficulty in starting; the motor will start of itself. There are these difficulties, however: the armature and field must both be thoroughly laminated to prevent eddy currents; the magnetism of such large masses of iron being rapidly reversed will cause losses unless both field and armature are far from saturation; that is, the mass must be great. Again, Mr. Kapp has shown that to obtain the maximum work we must have, approximately, the counter E. M. F. of the motor equal to the E. M. F. of self-induction—a condition almost impossible to realize in practice. The motor must be governed in some way, as it will not govern itself.

The alternating motor, with a field fed from the alternating circuit, the current being commutated, will start itself. If, in Fig. 286, A and B are poles and C one of the armature coils, the effects of the currents will be as shown by the two sets of signs. The effect of either arrangement is to move the armature in the direction of the arrow. When C gets opposite A the commutator changes the relative directions of the currents in C and A, and C is repelled by A and the armature continues to rotate. The maximum work will, according to Dr. Duncan, be done when the speed is such that a reversal takes place in a distance A to

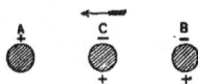

FIG. 286.

B. There is an advantage in this type in that it will start itself; a disadvantage is that the fields must be carefully laminated; there is loss in reversing them, and there will be for some speeds considerable sparking at the field commutator. It is probable, also, that the work obtainable from such a motor would not be as great as if its field were fed by continuous currents.

The motor obtained by reversing an ordinary alternating dynamo has advantages and disadvantages. It perfectly regulates itself, and the field magnets need not be laminated ; that is, it can be made cheaply. It will give a greater output from a given source of current than corresponding machines of either of the types already discussed. Its disadvantages are that it must be started independently. We must have a continuous current to excite the field, and if a load having any considerable inertia be suddenly applied the motor will stop. This last objection also applies to the type of motor mentioned above, provided the maximum work is obtainable when the counter E. M. F. has the same period as the applied E. M. F.

The motor might be started by passing the commutated alternating current through the field as in the second type of motor discussed, changing our connections when the proper speed is attained, so that a continuous current from some external source passes through the field, and the alternating current is shut off from it. Another way would be to have on the same shaft with the main motor a motor arrangement similar to that of Professor Elihu Thomson, described above. With this arrangement we can do more than start the motor. It was pointed out that when this auxiliary motor reached a certain speed it would produce a continuous current in the external circuit of its armature. This current could be used to excite the field of the main motor. By properly proportioning the number of coils in the main and auxiliary motors, this continuous current would be produced just when the motor had arrived at its proper speed, and it is evident that we can make this current operate an automatic device to make the circuit of the main motor at this moment. According to Dr. Duncan, however, a motor made in this way would be expensive and not particularly efficient.

Dr. Duncan considers the simplest, cheapest and most efficient means of running alternating current motors is this : Build the motor on the same general plan as the dynamo, with such modifications as the different conditions of working impose, and start it and excite the field magnets from a continuous current circuit, run with the alternating circuit, supplied with current by a dynamo at the central station. If it is desired to distribute 500 horse-power the continuous circuit should have a maximum capacity of about 50 horse-power. To start the motor the following arrangement would be used : There should be two breaks in the armature circuit, one between the regular brushes of the machine, the other as a commutator for the continuous current. At this second break the two ends of the circuit should be taken to alternate bars of the commutator, the number of bars being such that the direction of the current is reversed every time a coil passes a pole. The alternate bars are normally connected by a metal ring pressed against them by a spring ; in this case we will have the normal circuit just as if there were no continuous current commutator. On the motor would be a switch-board that would accomplish the following things : If we wish to start we turn the handle of the switch to a certain position; this will short-circuit our regular brushes and by the aid of a couple of levers will drop the continuous circuit brushes on the commutator, at the same time pulling away the metal ring. The motor will then start as a continuous current motor. When it has reached its proper number of revolutions, or is above it, turn the handle a little further ; the continuous current brushes will be raised, the metal ring will connect the commutator bars and the alternating circuit will be made, and the motor will continue to run and will do work.

In enumerating the various methods which might be employed in the construction of alternating current motors, Mr. Nikola Tesla, in a masterly paper read before the American Institute of Electrical Engineers in May, 1888, referred to those given by Dr. Duncan as stated above, and suggested two additional ones, viz.: (1) A motor with one of its circuits in series with a transformer and the other in the secondary of the transformer ; (2) a motor having its armature circuit connected to the generator and the field coils closed upon themselves. These, however, were only incidentally mentioned by Mr. Tesla, the paper relating to an entirely new class of alternate current motors, based upon the continuous rotation of the magnetic poles in a closed magnetic circuit, and which we will now proceed to describe.

In dynamo machines, it is well known, we generate alternate currents which we direct by means of a commutator. Now, the currents so directed cannot be utilized in the motor, but they must again be reconverted into their original state of alternate currents. The function of the commutator is entirely external, and in no way does it affect the internal working of the machines. In reality, therefore, all machines are alternate-current machines, the currents appearing as continuous only in the external circuit during their transit from generator to motor. But the operation of the commutator on a motor is two-fold; first, it reverses the currents through the motor, and secondly, it effects, automatically, a progressive shifting of the poles of one of its magnetic constituents. Assuming, therefore, that both of these operations in the system—that is to say, the direct-

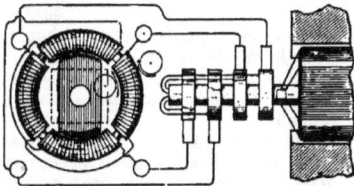

FIG. 287.

ing of the alternate currents on the generator and reversing the direct currents on the motor —be eliminated, it would still be necessary, in order to cause a rotation of the motor, to produce a progressive shifting of the poles of one of its elements, and the question presented itself to Mr. Tesla, How to perform this operation by the direct action of alternate currents? We will now proceed to show how this result was accomplished.

In the first experiment, a drum-armature was provided with two coils at right angles to each other, and the ends of these coils were connected to two pairs of insulated contact-rings as usual. A ring was then made of thin insulated plates of sheet-iron and wound with four coils, each two opposite coils being connected together so as to produce free poles on diametrically opposite sides of the ring. The remaining free ends of the coils were then connected to the contact-

rings of the generator-armature so as to form two independent circuits, as indicated in Fig. 287. The field of the generator being independently excited, the rotation of the armature sets up currents in the coils $C C$, Fig. 288, varying in strength and direction in the well-known manner. In the position shown in Fig. 288 the current in coil C is nil while coil C_1 is traversed

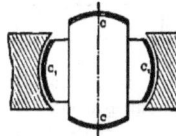 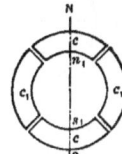

FIG. 288.   FIG. 288A.

by its maximum current, and the connections may be such that the ring is magnetized by the coils c, c_1 as indicated by the letters $N S$ in Fig. 288A, the magnetizing effect of the coils $c c$ being nil, since these coils are included in the circuit of coil C.

In Fig. 289 the armature coils are shown in a more advanced position, one-eighth of one revolution being completed. Fig. 289A illustrates the corresponding magnetic condition of the ring. At this moment the coil C_1 generates a current of the same direction as previously, but weaker, producing the poles n_1 s_1 upon the ring; the coil C also generates a current of the same direction, and the connections may be such that the coils $c c$ produce the poles $n s$,

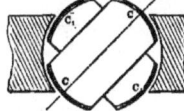 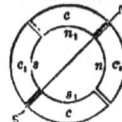

FIG. 289.   FIG. 289A.

as shown in Fig. 289A. The resulting polarity is indicated by the letters $N S$, and it will be observed that the poles of the ring have been shifted one-eighth of the periphery of the same.

In Fig. 290 the armature has completed one-quarter of one revolution. In this phase the current in coil C is a maximum, and of such direction as to produce the poles $N S$ in Fig. 290A, whereas the current in coil C_1 is nil, this coil being at its neutral position. The poles $N S$ in Fig. 290A are thus shifted one-quarter of the circumference of the ring.

Fig. 291 shows the coils $C C$ in a still more advanced position, the armature having completed three-eighths of one revolution. At

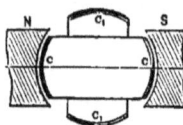 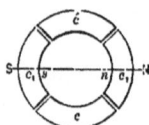

FIG. 290.   FIG. 290A.

that moment the coil C still generates a current of the same direction as before, but of less strength, producing the comparatively weaker poles $n s$ in Fig. 291A. The current in the coil C_1 is of the same strength, but of opposite direction. Its effect is, therefore, to produce

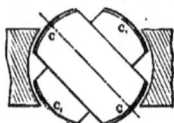 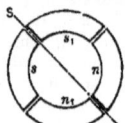

FIG. 291.   FIG. 291A.

upon the ring the poles n_1 and s, as indicated, and a polarity, $N S$, results, the poles now being shifted three-eighths of the periphery of the ring.

In Fig. 292 one-half of one revolution of the armature is completed, and the resulting mag-

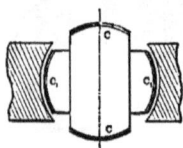 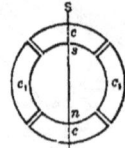

FIG. 292.   FIG. 292A.

netic condition of the ring is indicated in Fig. 292A. Now, the current in coil C is nil, while the coil C_1 yields its maximum current, which is of the same direction as previously; the magnetizing effect is, therefore, due to the coils $c_1 c_1$ alone, and, referring to Fig. 292A, it will be observed that the poles $N S$ are shifted one-

half of the circumference of the ring. During the next half revolution the operations are repeated, as represented in the Figs. 293 to 295A.

A reference to the diagrams will make it clear that during one revolution of the armature the poles of the ring are shifted once around its periphery, and each revolution producing like effects, a rapid whirling of the poles in harmony with the rotation of the armature is the result. If the connections of

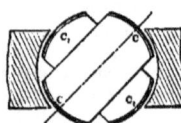 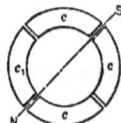

FIG. 293.   FIG. 293A.

either one of the circuits in the ring are reversed, the shifting of the poles is made to progress in the opposite direction, but the operation is identically the same. Instead of using four wires, with like result, three wires may be used, one forming a common return for both circuits.

This rotation or whirling of the poles manifests itself in a series of curious phenomena. If a delicately pivoted disc of steel or other magnetic metal is approached to the ring it is set in rapid rotation, the direction of rotation varying with the position of the disc. For in-

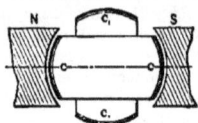 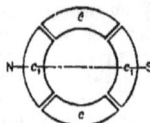

FIG. 294.   FIG. 294A.

stance, noting the direction outside of the ring it will be found that inside the ring it turns in an opposite direction, while it is unaffected if placed in a position symmetrical to the ring. This is easily explained. Each time that a pole approaches it induces an opposite pole in the nearest point on the disc, and an attraction is produced upon that point; owing to this, as the pole is shifted further away from the disc a tangential pull is exerted upon the same, and, the action being constantly repeated, a more or less rapid rotation of the disc is the result. As the pull is exerted mainly upon that part which is nearest to the ring, the rotation outside and inside, or right and left, respectively, is in opposite directions, Fig. 287. When placed symmetrically to the ring, the pull on opposite sides of the disc being equal, no rotation results. The action is based on the mag-

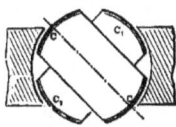 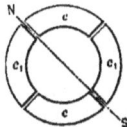

FIG. 295.   FIG. 295A.

netic inertia of the iron; for this reason a disc of hard steel is much more affected than a disc of soft iron, the latter being capable of very rapid variations of magnetism. To demonstrate the complete analogy between the ring and a revolving magnet, a strongly energized electromagnet was rotated by mechanical power, and phenomena identical in every particular to those mentioned above were observed.

Obviously, the rotation of the poles produces corresponding inductive effects, and may be util-

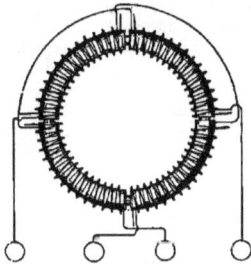

FIG. 296.

ized to generate currents in a closed conductor placed within the influence of the poles. For this purpose it is convenient to wind a ring with two sets of superimposed coils forming respectively the primary and secondary circuits, as shown in Fig. 296. In order to secure the most economical results the magnetic circuit should be completely closed, and with this object in view the construction may be modified at will.

The inductive effect exerted upon the secondary coils will be mainly due to the shifting or movement of the magnetic action; but there may also be currents set up in the circuits, in consequence of the variations in the intensity of the poles. However, by properly designing the generator and determining the magnetizing effect of the primary coils, the latter element may be made to disappear. The intensity of the poles being maintained constant, the action of the apparatus will be perfect, and the same result will be secured as though the shifting were effected by means of a commutator with an infinite number of bars. In such case the theoretical relation between the energizing effect of each set of primary coils and their resultant magnetizing effect may be expressed by the equation of a circle having its centre coinciding with that of an orthogonal system

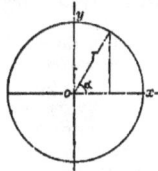 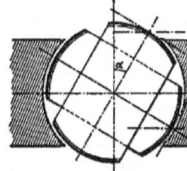

FIG. 297.   FIG. 298.

of axes, and in which the radius represents the resultant and the co-ordinates both of its components. These are then respectively the sine and cosine of the angle a between the radius and one of the axes ($o\,x$). Referring to Fig. 297, we have $r^2 = x^2 + y^2$; where $x = r \cos a$, and $y = r \sin a$.

Assuming the magnetizing effect of each set of coils in the transformer to be proportioned to the current—which may be admitted for weak degrees of magnetization—then $x = K\,c$ and $y = K\,c^1$, where K is a constant and c and c^1 the current in both sets of coils respectively. Supposing, further, the field of the generator to be uniform, we have for constant speed $c^1 = K^1 \sin a$, and $c = K^1 \sin (90° + a) = K^1 \cos a$, where K^1 is a constant. See Fig. 298.

Therefore, $x = K\,c = K\,K^1 \cos a$;
$$y = K\,c^1 = K\,K^1 \sin a, \text{ and}$$
$$K\,K^1 = r.$$

That is, for a uniform field the disposition of the two coils at right angles will secure the theoretical result, and the intensity of the shifting poles will be constant. But from $r^2 = x^2 + y^2$ it follows that for $y = 0$, $r = x$; it follows that the joint magnetizing effect of both sets of coils should be equal to the effect of one set when at its maximum action. In transformers and in a certain class of motors the fluctuation of the poles is not of great importance, but in another class of these motors it is desirable to obtain the theoretical result.

In applying this principle to the construction of motors, two typical forms of machines have been developed by Mr. Tesla. First, a form having a comparatively small rotary effort at the start, but maintaining a perfectly uniform speed at all loads, which motor has been termed synchronous. Second, a form possessing a great rotary effort at the start, the speed being dependent on the load.

These motors may be operated in three different ways: (1) By the alternate currents of the source only ; (2) by a combined action of these and of induced currents; (3) by the joint action of alternate and continuous currents.

The simplest form of a synchronous motor is obtained by winding a laminated ring provided with pole projections with four coils, and connecting the same in the manner before indicated. An iron disc having a segment cut away on each side may be used as an armature. Such a motor is shown in Fig. 287. The disc being arranged to rotate freely within the ring in close proximity to the projections, it is evident that, as the poles are shifted, it will, owing to its tendency to place itself in such a position as to embrace the greatest number of the lines of force, closely follow the movement of the poles, and its motion will be synchronous with that of the armature of the generator ; that is, in the peculiar disposition shown in Fig. 287, in which the armature produces by one revolution two current impulses in each of the circuits. It is evident that if, by one revolution of the armature, a greater number of impulses is produced, the speed of the motor will be correspondingly increased. Considering that the attraction exerted upon the disc is greatest when the same is in close proximity

to the poles, it follows that such a motor will maintain exactly the same speed at all loads within the limits of its capacity.

To facilitate the starting, the disc may be provided with a coil closed upon itself. The advantage secured by such a coil is evident. On the start the currents set up in the coil strongly energize the disc, and increase the attraction exerted upon the same by the ring, and currents being generated in the coil as long as the speed of the armature is inferior to that of the poles ; considerable work may be performed by such a motor, even if the speed be below normal. The intensity of the poles being constant, no currents will be generated in the coil when the motor is turning at its normal speed.

Instead of closing the coil upon itself, its ends may be connected to two insulated sliding rings, and a continuous current supplied to these from a suitable generator. The proper way to start such a motor is to close the coil upon itself until the normal speed is reached, or nearly so, and then turn on the continuous current. If the disc be very strongly energized by a continuous current motor it may not be able to start, but if it be weakly energized, or generally so that the magnetizing effect of the ring is preponderating, it will start and reach the normal speed. Such a motor will maintain absolutely the same speed at all loads. It has also been found that if the motive power of the generator is not excessive, by checking the motor the speed of the generator is diminished in synchronism with that of the motor. It is characteristic of this form of motor that it cannot be reversed by reversing the continuous current through the coil.

The synchronism of these motors may be demonstrated experimentally in a variety of ways. For this purpose it is best to employ a motor consisting of a stationary field magnet and an armature arranged to rotate within the same. as indicated in Fig. 290. In this case the shifting of the poles of the armature produces a rotation of the latter in the opposite direction. It results therefrom that when the normal speed is reached, the poles of the armature assume fixed positions relatively to the field magnet, and the same is magnetized by induction, exhibiting a distinct pole on each of the pole-pieces. If a piece of soft iron is approached to the field magnet it will at the start be attracted with a rapid vibrating motion produced by the reversals of polarity of the magnet, but as the speed of the armature increases the vibrations become less and less

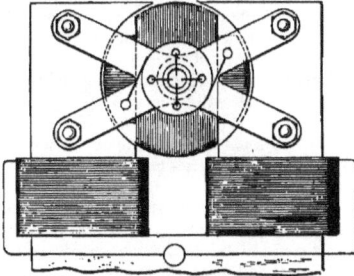

FIG. 299.

frequent and finally entirely cease. Then the iron is weakly but permanently attracted, showing that synchronism is reached and the field magnet energized by induction.

The disc may also be used for the experiment. If held quite close to the armature, it will turn as long as the speed of rotation of the poles exceeds that of the armature ; but when the normal speed is reached, or very nearly so, it ceases to rotate and is permanently attracted.

In motors of the synchronous type it is desirable to maintain the quantity of the shifting magnetism constant, especially if the magnets are not properly subdivided.

To obtain a rotary effort in these motors, it was necessary to make such a disposition that while the poles of one element of the motor are shifted by the alternate currents of the source, the poles produced upon the other element should always be maintained in the proper relation to the former, irrespective of the speed of the motor. Such a condition exists in a continuous-current motor ; but in a synchronous motor, such as described, this condition is fulfilled only when the speed is normal.

The object has been attained by placing within the ring a properly subdivided cylindrical iron core wound with several independent coils closed upon themselves. Two coils at right angles, as in Fig. 300, are sufficient, but a greater number may be advantageously employed. It results from this disposition that when the poles of the ring are shifted, currents are generated in the closed armature coils. These currents are the most intense at or near the points of the greatest density of the lines of force, and their effect is to produce poles upon the armature at right angles to those of the ring, at least theoretically so ; and since this action is entirely independent of the speed—that is, so far as the location of the poles is concerned—a continuous pull is exerted upon the periphery of the armature. In many respects these motors are similar to the continuous-current motors. If load is put on, the speed, and also the resistance of the motor, is diminished and more current is made to pass through the energizing coils, thus increasing the effort. Upon the load being taken off, the counter-electromotive force increases and less current passes through the primary or energizing coils. Without any load the speed is very nearly equal to that of the shifting poles of the field magnet.

According to Mr. Tesla, the rotary effort in these motors fully equals that of the continuous current motors. The effort seems to be

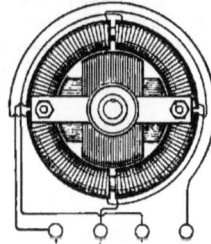

FIG. 300.

greatest when both armature and field magnets are without any projections ; but as in such dispositions the field cannot be very concentrated, probably the best results will be obtained by leaving pole projections on one of the elements only. Generally, it may be stated that the projections diminish the torque and produce a tendency to synchronism.

A characteristic feature of motors of this kind is their capacity of being very rapidly reversed. This follows from the peculiar action of the motor. Suppose the armature to be rotating, and the direction of rotation of the poles to be reversed. The apparatus then represents a dynamo machine, the power to drive this machine being the momentum stored up in the armature, and its speed being the sum of the speeds of the armature and the poles. If we now consider that the power to drive such a dynamo would be very nearly proportional to the third power of the speed, for this reason alone the armature should be quickly reversed. But simultaneously with the reversal another element is brought into action, namely, as the movement of the poles with respect to the armature is reversed, the motor acts like a transformer in which the resistance of the secondary circuit would be abnormally diminished, by producing in this circuit an additional electromotive force. Owing to these causes the reversal is instantaneous.

If it is desirable to secure a constant speed, and at the same time a certain effort at the start, this result may be easily attained in a variety of ways. For instance, two armatures, one for torque and the other for synchronism, may be fastened on the same shaft, and any desired preponderance may be given to either one; or an armature may be wound for rotary effort, but a more or less pronounced tendency to synchronism may be given to it by properly constructing the iron core; and in many other ways.

As a means of obtaining the required phase of the currents in both the circuits, the disposition of the two coils at right angles is the simplest, securing the most uniform action; but the phase may be obtained in many other ways, varying with the machine employed. Any of the dynamos at present in use may be easily adapted for this purpose by making connections to proper points of the generating coils. In closed circuit armatures, such as used in the continuous-current systems, it is best to make four derivations from equi-distant points or bars of the commutator, and to connect the same to four insulated sliding rings on the shaft. In this case each of the motor circuits

is connected to two diametrically opposite bars of the commutator. In such a disposition the motor may also be operated at half the potential and on the three-wire plan, by connecting the motor circuits in the proper order to three of the contact-rings.

In multipolar dynamo machines, such as used in the converter systems, the phase is conveniently obtained by winding upon the armature two series of coils in such a manner that while the coils of one set or series are at their maximum production of current, the coils of the other will be at their neutral position, or nearly so; whereby both sets of coils may be subjected simultaneously or successively to the inducing action of the field magnets.

Generally, the circuits in the motor will be similarly disposed, and various arrangements may be made to fulfill the requirements; but the simplest and most practicable is to arrange

FIG. 301.   FIG. 302.   FIG. 303.

primary circuits on stationary parts of the motor, thereby obviating, at least in certain forms, the employment of sliding contacts. In such a case the magnet coils are connected alternately in both the circuits; that is, 1, 3, 5 ....in one, and 2, 4, 6....in the other, and the coils of each set of series may be connected all in the same manner, or alternately in opposition; in the latter case, a motor with half the number of poles will result, and its action will be correspondingly modified. The diagrams, Figs. 301, 302 and 303, show three different phases, the magnet coils in each circuit being connected alternately in opposition. In this case there will always be four poles, as in Figs. 301 and 302, four pole projections being neutral, and in Fig. 303 two adjacent pole projections will have the same polarity. If the coils are connected in the same manner there will be eight alternating poles, as indicated by the letters n' s' in Fig. 301.

The employment of multipolar motors secures in this system an advantage much desired, and that is, that a motor may be made to run exactly at a predetermined speed, irrespective of imperfections in construction, of the load, and, within certain limits, of electromotive force and current strength.

In a general distribution system of this kind, according to Mr. Tesla, the following plan should be adopted : At the central station of supply a generator should be provided, having a considerable number of poles. The motors operated from this generator should be of the synchronous type, but possessing sufficient rotary effort to insure their starting. With the observance of proper rules of construction, it may be admitted that the speed of each motor will be in some inverse proportion to its size, and the number of poles should be chosen accordingly. Still, exceptional demands may modify this rule. In view of this, it may be advantageous to provide each motor with a greater number of pole projections or coils, the number being preferably a multiple of two and three. By this means, by simply changing the connections of the coils, the motor may be adapted to any probable demands.

If the number of the poles in the motor is even, the action will be harmonious, and the proper result will be obtained ; if this is not the case, the best plan to be followed is to make a motor with a double number of poles and connect the same in the manner before indicated, so that half the number of poles result. Suppose, for instance, that the generator has twelve poles, and it would be desired to obtain a speed equal to ¹⁄₇ of the speed of the generator. This would require a motor with seven pole projections or magnets, and such a motor could not be properly connected in the circuits unless fourteen armature coils would be provided, which would necessitate the employment of sliding contacts. To avoid this the motor should be provided with fourteen permanent magnets, and seven connected in each circuit, the magnets in each circuit alternating among themselves. The armature should have fourteen closed coils. The action of the motor will not be quite as perfect as in the case of an even number of poles, but the drawback will not be of a serious nature.

However, the disadvantages resulting from this unsymmetrical form will be reduced in the same proportion as the number of the poles is augmented. If the generator has, say, n, and the motor n_1 poles, the speed of the motor will be equal to that of the generator multiplied by $\frac{n}{n_1}$.

The speed of the motor will generally be dependent on the number of the poles, but there may be exceptions to this rule. The speed may be modified by the phase of the currents in the circuits, or by the character of the current impulses, or by the intervals between each, or between groups of impulses. Some of the possible cases are indicated in the diagrams, Figs. 304, 305, 306 and 307, which are self-explanatory. Fig. 304 represents the condition generally existing, and which secures the best result. In such a case, if the typical form of motor illustrated in Fig. 287 is em-

Fig. 304.   Fig. 305.   Fig. 306.   Fig. 307.

ployed, one complete wave in each circuit will produce one revolution of the motor. In Fig. 305 the same result will be effected by one wave in each circuit, the impulses being successive ; in Fig. 306 by four, and in Fig. 307 by eight waves.

By such means any desired speed may be attained ; that is, at least within the limits of practical demands. This system possesses this advantage besides others, resulting from simplicity. At full loads, according to Mr. Tesla, the motors show an efficiency fully equal to that of the continuous current motors. The transformers present an additional advantage in their capability of operating motors.

We may add that shortly after the publication of Mr. Tesla's work, Professor Galileo Ferraris, of Turin, Italy, published as the result of independent study an essay, in which, under the title of "Electro-Dynamic Rotation," he described phenomena similar to those already employed by Mr. Tesla in his motors.

CHAPTER XVI.

THERMO-MAGNETIC MOTORS.

WHILE the class of motors described in the last chapter occupies the attention of many electricians at the present time, still another type, depending upon an entirely different principle, has recently come into prominence, and although such motors are still largely in an experimental stage their possible future practicability, as well as their theoretical interest, makes some mention of them in this work necessary.

These motors, which may be termed "thermomagnetic," are based upon a principle which has long been known, and which, indeed, was first announced by that pioneer in electricity and magnetism, Dr. William Gilbert. He first showed that when a loadstone or iron bar was heated to redness it lost its magnetism. Since then the influence of heat on magnetic metals, such as iron, nickel, cobalt, etc., has been well recognized, and the experiment has often been made, showing that a soft iron armature strongly attracted by an electro-magnet when cold, is easily withdrawn when the armature is heated to redness. The explanation of this phenomenon is referred to the fact that at red heat the molecules of iron lose their coercitive force and the iron becomes magnetically inert, exhibiting no greater magnetic properties than other metals.

Although, as stated, these phenomena have long been known, they were not for a long time applied to the construction of machines either for the generation of electricity or the conversion of heat into mechanical work through the medium of magnetism. Among the first to undertake experiments in this direction was Dr. G. Gore, who in 1868 succeeded in generating a current by the heating and cooling of a magnet placed in inductive proximity to a coil of wire.

Dr. Gore's apparatus[*] consisted of a coil of wire through the hollow of which passed an iron wire 1.03 mm. in diameter. This iron wire was kept magnetized by contact with the ends of a permanent magnet which passed around the outside of the coil, and the arrangement was such that the iron wire forming the core of the helix could be heated by the passage of an electric current from a battery. The helix being connected with a galvanometer the current was alternately made and broken through the iron wire, causing alternate heating and cooling and a corresponding change in its magnetic strength. This variation in magnetism acting on the surrounding helix caused the generation of induced currents in the well-known manner, which currents were made manifest in the galvanometer.

In a subsequent paper, Dr. Oliver Lodge showed the influence of heat upon the action of two spirals, one of magnetic and the other of non-magnetic metal, placed in proximity to a magnetic needle.[†]

These experiments, though crude, nevertheless gave some clue to the methods to be employed; the latter experiment especially indicating a means of obtaining motion through the medium of magnetism and heat.

While engaged in investigations concerning the increase in the coercitive force of steel by changes of temperature, Professors Elihu Thomson and E. J. Houston, in 1878, devised an elementary form of thermo-magnetic motor,[‡] shown in the illustration, Fig. 308.

[*] For a full description, see Dr. Gore's paper, "On the Development of Electric Currents by Magnetism and Heat." Proc. Roy. Soc., Vol. XVII., 1868-1869, p. 261. Also, Phil. Mag., Vol. XXXVIII.. 1869, p. 60.

[†] See Phil. Mag., Vol. XL., 1870, p. 170. "On the Magnetism of Electro-Dynamic Spirals."

[‡] See Journ. Franklin Inst., 1879, p. 39.

As will be seen, a disc or ring of thin steel D is mounted on an axis, so as to be quite free to move. The edges of the wheel are placed opposite the poles N and S of a magnet, and in this position the wheel of course becomes magnetized by induction. If, now, any section of the wheel, as H, be sufficiently heated, the disc will move in the direction shown by the arrow. The cause of the motion is as follows:

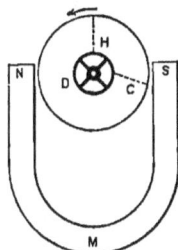

FIG. 308.—EXPERIMENTAL PYROMAGNETIC MOTOR.

The section H, when heated, has its coercitive force increased thereby, and, being less powerfully magnetized by the induction of the pole S than the portion C immediately adjacent to it, the attraction exerted by the pole S on the latter portion is thereby sufficient to cause a movement of the disc in the direction shown by the arrow. If a constant source of heat be placed at H, a slow rotation in the direction shown is maintained.

To insure success, the steel disc must be sufficiently thin to prevent its acquiring a uniform temperature. If the source of heat be at the same time applied at diametrically opposite portions of the disc, as at H and D, adjacent to the poles, the same effect will be produced. The experimenters remark truly that since the amount of heat expended in producing motion of the disc is so enormous when compared with the force developed, it will be readily understood that this motor is of no value as such, but must be regarded as an interesting example of the interconvertibility of force.

A somewhat similar experiment was described by C. K. McGee in 1884,[*] and is illustrated in Fig. 309.

* Science, March 7, 1884, p. 274.

Here a b c represents a ring thirteen centimetres in diameter, and supported horizontally upon radial arms and an axis of some non-magnetic metal. This ring was made of one or more turns of iron wire of about a millimetre diameter, and N S was either a permanent or an electro-magnet. The axis was furnished with a driving pulley, cord and weight, as shown in the figure.

That part of the ring which lay between a and c was heated to bright redness by means of two or three Bunsen burners. The magnet then exerted a preponderating attraction upon the farther or cool side of the ring, and the latter revolved as indicated by the arrow. As fast as the ring entered the space a b c, it became red hot and non-magnetic, and a lack of equilibrium was thus maintained which resulted in a continuous rotation.

The motion was necessarily quite slow, on account of the considerable time required to heat the iron ring. In the actual experiment, considerable difficulty was experienced from the distortion which the ring underwent when softened by the heat, in consequence of which the speed of rotation became very irregular. With a permanent steel magnet a speed of

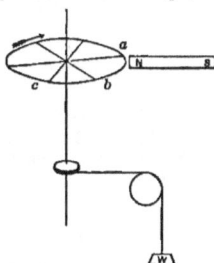

FIG. 309.—McGEE'S MOTOR.

about one revolution in two minutes was obtained; and with a powerful electro-magnet a weight of six grams was raised fifty centimetres in six minutes; and, in a second experiment (the ring having become quite distorted), ninety centimetres in thirty minutes.

Still another experiment recorded, of a similar character, is that of Schwedoff. As shown in the accompanying illustration, Fig. 310, an iron ring is provided with four brass spokes, by which it is pivoted at the centre so as to be free to rotate in its own plane. A steel magnet is brought near one side, and at a distance of about 45° a Bunsen burner is placed. The magnetic permeability of the iron being greatly weakened by the heat from the flame, the effect is the same as if this part of the ring were made of brass or some other non-magnetic material; consequently the magnet tends to pull the ring round in the direction of the arrow, so as to restore the symmetry of the field. But as the hot part rapidly cools when removed from the flame, and the adjacent section is heated, the distribution of temperature, and therefore of magnetic permeability, remains fixed in space, and a continuous rotation is therefore set up.

These experiments, however, attracted but little notice, and it was not until recently that general attention was arrested on the subject,

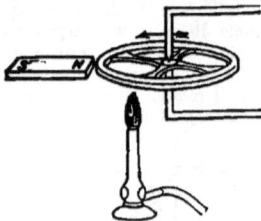

FIG. 310.—SCHWEDOFF'S THERMO-MAGNETIC MOTOR.

by a paper read by Mr. Thomas A. Edison, before the American Association for the Advancement of Science, in August, 1887.* Mr. Edison introduced an adjunct in his "pyromagnetic motor," in the shape of a refrigerator or cooling arrangement, which greatly increases the efficiency of the apparatus, and his machine included a better arrangement for obtaining the maximum effect due to the magnetic action.

In order to explain the action of this interesting motor, let us suppose a permanent magnet having a bundle of small tubes made of thin iron placed between its poles and capable of rotation about an axis perpendicular to the plane of the magnet after the fashion of an armature. Suppose further, that by suitable

* See The Electrical World, August 27, 1887.

means, such as a blast or a draught, hot air can be made to pass through these tubes so as to raise them to redness, and that by a flat screen symmetrically placed across the face of this bundle of tubes, and covering one-half of them, access of the heated air to the tubes beneath it is prevented. Then it follows that if this screen be so adjusted that its ends are equidistant from the two legs of the magnet, the bundle of tubes will not rotate about the axis, since the cooler and magnetic portions of the tube-bundle (i. e., those beneath the screen) will be equidistant from the poles and will be equally attracted on the two sides. But if the screen be turned about the axis of rotation so that one of its ends be nearer one of the poles and the other nearer the other, then rotation of the bundle will ensue, since the portion under the screen, which is cooler, and therefore magnetizable, is continually more strongly attracted than the other and heated portion. This device acts, therefore, as a magnetic thermo-motor, the heat now passing through the tubes in such a way as to produce a dissymmetry in the lines of force in the iron field, the rotation being due to the effort to make these symmetrical. The guard-plate in this case has an action analogous to that of the commutator in an ordinary armature. The first experimental motor constructed on this principle was heated by means of two small Bunsen burners, arranged with an air blast, and it developed about 700 foot-pounds per minute.

A second and larger motor embodying these principles has also been constructed, weighing about 1,500 pounds and calculated to develop about 3 horse-power. This machine is shown in diagrammatic perspective and in section, respectively, in Figs. 311 and 312. As will be seen, the armature placed between the poles of an electro-magnet consists of a bundle of small iron tubes $\frac{1}{100}$ of an inch thick. The armature is fixed to a hollow upright rod R, through which a blast of cool air is brought to the machine. The cool air issuing through the openings at a enters the screen S, and is directed downward through the tubes directly under the screen, as shown by the arrows. At the bottom of the armature it meets a similar screen S', which forces the air back into the tube R through the openings at o. The air then passes down the tube and passes out of the openings d d under the fire, acting as a draught for the same. Continuing on as hot furnace gases, it passes up through the armature tubes, which are not screened, heating them up, and finally passing out at the chimney c.

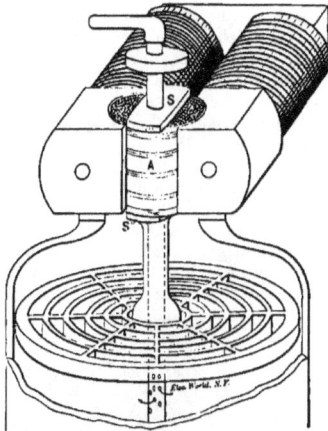

FIG. 311.—EDISON'S PYROMAGNETIC MOTOR.

Now, since the screens S S¹ are placed unsymmetrically, and from what has been said before, it follows that the cool tubes between the screens will be magnetic, while the tubes heated by the furnace gases will be non-magnetic. The result is that an attraction will take place between the magnetic tubes and the pole-pieces of the electro-magnet, causing a rotation. This rotation is maintained continuously by the blast of cool air, which converts the inert magnetic iron into magnetic as soon as it comes under the screen, the cooled portion remaining fixed in space while the armature revolves.

Like all heat-engines, the efficiency of this motor, other things being equal, will depend upon the temperature difference in working, the rate of temperature variation, and upon the proximity to the points of maximum magnetic effect. No advantage will be gained, of course, by raising the temperature of the heated por-

tion of the armature above the point at which its magnetizability is practically zero; nor, on the other hand, would it be advantageous to cool the part between the screens below the point where its magnetic strength is practically a maximum. The points of temperature, therefore, between which for any given magnetic metal it is most desirable to work, can be easily determined by an inspection of the curve showing the relations between heat and magnetism for this particular metal. Thus the points of temperature at which the magnetizability is practically zero, as above stated, are a white heat for cobalt, a bright red for iron, and 400° C. for nickel. On the other hand, while, according to Prof. Rowland,* at ordinary temperatures iron has a maximum intensity of magnetization represented by 1,390, its intensity at

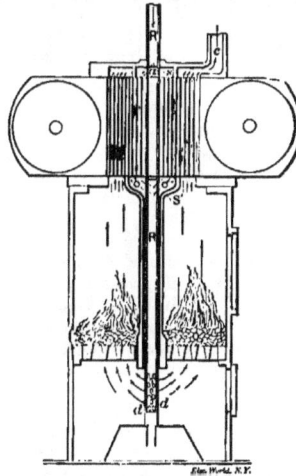

FIG. 312.—EDISON'S PYROMAGNETIC MOTOR.

220° C. is 1,360, and hence no practical advantage is gained by cooling the iron below this temperature. Nickel, however, whose maximum intensity of magnetization at ordinary temperature is 494, has an intensity of only 380 at 220° C. Hence, while this metal requires a lower maximum temperature, it also requires a

* See Phil. Mag., 1873, Vol. XLVI., p. 157, and Nov., 1874.

lower minimum one; but it may be worked with much less heat.

In the paper referred to above, Mr. Edison also described a pyromagnetic dynamo,* which, though based on the same principle, differed somewhat in construction from the motor. We mention it here, however, on account of the close connection between the two, and the ready convertibility of the dynamo into a motor.

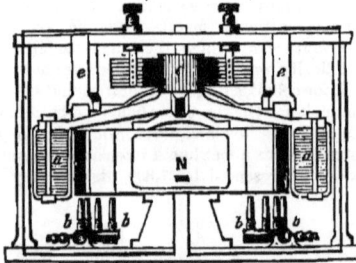

FIG. 313.—MENGES' PYROMAGNETIC GENERATOR-MOTOR.

A dynamo and a motor analogous in principle to those of Mr. Edison have also been designed by M. Menges, of the Hague, Holland. In this motor the inventor has tried to overcome the difficulty encountered in the heating and cooling of large masses of metal and in the time required to effect that purpose. The action of a magnetic pole diminishes so rapidly with the increase of distance that it may suffice to remove the armature to a distance relatively small compared with its own dimensions, or with those of the magnet, in order to reduce the action to a negligible value. But if, as in the accompanying illustrations, Figs. 313 and 314, the magnet N S and the armature a being at a certain distance, we bring between them a piece of iron or nickel d, then the magnetic force upon a is immediately and very considerably increased. In modern terms, the resistance of the magnetic circuit has been reduced by the introduction of a better magnetic conductor, and the number of lines of force passing through a is proportionately increased. The mass of the piece d may, moreover, be relatively small, compared with

* See *The Electrical World*, August 27, 1888.

that of N S and a. If d be again withdrawn the magnetic resistance is increased, and the lines through a are again a minimum.

Now it is evident that we can also obtain the same effect by sufficiently heating or cooling the intermediate piece d, and, again, with a broad field, we can alter the distribution of the lines at will by heating or cooling one side of this piece or the other. For this reason we will call the piece, d, the thermo-magnetic distributor, or, briefly, the distributor.

We will now describe the manner in which this principle has been realized in the practical construction of both a thermo-magnetic generator and a motor. Fig. 313 shows an elevation and part section of one of the arrangements employed. Fig. 314 is a plan of the same machine (in the latter the ring a a appearing on a higher plane than it actually occupies). N S is an electro-magnet, a a the armature, wound as a Gramme ring, and fixed to a frame with four arms, which can turn freely upon a pivot

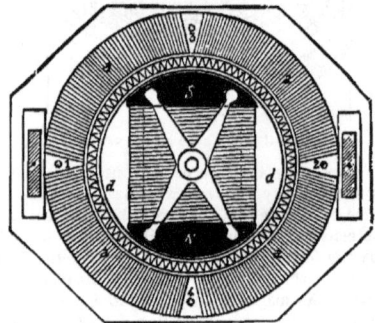

FIG. 314.—MENGES' PYROMAGNETIC GENERATOR-MOTOR.

midway between the poles. The cross arms of the frame are attached at 1, 2, 3, 4, Fig. 314. Between the magnets and the armature is placed the distributor d d, where it occupies an annular space open above and below. Both the magnets and the armature are coated with mica, or some other non-conductor of heat and electricity on the sides facing the distributor. The distributor is attached to and supported by the cross-arms, so that it turns with the armature.

The distributor is composed of a ribbon of iron or nickel bent into a continuous zig-zag. This form has the advantage of presenting, in the cool part of the distributor, an almost direct path for the lines of force between the poles and the armature, thus diminishing the magnetic resistance as far as possible. At the same time the Foucault currents are minimized. To the same end it is useful to slit the ribbon,

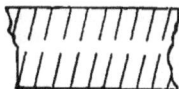

FIG. 315.—MENGES' PYROMAGNETIC GENERATOR-MOTOR.

as in Fig. 315; this also facilitates the folding into zig-zags.

The distributor is heated at two opposite points on a diameter by the burners b b, above which are the chimneys e c. The cooling of the alternate sections is aided by the circulation of cold air, which is effected by means of the draught in the chimneys e e. At the points of lowest temperature a jet of air or water is maintained. The cross-arms are insulated with mica or asbestos at the points where they extend from the armature to the distributor.

It will now be evident that while the distributor is entirely cool, many of the lines of force pass from N S without entering the armature core; but if heat is applied at the points 1 and 2 in the figure, so as to increase the magnetic resistance at these points, then a large portion of the lines will leave the distributor and pass through the armature core. Under these conditions, so long as heat is applied at two points equidistant from N and S, we might, if we so pleased, cause the armature to be rotated by an external source of power, and we should then have an E. M. F. generated in the armature coils; that is to say, the machine would work as an ordinary dynamo, and the power expended in driving the armature would be proportionate to the output.

Suppose next that the points of heating, and with them the alternate points of cooling 90 degrees apart, are shifted round about 45 degrees, so that the two hot regions are no longer

symmetrically situated with respect to each pole of the field. The distribution of the magnetization has therefore become unsymmetrical, and the iron core is no longer in equilibrium in the magnetic field. We have, in fact, the conditions of Schwedoff's experiment upon a larger scale, and if the forces are sufficient to overcome the frictional resistance a rotation of the ring ensues in the endeavor to restore equilibrium. The regions of heating and cooling being fixed in space, this rotation is continuous so long as the difference of temperature is maintained. The ring in rotating carries with it the armature coils, and, of course, an E. M. F. is generated in the same way as if the motive power came from an external source. In this respect the machine, therefore, resembles a motor-generator, and the rotation is entirely automatic. The armature coils are connected with a commutator in the usual way, and the field may, of course, be excited either in shunt or in series. M. Menges states that the residual magnetization is sufficient in his machine to start the rotation by itself. When the machine is to be

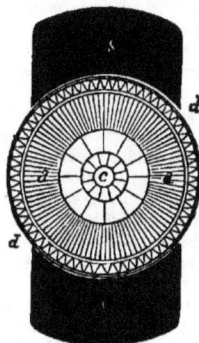

FIG. 316.—MENGES' PYROMAGNETIC GENERATOR-MOTOR.

used as a motor it is evident that the windings on the armature core need only be sufficient to supply current to excite the field, or by the use of permanent magnets they may be dispensed with altogether.

M. Menges has further designed a large number of variations on the original type, varying the arrangement of the several parts, and employing armatures and fields of many different types, such as are already in use for dynamos. In Fig. 316 a machine is represented in which the field is external to the armature.

It is evident that what has been said in reference to the magnetic properties of the iron in the Edison motor holds true also in the machine just described. In the *Journal of the Franklin Institute* for October, 1887, Mr. Carl Hering discusses the theory of pyromagnetic generators of this form. His remarks are, to a large extent, applicable also to motors of this class. A study of the principles involved, says Mr. Hering, will show that there are practical limits to the development of such machines, some of which have been very nearly, if not quite, reached in those of the type described above. The chief one of these is the limit to the speed of cutting of the lines of force, and therefore a limit to the electromotive force, or electrical pressure, which can be generated with a given field-magnet. It is well known that it is the electromotive force or electrical pressure, and not the current, which is primarily generated by any electrical generator, the current being dependent on this electromotive force, the size of the wire and the external resistance. This electromotive force is proportional to two factors: first, the amount of magnetism, or the number of lines of force, which is practically unlimited, as the field-magnets may be made of any size; second, the speed with which these lines are cut, which in these machines is dependent on the speed with which the magnetic qualities of the iron core may be destroyed and restored by the heat. This latter has a comparatively low limit in practice, 120 heatings per minute being, according to Mr. Edison, the fastest rate at which these changes can take place. This means that the useful lines of force of the field-magnet can be cut only 240 times a minute, while in the ordinary dynamo, having a speed, say, of 1,200 revolutions, the lines are cut 2,400 times a minute; from this alone it would follow that the amount of magnetism of a pyromagnetic generator of this kind, neglecting all other factors, would have to be ten times as great as in a dynamo generating the same potential. Such a generator is, therefore, allied

to a dynamo having a very low speed, and must, therefore, be quite large as compared to an ordinary high-speed dynamo for the same output. Other considerations will modify these proportions somewhat, but the machines of this kind must necessarily be quite large and heavy. Fortunately, however, magnetism is cheap, and the large size and heavy weight of a machine is not always a very objectionable quality; it is outweighted many times by the fact that such a machine is stated to require no more attendance than that required for an ordinary furnace for heating houses.

Another feature of these machines, and one which will no doubt present many serious difficulties, is that the iron cores which are to be heated and cooled so rapidly, must necessarily be made of very thin metal, and as it has to be heated to redness to destroy the magnetic qualities, it is evident that rapid oxidation and disintegration of the metal will take place, which will seriously affect the life of those parts of such a machine. This feature, may, however, not be an insurmountable obstacle, and opens a sphere for inventive genius or for discovery.

In order to complete the historical record of this class of machines, we would mention that, shortly after the publication by Mr. Edison of his new form of pyromagnetic generator and motor, Mr. Emile Berliner, of Washington, D. C., announced that he had some time before already applied for a patent on a thermo-magnetic generator embodying similar principles.[*] We would also draw attention to the investigations of Mr. E. G. Acheson in this field.[†]

The limits of this work do not permit of a more detailed treatment of the thermo-magnetic properties of iron and other magnetic metals, an important factor in this class of motors, but the reader will find the subject treated at considerable length in the researches of Rowland—already referred to above—Hopkinson, Ewing, Tomlinson, and others. A valuable and succinct account will also be found in the *Encyclopædia Brittanica*, 9th Ed., under the head of "Magnetism."

[*] For the full patent specification and drawings, see *The Electrical World*, Sept. 10, 1887.

[†] See *The Electrical World*, Feb. 25, 1888.

APPENDIX.

THE DEVELOPMENT OF THE ELECTRIC MOTOR SINCE 1888.

THE past two years have made an immense change in the status of the electric motor. At the time when the last edition of the present volume was published the art of the electrical transmission of energy, and the mechanisms for accomplishing it, were in a state wonderful, to be sure, when viewed from the standpoint of a decade ago, but far behind what they are at present. In particular the electrical street railway, perhaps the most interesting and important development of recent years, was in its infancy. Inventors were struggling with the practical problems involved in that special

so, giving an account of every inventor who has toiled or dabbled in this particular field, would occupy more space than is appropriate. We shall simply give a brief account of the systems in use to-day, referring our readers for details of their performances to the current technical journals.

The first electric road on a considerable scale was that at Richmond, Va., the scene, a little more than two years ago, of Mr. F. J. Sprague's experimental work. It was an unqualified success in spite of the crudeness of the apparatus employed, and since then the number of

FIG. 317.—EARLY SPRAGUE MOTOR TRUCK.

task, but the electric roads, even in an experimental stage, could be counted on the fingers of one hand. To-day, electric traction is so familiar that it may seem almost superfluous to indulge in a description of the apparatus that is in everyday use; especially as in two years more much of it will have only the same historical value that the early experiments detailed in this volume have to-day. We shall, therefore, make no attempt to give a complete description of the advances in electrical traction, or of the apparatus employed, for to do

roads has run up into the hundreds, the number of cars into the thousands. Mr. Sprague's first practical form of motor truck bears little resemblance to that described in Chapter XII. The principles involved are much the same, the details, however, widely different. In the early model used on the Richmond road two 7½ horse-power motors were employed; one end of each motor frame was pivoted on the car axle, the other end supported flexibly by a spring. The magnetic circuit was of the consequent pole type, and the armature was drum wound, with a commutator at each end, the latter device mainly for the purpose of enabling the brushes to be very easily accessible. The armature was wound in 24 sections, three layers

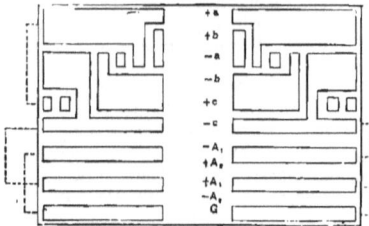

FIG. 318.—SPRAGUE SWITCH BOX.

of wire being in parallel. The armature core was about 10 inches in diameter, and perfectly smooth.

Fig. 317 gives a view of the early Sprague truck, equipped with two such motors. These,

turn current on and off, but to regulate the speed of the motor by changing the effective number of amp re turns around the field mag nets. The motors were series wound, each field coil being a triple one. All the coils in these fields were brought to the switch box, so that the individual coils could be thrown in series or in parallel and combined in different ways. The switch box had seven steps, each representing a combination or a step from one combination to another. The actual number of combinations was five, of which only three represented any considerable change in magnetizing force, the other two being very slight modifications. Fig. 318 gives a view of the somewhat complex cylindrical switch, represented spread out on a surface so that its details can be more easily seen. On the first step of the switch all three coils were in series; on the second step the coil of highest resistance was short-circuited; the third step cut it out ready to be thrown in parallel with one of the other coils; on the fourth step the low resistance coil

FIG. 319.—RECENT SPRAGUE MOTOR TRUCK.

like all other street railway motors since, were wound for relatively high voltages, at first for 400 and afterwards for 500 volts. The two machines were placed in parallel, and controlled from a switch box at either end of the car. These switch boxes served not only to

was in series with the other two combined in multiple arc; the fifth step threw two coils in parallel with each other, short-circuiting the third; the sixth step cut out the latter as a preliminary to the seventh, where all three coils were placed in multiple arc.

The armature speed under ordinary conditions was about 1,200 revolutions per minute, geared down in transmission to the axle very nearly 12 times. This was accomplished by a double set of gears. The armature pinion engaged the larger of a pair of intermediate gears, the smaller of which drove the car axle through an interior gear wheel, split so that it could be readily placed in position. These gears were originally of cast iron or steel, but later exhaustive experiments were carried on with a view of obtaining a better wearing material. The armature pinion has been made of nearly every substance ever used for gear wheels; compressed rawhide and vulcanized fiber being favorites. No material, however, can stand the wear and tear of exposure to dirt and mud for any length of time. After a num-

was but slightly altered. The system of supplying the current to the moving motors has been in every case the overhead bare trolley wire, strung over the center of the track from cross wires carried on poles arranged in pairs on opposite sides of the street. Sometimes, however, a modification has been introduced by carrying the trolley wire from brackets extending from a single pole line. The efficiency of the later motors, reckoning it as the ratio between the power supplied and that applied at the car axle, amounts to very nearly 65 per cent. A very large proportion of the loss occurred in the somewhat cumbersome gearing, for the motors themselves have an efficiency of nearly 90 per cent.

During the period in which the Sprague system was developed, the Thomson-Houston elec-

FIG. 320.—THOMSON-HOUSTON MOTOR TRUCK.

ber of roads had been equipped with this form of Sprague truck, it became very evident that the motors were not powerful enough to do the heavy work required of them, and a second truck carrying two 15 horse-power motors was brought out. This later motor had a single U-shaped magnetic circuit with an armature slightly larger and considerably more powerful than before, and field coils of lower resistance. The efficiency of the machine was thus decidedly improved. The method of hanging the motors on the truck, and the character and combinations of the regulating mechanism were practically the same as before. Fig. 319 gives a view of the later Sprague truck, equipped with its two motors. The gearing remained virtually unchanged, and the armature speed

tric milway appeared, and very soon made a considerable place for itself. It employed the same system of supply, virtually the same method of suspending the motors, and nearly the same gearing. As will be seen from Fig. 320, which shows the standard Thomson-Houston truck, the motors have the same general appearance as the Sprague motors, but differ somewhat in construction. Ordinarily two 15 horse-power motors are supplied to each truck, and are run in parallel at a pressure of about 500 volts. While the Sprague system had depended entirely on the high resistance of the three field windings in series to choke down the initial rush of current while the motor was starting from rest, the Thomson-Houston engineers preferred to employ an external rheostat for this purpose. The motor fields are wound with what are practically double coils, one or both being employed, as occasion demands. On starting, the rheostat, semi-circular

FIG. 321.—THOMSON-HOUSTON RHEOSTAT.

in form, and controlled by a sprocket wheel, generally operated by a handle on the car platform, offers sufficient resistance to check the initial current. Afterwards more or less of this rheostat is cut out, and finally the motor coils alone are in series. The current is then said to be "on the end," that is, both the motor coils are in circuit. A further operation of the switch throws the motors "on the loop," that is, only one of the magnet coils is in use. It will thus be seen that the two systems differ principally in minor points of mechanical construction and in the means employed for regulation. In the method of distributing the current supply the two systems vary somewhat. The Sprague supply is from a line of feed wire extended parallel to the trolley wire and tapped into it at short intervals; the Thomson-Houston distribution is effected by using a rather larger trolley wire and feeding in less frequently. The rheostat (Fig. 321) permits, if necessary, a somewhat nicer regulation than is obtained by simply commutating the field coils, but in practice the rheostat is for the most part cut entirely out and the motors run either "on the loop" or "on the end." The commercial efficiencies of the two are not widely different; the Sprague having perhaps a few per cent. the advantage, gained, however, at the expense of a more complicated system of field winding.

Meanwhile the Short railway motor had undergone development. The considerable difficulties and questionable advantages of the series system of distribution has led to a practical abandonment of the original device, and the system is now operated with two 15 horse-power motors per car, run in parallel. The arrangement of the Short field magnets is—as will be seen by a glance at Fig. 322, which shows the motor complete—quite similar to that of the Brush arc dynamo, and the armature has the same characteristic form. It is a closed coil armature, however, with a commutator having a considerable number of segments.

FIG. 322.—SHORT RAILWAY MOTOR.

The motors are swung from the truck by virtually the same arrangement employed in the systems just mentioned, with the difference, however, that from the form of the magnets it has proved advisable to swing clear of the axle on a massive wooden frame pivoted at one end and supported flexibly at the other. The gearing is of the same character as in the Sprague and Thomson-Houston systems, and the armature speed somewhat lower. The construction of the motor makes the armature singularly easy to get at for repairs, or to take out entirely. Fig. 323 gives a good view of the Short motor truck. No efficiency tests on the Short motors have been published, but their commercial efficiency probably does not differ essentially from the machines just mentioned. The Short motors are governed by a rheostat, without the intervention of commutating the

there are a considerable number of minor improvements in the mechanical details that are worth describing at some little length.

The body or skeleton of the motor consists of only five parts; the cast iron frame, the keeper, the two pole pieces, and the brass casting joining the upper and lower pole pieces; forming a mechanical framework of a very strong and simple character. The cast iron frame carries the car axle, the intermediate axle and the armature in perfect alignment and parallelism, thus enabling the gears to mesh with great exactness. The pole pieces, as can be readily seen, are hinged to the keeper, and both are firmly held in position by the retaining

FIG. 323.—SHORT MOTOR TRUCK.

field coils. The system of distribution employed is practically the same as in the two systems just mentioned, and the electro-motive force of supply is 500 volts.

In the summer of 1890 the Westinghouse Electric Company entered the field of electric traction with a motor that, while electrically very similar to the Sprague and Thomson-Houston machines, possesses some unique mechanical features. The motor proper, as will be seen from Fig. 324, is within a square iron frame that serves both to support it and to furnish bearings for the counter-shafts for gearing. It will be seen at once that there are no radical changes from the existing types, but

bolts through the brass casting that joins them at their extremities. When, therefore, it becomes necessary from any cause to remove the field coil, the retaining bolts are withdrawn and the entire pole piece swung back, when the coil can be at once slipped off and replaced. In the same way the armature can be removed by opening the box and lifting it out after the pole piece is swung out of the way. In fact, one man with the aid of a differential pulley and support which can be attached to the motor frame, will have no difficulty in accomplishing these operations. When the lower pole piece is swung back, the armature can be passed through into the pit if it is desirable to take it out in that direction. This hinged construction renders it very easy to get at any portion of the motor for repairs.

The gears are encased in cast iron boxes, oil tight, and partially filled with grease. They are thus entirely free from the access of dust and grit, and can be continually and thoroughly lubricated. This arrangement of the gear wheels is the secret of the fact that the motors

built up of plates, each of which is cut with a key way, so that the entire inner structure of the armature can be locked firmly upon the shaft. The armature is wound with due reference to the fact that street car motors have occasionally to support heavy discharges of current, beyond their carrying capacity. The double wires of the armature are equivalent in conductivity to No. 7 wire, so that there is

FIG. 324.—WESTINGHOUSE STREET RAILWAY MOTOR.

make an unusually small amount of noise. At the same time the life of the gears is improved by the thorough lubrication, and their wide faces, five inches, give them additional durability. The entire motor is protected beneath by a sheet iron pan and on the sides by waterproof sail-cloth.

As to the details of construction, the armature is of the usual drum type; the core is

little danger of undue heating under the severest strain of service. The wires from the armature are brought out straight to the commutator without crossing or overlapping, and the surface of contact between the wire and commutator is three or four times larger than that used in the earlier street car motors, thus avoiding what was, at one time, not an uncommon fault, the destruction of the connections at that point. Particular care is taken in binding the armature wires to the core, so that there is no danger of any play, even under the severe torque brought upon the windings. Between the bearings and winding at the head of the armature is placed a little brass ring so arranged as to throw out any oil which creeps along the shaft, so that it cannot affect the commutator.

The armature shaft is tapered, having the greatest diameter where the core is keyed upon it, and being tapered at the ends to receive the armature pinion. This is cut with a corresponding conical hole, an arrangement that enables it to be very readily removed and replaced. The commutator has been constructed with the severe conditions of ordinary service in mind, and experience has shown that it is thoroughly insulated, unlikely to get out of line from any cause, and quite free from any injurious amount of heating.

In the insulation the greatest pains is taken to render the coils as nearly water-proof and fire-proof as possible, and where the wires are bent around under the coils, double insulation is used, so as to avert any probability of short-circuiting. The wires are not brought out direct from the coils to the terminals, but through special castings, averting breakage of the wires from continual jarring at the point where they leave the coils.

As to the running gear, comparatively little need be said. The wheels are all of five-inch face, and are dressed down and tested in position before leaving the shops, so that the meshing between the teeth is as precise as it can well be made. The armature pinions are made of forged steel, and it is believed that when running in oil they will show a life far above the ordinary. The axle gears are made of selected iron, with solid webs. Both pinions and axle gears are tested with circular jigs, so that the tapered hole brings each wheel to exactly the proper position when the gearing is set up. Circular jigs were sent to the principal truck builders, with the request that car axles be made to conform with it, so that every portion of the driving gear will fit accurately when it is first placed in position.

The distribution is by overhead trolley wires

and poles, as usual. It is evident from what has been said that the Sprague, Thomson-Houston, and Westinghouse systems possess very many features in common. The motors are of about the same size and weight, have, roughly speaking, similar forms of magnetic circuits, and the same sized drum armatures. The design of the first has been from the standpoint of the electrician; that of the last from the standpoint of the mechanician; while the Thomson-Houston motor may be said to be midway between the two. The Sprague motor has been made even more similar to the others by the recent introduction of a small resistance coil, a rheostat, external to the motor and thrown in at the moment of starting, to be immediately cut out as the switch is turned to the other steps. The Sprague motor, then, is governed by commutating the field coils with slight assistance from a small rheostat. The Thomson-Houston motor is governed by a rheostat of many sections, with an additional modification introduced by cutting out part of the field winding. The Westinghouse employs a rheostat of only four sections, but otherwise follows the Thomson-Houston practice of a double field winding. All the three, and with them the Short motor, employ two 15 horse-power motors per car, geared down twice, the gear ratio being in each 10 or 12 to one. It has been found by experiment that two motors thus placed on a car never pull together in perfect harmony, for one is apt to take more than its share of the load. The result of this and the consequent uneven wear on the gears produces a serious loss of efficiency. The double gear itself wastes at least 20 per cent, of the power employed, and where two motors are used even more. The first real improvement in the matter of gearing efficiency was made by the introduction of gears running in oil-tight cases, and therefore free from dirt and thoroughly lubricated.

In connection with the unequal load, almost certain to be found in the case of using two motors to a car, a particular point of weakness was found in the connections employed in the Sprague system. At a certain stage of the field commutation an unstable state was reached, and one motor was almost sure to take more than its share of current. Consequently, in the latest motor equipments, manufactured since the Sprague Company was merged with the Edison Company, equalizing coils are employed, consisting of series windings so arranged that an extra rush of current through a single motor will pass around the fields of the other motor, and tend to restore equilibrium.

The experience of a couple of years has rendered it evident that an armature speed so high as to require a double gear reduction causes serious loss of efficiency, as has been

The Rae electric railway system presents some radical differences from any of the others heretofore mentioned. In the first place, only a single motor is used. ·It is rigidly attached to the truck, and the armature spindle is parallel to the length of the car. The power is transmitted to both axles from the same motor through beveled gearing. Fig. 325 gives an excellent idea of the principal characteristics of the system. As will be seen at once, the motor—of the consequent pole type—is placed crosswise of the car midway between the

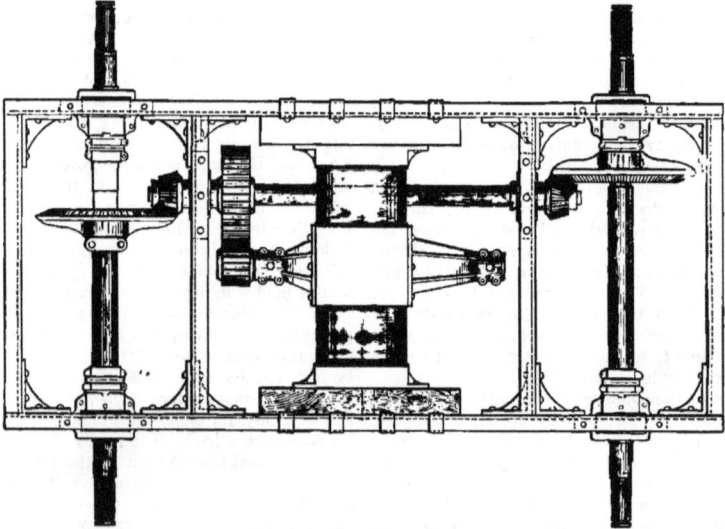

Fig. 325.—Rae Motor Truck.

mentioned before, and of late the efforts of all the railway companies have been bent towards securing lower speed machines. In addition, there has been a growing tendency towards the use of a single motor, which may, perhaps, be best illustrated by the Fisher-Rae system, before going on to describe the evolution of the slow speed motor, and the consequent abolition of one or both sets of the gearing that had been so productive of waste of power and frequent repairs.

wheels, and fastened rigidly to the frame-work of the truck. The armature pinion drives an intermediate gear that through bevel wheels turns the axles. The motor is of 30 horse-power, with a Siemens armature; it is thoroughly insulated at the sides by oak bars saturated with asphalt, and the employment of rawhide or fibre armature pinions still further frees the machine from danger of a ground. The whole truck is put together as rigidly as possible, no attempt whatever being made to secure the usual flexibility, which in fact would interfere with the action of the system. The motor is series wound, strongly made, and of creditable efficiency. The regulation of speed is through the interposition of a rheostat consisting of four coils that are successively thrown in parallel arc with each other, and finally short-circuited. The rheostat with its switch is placed under the car, as in the Thomson-Houston, Short, and Westinghouse systems, and is operated from the car platform by a simple handle. The use of the single motor is an advantage, as it somewhat reduces the weight of driving gear, simplifies the connections, and is cheaper. The use of beveled gears is of questionable value, although the system is in use on a number of roads, and is reported to work very well. The mechanical details of the apparatus are thoroughly worked out. The beveled gears run in oil, and all the bearings are bushed with graphite. These good mechanical features probably go far to offset some of the apparent disadvantages. The method of distribution and the minor details of the system present no striking peculiarities, although the electro-motive force is intended to rise as high as 550 volts, rather more than is employed in general electric railway practice. The use of a single large motor, as employed by Mr. Rae, enables the armature speed to be somewhat reduced, and, as a matter of fact, it does not rise above 900 revolutions per minute at the full speed of the car.

But, in spite of the advantage of the single motor, the Rae system, like all the others previously mentioned, employs a double set of gearing, and consequently loses nearly 25 per cent. of the power applied. To secure a high commercial efficiency for any electric railway system it is necessary to diminish the amount of gearing employed in transmitting the power from the armature spindle to the axle, and after about two years' experience with the modern electric railroad the attention of a number of inventors was simultaneously drawn towards the problem of producing a low speed motor, the armature spindle of which should carry a pinion to engage directly the gear wheel upon the car axle. No practical machine of this sort appeared until the summer of

1890, when the Wenstrom Company of Baltimore, Md., brought out a very efficient and well designed motor that possessed this valuable property of slow speed. In addition several ingenious new devices were embodied in the car equipment.

The Wenstrom system is in some respects a very radical departure from previous street railway practice, and, as such, is worth more than a passing notice. It has not yet been put into operation on a road of its own, although the preliminary experiments have met with very gratifying success.

To begin at the beginning, it is worth while saying something of the Wenstrom generator. Although the machine is by no means new, yet there have been recent and important improvements. The Wenstrom machine has for its fundamental point in design a short magnetic circuit as nearly closed as the mechanical exigencies of the case will allow. The type of field magnet is that which has come to be known as the "iron clad," where the magnetic circuit completely shuts in the armature and field coils, so that, as one looks at the machine from the outside, nothing is visible except smooth, easily curved surfaces of iron. The merits of the type are two: Absence of external magnetism, and economical and convenient arrangement of the magnetizing coils.

Perhaps the most extraordinary feature of the Wenstrom machine is the armature, which does not present the familiar appearance of a core wound with wire, but rather seems to be a smooth cylindrical mass of iron turned and polished on the exterior and barely out of contact with the pole-pieces. The winding consists of wires passed through the apertures cut lengthwise through the core just under the surface. These are actually stamped in the iron discs of which the core is built up. We may then imagine the Wenstrom armature as a modified drum armature furnished with Pacinotti projections, closed over the top, however, by a thin casing of iron. As a matter of fact, the apertures which receive the windings come within so short a distance of the external surface of the armature that only a very small amount of magnetism is lost by leaking around the armature on the outside of the wires, while the complete freedom from external winding allows this type of armature to be run with a gap in the magnetic circuit enormously less than can be found in any other construction. As a natural consequence the magnetism is cheaply obtained, and the armature wires are subjected to a very powerful induction. The street car motor shown in diagram (Fig. 326) has all the characteristic features of the generator, although the frame does not retain the cylindrical form, but is flattened into a somewhat more compact shape. It is, like the generator, a four-pole machine, the magnetic circuit being cast of mitis metal in one piece, with two consequent and two salient poles.

The standard street car motor is rated at 25 horse-power and weighs, complete, very nearly one ton. Owing to the powerful magnetic field practicable with the Wenstrom construction, and to the fact of the new motor being a four-pole machine, its speed is only 400 revolutions per minute. The armature can consequently be geared directly to the car axle without the intermediate countershaft that has been the subject of frequent objurgations from every electric railway man who has been in the business long enough to have gearing give out. A small convenience which should be mentioned here is that the brush holder is fitted to the outside of the bracket that supports the

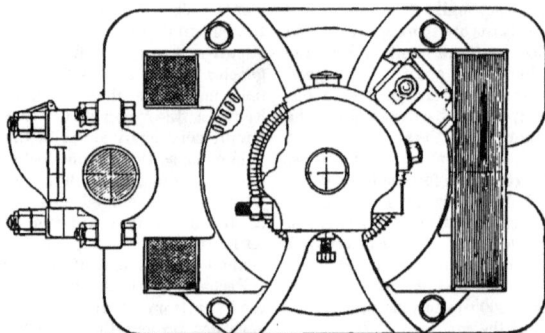

FIG. 326.—WENSTROM RAILWAY MOTOR.

The frames that sustain the armature are bolted to the sides of the field magnets and are readily removable, permitting, therefore, the removal of the armature and consequently of the field coils, which, when the armature is taken out, can be slipped off the pole-pieces with great readiness. The armature is of the same type as that of the generator and runs within one-sixteenth of an inch of the pole-pieces. The commutator is provided with two brushes only, 90 degrees apart, the armature being cross-connected so that this arrangement is possible. For convenience in taking off the armature the pinion seat is tapered, so that the armature pinion can be very readily slipped off.

armature on the commutator end, so that it can be adjusted or taken off without loosening the armature frame at all. The absence of intermediate gear in the Wenstrom motor is a very considerable advantage, as it has already been mentioned that losses in gearing may rise to a serious amount in the motors now in use. The standard construction adopted for the Wenstrom street car system is a single 25 horse-power motor geared to one axle, the split gear being preferably of the wooden tooth construction recently introduced. Perhaps, however, the most ingenious modification of present street car practice is to be found in the hydraulic gear which forms the connection between the split gear and the driven axle. This has been but recently worked out, and its purpose is to furnish a variable clutch between the driving and the driven axle, so that in starting the motor it may be allowed to run

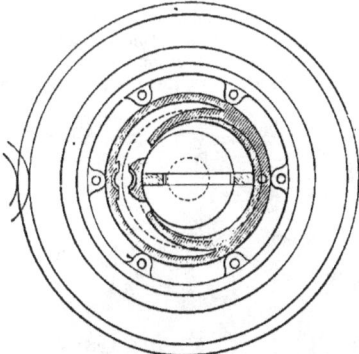

FIG. 327.—SECTION OF HYDRAULIC CLUTCH.

free and its power applied gradually to start the car, and in addition to provide a sort of mechanical safety-valve, so that when there is a severe overload the hydraulic clutch will slip and allow the armature to rotate fast enough to save it from the excess of current, instead of subjecting it to the dangerous overloading which would otherwise follow.

Fig. 327 shows a section of this hydraulic gear. It consists of a cylindrical cavity placed eccentrically to the driving shaft and turned up true and smooth within. It is fitted with a tight cover firmly bolted on. Through a disc fast to the axle slides a brass key, forming a partition across the eccentric box free to slide as the shaft turns, and forcing the oil with which the eccentric box is filled through a port in the rim connecting one side of the cavity with the other. The arrangement thus forms a rotary circulating pump of which this brass slide is the piston and the eccentric box the pump cylinder. So long as the port between the two sides of the piston is open the eccentric box can revolve freely, forcing the oil around through the port as it turns. If, however, the port is closed, the oil can no longer flow and

forms an incompressible mass through which the power is transmitted to the axle. The arrangement is equivalent, as before remarked, to a rotary pump, the piston of which is free to move within the cylinder so long as the valves are open. When they are closed, piston and cylinder must move together, if at all. The result of this arrangement is that, the port being open, the split gear to which the apparatus is applied rolls freely around on the axle without communicating any motion to it. If, however, the valve be closed, the gear and axle act as a rigid body and the motor can exert its full force. Suppose the car to be at rest and the valve and hydraulic gear open, the motor on starting will revolve freely. Now gradually close the valve; as the passage for the oil is more and more contracted a greater and greater pressure will be exerted tending to turn the axle, and, although during this period there will be some slip, finally, as the valve closes, the pressure is sufficient to start the car, and when the by-passage is completely closed, gear and axle act as a rigid body, and there is no appreciable amount of slip. It will be seen that this hydraulic gear enables a flexible connection to be made between the axle and the gear, so that the latter will run freely or with

FIG. 328.—HYDRAULIC CLUTCH IN POSITION.

varying amounts of resistance and slip up to the point where, when the valve is completely closed, the two are to all intents and purposes rigidly connected. It is worth while thus to go into details concerning this unique contrivance, for the reason that, while it is simple, its exact action might not be at first sigh' obvious. Fig. 328 shows the hydraulic gear as attached to the corresponding wheel. It is not a bulky contrivance, and is under perfect control from the platform of the car, from which point the valve can be regulated. It allows the armature to run, if desired, at a constant speed, whether the car is going at its full rate or standing still. Perhaps, however, its greatest convenience is

fitted to any of the eight-wheeled forms of truck, even above the car floor, if desirable.

The next slow speed motor to appear was the Baxter, still another development of the multipolar type of machine.

The armature in the Baxter street railway motor is 18 inches in diameter, and at each end of the armature shaft are 4½-inch pinions which gear in the 18-inch wheels on the car axle; the speed of the armature, therefore, is four times

FIG. 329.—BAXTER MOTOR TRUCK.

in the matter of applying the power gradually at the start, so as to avoid straining the armature or producing an unpleasant jar. In applying the Wenstrom motor and the hydraulic driving gear to a car equipment, any of the usual arrangements can be employed. One motor can be geared to one or both axles; two motors can, in the rare cases when it may be necessary, be employed, or the motors can be

as great as that of the wheel, so that, when the car is running at about eight miles per hour, the armature will turn at something like 335 revolutions per minute. With the cars ordinarily in use the armature speed would be something like 1,000 per minute for the same rate. The motor of course is multipolar, having eight poles formed by four separate magnets, each having its own magnetizing coil, placed so that the poles alternate around the armature. The core of the armature and the hub of the motor are insulated with particular care to avoid the short circuits which sometimes occur through grounding on the frame of the motor. The motor is series wound and has the various segments of the armature coils connected in series by a special method devised by Mr. Baxter. The conductors lie in grooves on the exterior of the armature, so that the Pacinotti construction is really the one followed. The brush holders are composed of two sliding castings, so arranged that the upper can be pulled out of place in an instant with the

noise that is produced is much muffled and the car runs very quietly and easily. All bearings are dust-tight, and are self-oiling. The motors are suspended something after the usual method, but the suspension rod is some six or seven inches long and prevents the motor from slipping up and down, although it does not restrain lateral motion; therefore, if the car or truck should sway even an inch or two, no strain will be brought upon the motors or gearing. In addition to this the end of the motor next to the suspension moves with the car, and there is, therefore, little vibration at this point where the circuit wires are connected from

FIG. 330.—THOMSON-HOUSTON SLOW SPEED MOTOR.

fingers, and the brush removed and replaced again just as quickly. The solidity of construction in the brush holders prevents vibrations of the brushes, and aids them to run without any visible sparking. Both brushes are on the upper side of the commutator in full view. The illustration on the preceding page (Fig. 329) gives a good idea of the construction of the motor and its method of application to the truck.

The gears are inclosed in an oil-tight chamber, so that they are perfectly lubricated and protected from dust and grit which would tend to wear them loose. In this way the little

the motor to the switches, so that there is very little liability of bad connections. In addition a dust-tight casing is presented to cover over the entire machine.

The governing is by rheostat, and the car is operated by a single crank shaft. Turning it in one direction the car moves forward, in the other direction reverses, and the speed of the car is controlled by the distance through which the crank is turned. In addition to securing slow speed by the multipolar construction a high weight efficiency is also obtained, so that the two motors, each of 20 nominal horsepower, weigh together but 3,000 pounds. This is considerably lighter than the usual street car motors, which weigh from 2,000 to 2,500 pounds apiece. The rheostat and regulating apparatus is placed between the motors under the car floor, and therefore can be easily reached by opening the trap doors. The lightning arresters are placed in the same situation and are in series with the rheostat, which serves as a choking coil to afford additional protection.

As in the case of the Wenstrom, the Baxter motor has at the date of writing no extensive commercial use, but it is of interest as a meritorious endeavor to improve street railway practice, and as a forerunner of the slow running motor type. For once the larger electric

round the armature core. The magnetic circuit is completed on the front end of the motor through the face plate and at the back through the frame, on which are cast the axle boxes and arms that serve as a support for the armature shaft bearings. The armature is of the Gramme ring type, and the bobbins are wound close together around the entire rim. One great advantage of this construction is the fact that any coil can be easily rewound without disturbing its fellows, while with the drum armature in the type of motor formerly used by the company the winding all had to be removed down to the injured coil.

The brushes are placed exactly opposite and

Fig. 331.—Details of Thomson-Houston Slow Speed Motor.

companies were behindhand, but during the first months of 1891 the Thomson-Houston and Westinghouse companies brought out motors for street railway service having the same valuable property of slow speed. The first of these is notable as being a two-pole machine. It has been the subject of much experiment on the part of the Thomson-Houston company and has been named, from its construction, the "single reduction gear" motor, but is ordinarily called the "S. R. G." As will be seen by Fig. 330, the motor is very nearly iron clad, having two pole pieces of ample surface and carrying two field coils, which partially sur-

in a horizontal fixed position. There seems to be no sparking under the ordinary running conditions, and the brushes are easy of access.

The field spools are protected on all sides by the fields and frame. The gears are entirely inclosed in a dust and oil-tight case, which is provided with a hand-hole closed by a spring cover, permitting ready examination of gears and the introduction of lubricants.

A sheet-iron pan extending above the center of the armature shaft entirely incloses the bottom and sides of the motor and protects the armature and commutator from dust, snow and water. The advantage of this casing was strikingly demonstrated during several snow storms, the "S. R. G." coming through unharmed, while some of the fields on the old style of motor came to grief from the effects of excessive moisture. The pan has a sliding bottom, and is attached to the motor in such a manner as to permit of being readily removed for access to the various parts.

The motor when mounted on a truck with thirty-inch wheels is designed to clear the tops of the rails four inches. The spur gear on the armature shaft is of steel, four and one-half inches face, and has fourteen teeth. The split gear on the car axle is of cast iron, with the

displacement of coils, breaking of commutator connections, all insure a minimum amount of expenditure for repairs.

Fig. 331 shows the detailed construction of the working parts. A glance shows the great compactness that has been attained. The field castings fit snugly around the armature, and the flat magnetizing coils slipped over the poles embrace the ends of the armature. This is perhaps the only objectionable feature of the machine, for it necessitates bunching the wires that lead from the armature to the commutator in a way that demands the greatest care in insulation. The use of mitis iron for the fields

FIG. 332.—WESTINGHOUSE SLOW SPEED MOTOR.

same width of face, and has 67 teeth. The speed of the armature shaft relative to that of the car axle is nearly as 4.8 to 1 ; when the car is running ten miles per hour the armature makes 538 revolutions per minute, or the speed of the armature is 53.8 turns per minute when the car speed is one mile per hour. The gears are surrounded by an iron box, so that they may be run in oil.

The facility with which the armature can be removed simply by lifting the upper field, the ease with which an armature bobbin can be rewound, small liability to damage from centrifugal action, such as bursting of binding wires,

is a very important element in making possible the excellent efficiency that the motor has shown under test. It was a somewhat daring experiment to design a two-pole motor for so low a speed, but thanks to a good magnetic circuit and a powerful armature, the result has been satisfactory. The regulation of the "S. R. G." motor is through a rheostat like that of the old standard type of motor, so that the general car equipment presents no especially novel features. In fact, the number of useful innovations that can now be made in the governing of a series wound motor is comparatively limited.

In evolving a slow speed machine the West-inghouse Company adopted the four-pole con-struction and very ingeniously designed it to follow the general mechanical principles that had proved advantageous in the earlier standard motor of the same make. After considerable experimenting the multipolar machine was put upon the market in the shape shown in Fig. 332.

It will be seen that its general form is cylin-drical, giving both the shortest possible mag-netic circuit and very great strength with makes it possible to utilize four poles with great advantage, and they are, as will be seen at a glance, rather narrow, and consequently are capable of being magnetized by compar-atively short and small windings. One of these coils, together with a brush-holder, is shown in Fig. 334. The brush-holder is a solid-looking casting bolted on to the lower side of the main frame of the motor, and lift-ing its brushes quite up to the top of the com-mutator, where they rest 90 degrees apart. The field coils are of coarse wire, and, by rea-

FIG. 333.—FIELD MAGNETS OF WESTINGHOUSE SLOW SPEED MOTOR.

minimum amount of material. Besides this, all the sharp corners that tend to leak magnetism are eliminated, and the machine is rendered thereby slightly more efficient. The details of the magnetic circuit are best shown by examin-ing Fig. 333, which shows the casting freed from armature and coils and opened up to ex-hibit its arrangement. The motor has the same square form of frame that is already familiar in the older Westinghouse motor. But the change in the shape of the magnetic circuit son of their small length and low resistance, give the necessary magnetization without a serious loss of energy. The castings are of a specially soft grade of iron that has proved to have excellent magnetic properties.

Fig. 332 gives an excellent idea of the gen-eral arrangement of the motor, showing the gear casing—exhibited by itself in Fig. 335—the arrangement of the fields and the disposi-tion of the frame, supported as usual, on the axle at one end and flexibly at the other. The gearing, inclosed as it is in an oil-tight case, is always thoroughly lubricated and free from dirt. All the bearings are bushed with metal, and the armature shaft is slightly tapered to facilitate the removal of the pinion. The gear ratio is 3.3 to 1. The iron clad form of the motor enables it to be completely shut in by applying side plates, so that in actual practice it is inclosed so tightly as to be quite free from the numerous difficulties so often experienced from dirt and moisture finding their way into the working parts of a machine. As the lower surface of the motor presents only a solid cast-

cent., the abolition of the intermediate gear ought certainly to be good for more than 10 per cent. increase in efficiency. The normal speed of the armature at a car speed of about 10 miles per hour is 380 revolutions per minute. Thus it will be seen that the machine in question is really a very slow speed motor, the result of good magnetic circuits and the four-pole construction. The commutator is designed with special reference to obviating the heating that is sometimes so disastrous in street car motors. Each segment has a bearing along its entire lower edge, so that even if there should be any

FIG. 334.

BRUSH HOLDER OF WESTINGHOUSE MOTOR.   MAGNETIZING COIL OF WESTINGHOUSE MOTOR.

ing, it cannot be injured by casual blows from projecting rubble, a source of difficulty with which electric street railway men are only too familiar. The cut also gives a perspective view of the motor, showing its arrangement in the frame and connection to the gears. The armature is of the drum type, and the core is built up of grooved iron plates, so that the windings are inside slots upon its surface, being completely imbedded in insulating material. The surface of the finished armature is therefore entirely smooth and the clearance space very small. Even should the bearings become worn so that the armature would brush against the pole pieces, no serious damage would be done because no wire is exposed.

The electrical efficiency of the motor is said to rise to 95 per cent., and the commercial efficiency to 75 or 76 per cent. This is an excellent showing, and displays the slow-speed motor with a single gear reduction in a very favorable light. It is about the figure that would be expected from a machine of this construction. Inasmuch as the efficiency of the two-pole motors of various forms with the complicated gear is generally held to be a little over 60 per

slipping the symmetry of the commutator would not be destroyed. The winding of the armature enables the two brushes, as before mentioned, to be placed 90 degrees apart, and both upon the top of the commutator, where they can be readily inspected or replaced if necessary.

Not content with the reduction in speed

FIG. 335.—GEAR CASING OF WESTINGHOUSE SLOW SPEED MOTOR.

gained in this promising motor, the Westinghouse Company set about the task of devising a motor the armature of which should be directly upon the car axle, thus doing away with all gearing whatsoever. The abolition of the remaining gear would mean a still further increase of efficiency, and besides, would lessen the moving parts that require repair. In less than sixty days after the single reduction gear Westinghouse motor appeared the gearless motor was in experimental operation.

The general appearance of the Westinghouse gearless motor is roughly shown in Fig. 336. It is a four-pole machine, completely iron-clad, and with the same hinged arrangement of fields that has proved so convenient in the other types of Westinghouse motor. The armature is built directly on the car axle without any attempt at flexible connection; it is of

ance between the bottom of the motor and the tread of the 30-inch wheel. The motor has not, at the time of writing, been used in anything but an experimental way, and the details of its winding have been carefully kept secret. A high efficiency is claimed for it, however, as great as 90 per cent., and it is said that after a two hours' run at a load of over 20 horse-power the rise in the temperature of the armature and field coils was only 30 degrees centigrade above the surrounding air, showing at least a tolerably efficient electrical design.

Almost simultaneously with the appearance of the Westinghouse gearless motor came the Short machine of the same type. In this motor the same style of armature is employed as

FIG. 336.—WESTINGHOUSE GEARLESS RAILWAY MOTOR.

the drum type, 16 inches in diameter, and instead of having a smooth surface is grooved to receive the wires, thus holding them rigidly in place and, of course, lessening the magnetic resistance of the air space. The brush holder is rigidly fastened to the magnet frame, and is easily accessible through the openings shown on the top of the casting, which are closed when the car is running by water-tight lids. The weight of the magnet frame is counterbalanced and cushioned on heavy spiral springs resting on the cross-bars of the truck ; these prevent the field from rotating, and give the motor the necessary flexibility needed for easy starting. The total depth of the field magnets over all is but 20 inches, giving 5 inches clear-

in the ordinary Short motor—that is, a flat Gramme ring of many sections, with a magnetic circuit arranged like that of the Brush dynamo. The motor and its connections are admirably shown in Fig. 337. The armature itself is not mounted, as in the Westinghouse motor, directly upon the axle but on a hollow shaft concentric with it, with plenty of inside clearance. The armature proper consists of a laminated iron core of the usual Short type wound in a large number of independent segments. The style of construction obviously allows excellent ventilation and very free rewinding. The commutator is mounted on the same hollow shaft as the armature and close to it. The motor is really a four-pole machine. The clearance allowed is very small and the magnetic field most intense. The field coils are bolted to a circular frame at each side of the motor, in the center of which are the bearings that carry the hollow armature shaft. The spring connections for easy starting are shown in the cut. A double arm running out from the frame-work to the cross-girders of the truck makes provision for supporting the entire motor. The insulation between these brackets and the girders is by means of heavy rubber bushings through which the bolts pass. By removing the bolts attaching the fields to the supporting frame-work the coils may be readily taken out for repairs, or for access to the armature.

Fig. 338 gives a plan of the truck equip-

is difficult to design for small electrical losses, but there is every reason to think that the working efficiency, all things considered, is decidedly higher than most of the machines at present in use, to say nothing of the gain in wear and tear from the absence of gearing. One of the very convenient features of the Short motor is the ease with which it can be repaired, for by loosening the four bolts supporting the motor on the truck and taking off the iron strips below the wheel-box, one end of the car may be jacked up and the axle-wheel and motor run out from under the car where they may be easily reached. Armature repairs may be made by removing two field coils; the armature coil can then be rewound as it stands. The field coils can be as easily re-

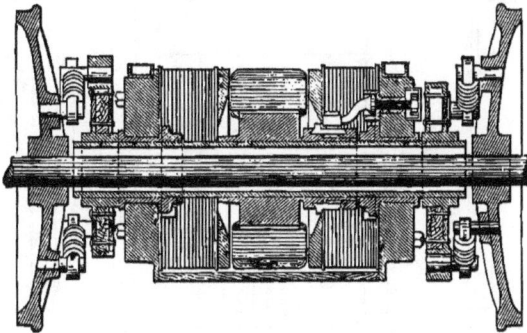

FIG. 337.—THE SHORT GEARLESS MOTOR.

ment, showing a single motor. From the center of the axle to the bottom of the casing is 12¼ inches; a 36-inch wheel is generally employed, giving a clearance of 5¼ inches over the track. At a speed of ten miles per hour the armature drives a 36-inch car-wheel ninety-four revolutions per minute; the equivalent speed of a single reduction motor would be about four hundred, showing clearly enough the advantage of the gearless form. The efficiency of the motor is not stated, but it is evident enough from what has already been said that the gain by the abolition of gearing is sufficient to compensate for no small loss of electrical efficiency. So slow running a motor

paired, while the commutator may be reached and sandpapered while the machine is running, as in the ordinary forms of geared motor. This Short motor is of special interest as being by a few days' priority the first of the direct connected motors to appear. It is certainly a very ingenious and interesting machine, and may be expected to give a good account of itself in actual service, although up to the present it has been used only in an experimental way, and almost nothing can be told of what will be its performance in commercial service.

Several other gearless motors are known to be under way, some of them possessing very remarkable characteristics. One of these is worth especial attention, although it does not properly belong in the same category as the two just mentioned. This is the Eickemeyer gearless motor, really earlier than either the Short or Westinghouse machines. In its first form, however, an attempt was made to dodge the difficulties inherent in low speed by using a very small driving wheel, hence allowing a higher armature speed for a given number of miles per hour. In machines now under construction the driving wheels are of about the ordinary size. The peculiarity of the Eickemeyer construction is the use of a motor not connected to the axle, but operating through

singularly compact. It is mentioned somewhat by itself as a machine of radically different construction from the other gearless motors described, and as having in its early form possessed the extraordinary characteristic of exceedingly small drivers. In its present type, however, many improvements have been introduced and the machine ranks as one of the promising solutions of the slow speed problem that confronts the designer.

None of the gearless motors have been as yet put to any extensive trial, hence any account of them must necessarily be imperfect and unsatisfactory. In the course of the coming year

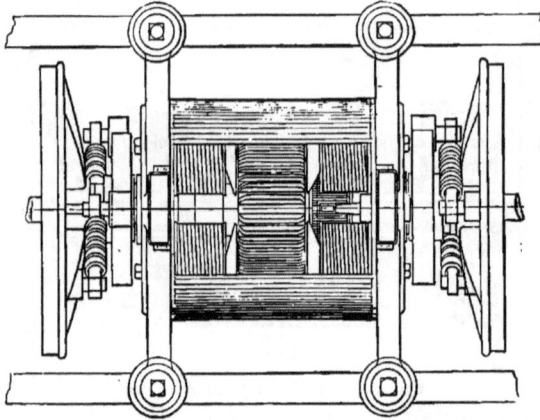

Fig. 338.—Plan of Short Gearless Motor in Position on the Truck.

the medium of a connecting rod. A heavy disc on the armature spindle is attached to the car wheel much as the drivers of a locomotive are connected. The possible advantage of this form of construction is freedom from injury to the armature by jarring of the axle. As neither this nor the previous form of machine have been in anything but experimental use, how much of real value the Eickemeyer construction has cannot yet be told. The machine is a thoroughly well designed and efficient one, like all that have thus far been elaborated by Mr. Eickemeyer. The motor is iron-clad and

they will be severely tested, and undesirable features will be gradually eliminated. The advantages possessed by them in common are extreme simplicity and a high degree of mechanical efficiency. The difficulties that may be met are breaking down of the armature from vibration, and a very severe strain upon it when starting the motor, or mounting heavy grades.

This account of modern types of motors for electrical traction would be notably incomplete without mention of the remarkable City and South London Railway, inaugurated in November, 1890, both as the first deliberate attempt to handle a large suburban traffic exclusively by electric locomotives, and as the first thoroughly successful experiment at placing the motor armature directly on the car axle.

As is well known, the line is an underground one, and is three miles in length, running from the heart of the city of London, near the "Monument," beneath the Thames, to within a mile of Clapham Common. Including the two termini of the road, there are along the line six stations, placed at about equi-distant points. The track is double, and is laid in two separate steel tubes, each 11 feet in diameter. The sleepers are laid directly, without ballast, upon the bottom of the tube, and the light weight of the rolling stock has made it possible not only to dispense with track chairs altogether, but to use rails of little more than half the usual weight.

The cost of the line in round numbers has been a little more than $1,100,000 per mile. Messrs. Mather & Platt, the constructors, have guaranteed that the cost of motive power for the first two years shall not exceed 7 cents per train mile. As each train is made up of three cars and a locomotive, and is able to carry 100 passengers, this compares very favorably with the cost of working the underground railway, which expends 20 cents per train mile, on trains capable of carrying about 450 passengers, while its maintenance expenses are said to be something formidable. Messrs. Mather & Platt are sanguine that the net efficiency of the line will amount to at least 60 per cent.

The central generating station is situated at Stockwell, the suburban terminus of the line. The plant consists of three large dynamos of the Edison-Hopkinson type, each worked independently by a vertical compound engine, designed and constructed by Messrs. John Fowler & Co. Steam is furnished from six 250 horse-power Lancashire boilers.

The engines work at a steam pressure of 140 pounds per square inch, and are of very massive proportions. They run at 100 revolutions per minute, giving a piston speed of 450 feet per minute. They are fitted with automatic expansion gear on both the high and low pressure cylinders, the governor being driven direct from the crank shaft by cotton ropes. The engines will indicate up to 375 horse-power each.

The dynamos used embody the latest improvements of Messrs. Mather & Platt, and are, as well as the engines, built with extra heavy parts. The weight of the entire machine is something over 17 tons, the armature alone weighing about 2 tons, the yoke of the machine about 3 tons, and each magnet limb with its pole piece, about 4 tons. The capacity of the machine is 450 ampères at a pressure of 450 volts. The commutators are of hard copper, insulated with mica, each rocker arm carrying three brushes, which are separably adjustable. The machines can be run either as shunt or compound, as required. The total weight of copper wire on the magnets of each machine is nearly one and one-half tons. The machines are said to have an electrical efficiency of 96 per cent., or slightly more, and the measured efficiency of the engine and dynamo, that is, the ratio of the electrical power available outside the dynamo, to the indicated horse-power of the engine, is said to be over 75 per cent.

The current from the dynamo is conveyed to a general distributing and testing switchboard fixed in a recess of the engine house. From this board the main circuits are taken to various parts of the line, and the current passing through each circuit is measured by suitable ammeters, while arrangements are provided by means of which the current may be switched over from one circuit to another. Sir William Thomson's multicellular electrostatic voltmeters are used for measuring the electromotive force.

The working conductor is of channel steel, carried on glass insulators, the joints being fished, and also connected with copper strips. The general arrangement of the working conductor is exactly the same as that employed by Dr. Edward Hopkinson on the Bessbrook & Newry line. The steel employed is of very high conductivity. The working conductor is divided into sections for convenience of testing and the making of necessary repairs. When the full pressure of 500 volts is on the complete system of working and feeding conductors, the leakage current is said to be not more than one ampère, so that the total loss by leakage is less than one horse-power, a small fraction of one per cent. of the total power required for working the line to its full capacity. The current is collected from the working conductor by sliding shoes of iron or steel arranged much like those employed on the Bessbrook line.

Fourteen 10-ton electric locomotives like that shown in section in Fig. 339, have been supplied by Messrs. Mather & Platt for working the line, each capable of developing 100 effective horse-power, and of running up to 25

tives are fitted with Westinghouse automatic air brakes and also screw hand brakes, and they are lighted by electric lights, the current being derived from the motor circuit. The train, when loaded, will weigh about 30 tons, and it is intended ultimately that ten trains shall be worked on the line at one time, these being run at three-minute intervals.

Up to date the operation of this unique road has been a complete success.

Before leaving the subject of motors for electric traction some of the minor but very important improvements in street railway

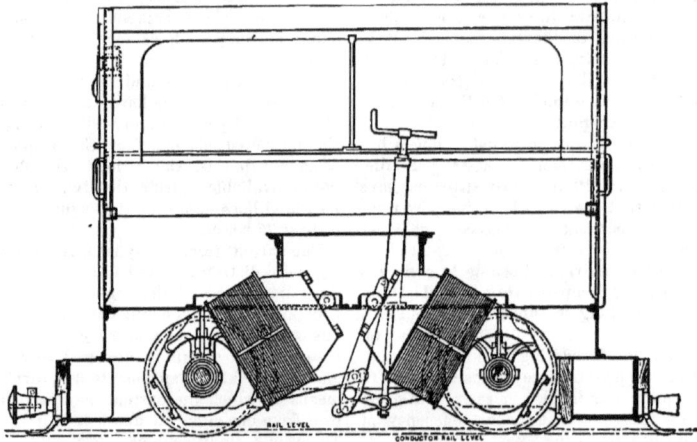

FIG. 339.—ELECTRIC LOCOMOTIVE ON CITY AND SOUTH LONDON RAILWAY.

or 26 miles per hour. The armatures of the locomotives are constructed so that the shaft of the armature is the axle of the locomotive; in this way all intermediate gear and all reciprocating parts are entirely avoided. A motor is fitted on each axle, as shown in the cut, the axles not being coupled, but working independently. The current is conveyed from the collecting shoes, through an ammeter, to a regulating switch, then to a reversing switch, thence to the motors and back through the framework of the locomotive to the rails, so completing the electrical circuit. The locomo-

practice are well worth mentioning. Perhaps the most valuable single advance in apparatus has been the introduction of the carbon brush for railway motors. This improvement really made the difference between the success and failure of the electric street car. It was of course quite practicable to operate motors with copper brushes, but they were continually turning when the motors were reversed, and were a never-ceasing source of delay, annoyance and vexation to the car operators and to the public. The present carbon brush is a mere bar, closely resembling electric light carbon. It is in general from two to three inches wide by a quarter to one-half inch thick and from two to three inches long. It is usually copper-plated, and held by spring pressure in a simple holder radially against the commutator. It gives an excellent contact, very seldom breaks, stands reversal of the armature direction without causing the slightest trouble, and never needs trimming unless accidentally nicked. Sparking is reduced to a very small amount by this device if the commutator is kept anywhere nearly clean. Its only disadvantage is the free way in which it distributes small particles of carbon during the progress of wearing. With proper care, however, it simply forms a dark glaze over the surface of the commutator without either short-circuiting the sections or causing any tendency to sparking. If, however, the commutator is allowed to get dirty and accumulate carbon dust, there is immediate trouble from the commutator heating through short circuiting, and also from the tendency of sparks to flash around the commutator in the conducting layer thus formed. With a reasonable amount of care no trouble of this kind need be experienced.

Another source of great difficulty that has been partially eliminated at the present time is the trolley that collects current from the working conductor. The main trouble in this case comes not from the trolley wheel proper but from the difficulty of getting a smooth upward pressure by springs without the danger of frequent breakage. The early trolleys were continually getting out of order ; the wheels would get jammed and refuse to turn ; their bearings would give way, and occasionally the trolley and pole would come down together into the street with a crash. The principal difficulty with the wheel is due to the impracticability of using much lubrication, and a long series of experiments was necessary before any ways were found of constructing bearings which should give good conduction for the current, and at the same time good mechanical properties. By the free use of anti-friction material and more careful workmanship this difficulty has been largely eliminated. Experience, too, has taught ways of giving a spring pressure to force the trolley

pole upward against the trolley wire without continual breakage. This perhaps has been best attained by the use of spiral springs strictly in tension. Very recently aluminium has been occasionally substituted for brass and bronze as the material for the trolley wheel and its immediate support. The result of this is very much to reduce the weight that has to be held against the trolley wire and thus simplify the mechanical process of obtaining even pressure. With a trolley wire properly lined up a continuous contact is the rule and only rarely does the collector jump from the wire. Some experiments have been tried with sliding contacts instead of wheels, with a fair degree of success.

In line construction a vast number of clever and ingenious devices for facilitating various portions of the work have been introduced. The substitution of iron for wooden supporting poles for the trolley line has been a very great advantage, as it enables the line to be kept taut and true without the continual annoyance from sagging that was the experience of a couple of years ago. The general construction of electric railway lines is steadily improving and consequently the electric motor is more thoroughly appreciated. Most of the difficulties that have to be met are mechanical ones, and when it was thought practicable to operate heavy motor cars on old horse-car tracks with light rails on lighter stringers, trouble of every kind was incessant. But now that engineers have come to appreciate the importance of solid track construction and careful line work the motors are found to perform better than was ever supposed possible.

During the last two years underground and storage battery systems of supply have remained practically at a standstill. The former is not now in use except in an experimental way in the United States, although some of the foreign roads have met with success. A steady effort has been made to bring the storage battery to the front but the inherent difficulties have proved too much for it. It has been tried thoroughly and carefully, but with very indifferent success. Two years' experience in Philadelphia with half a dozen cars in active service has led to the abandonment of the scheme. Two years' trial on a considerable scale in New Orleans has just terminated in a similar failure. Cars have been running spasmodically on the Fourth Avenue line in New York, with fair success, when they have been operated at all, but nothing has come of it. There is a single small storage battery road in Massachusetts; a few cars in Dubuque, Ia., a few in Washington, and scattered experimental cars here and there. That is all that has come of the storage battery so far. The case is by no means a hopeless one, however, and there is every reason to believe that eventually the storage battery may take a prominent part in city electrical traction. Up to the present then there is very little to report. One promising series of experiments, however, was recently carried out in Philadelphia with the Waddell-Entz storage battery, a modification of the alkaline zincate type. To sum it all up, storage battery traction may be a success but it generally is not.

Of stationary motors there has been little less than a deluge during the two years since the second edition of this book was published. Some of this numerous list of machines have interesting and useful peculiarities; others are simply imitations of some well-known type, or designed with more or less skill on general principles, but without possessing any striking merits or demerits. Within the scope of this volume it is simply out of the question to describe, or even mention, the majority of the stationary motors that are on the market. Most of them are good; the general character of their designs follows a comparatively small number of models, and the running speed, efficiency, and price are for the most part not widely different. Hence, in the brief space that must necessarily be allotted to description, no attempt will be made at completeness, but a few characteristic machines that are in especially wide use, or that possess some peculiarly interesting features of design or construction, will be described. For the further information of those who desire details of particular machines, we can do no better than to refer them to the current files of THE ELECTRICAL WORLD, wherein new motors are generally described very soon after their appearance before the public. As this chapter is not intended for a catalogue of all known machines, but for the intelligent information of those who desire to make themselves acquainted with good modern practice, attention will be confined to a limited number of thoroughly well-known and practical machines.

A couple of years ago the Sprague stationary motor was the best known machine of its class, and continued a general favorite up to the time that the Sprague Company was swallowed up by the Edison General Electric Company. The Sprague type was then immediately

FIG. 340.—TWENTY-FIVE KILO-WATT EDISON MOTOR.

and completely discarded, and in its place was put the Edison motor, that is at once an excellent specimen of modern design and a capital lesson in the reversibility of the dynamo. It is nothing more nor less than the well-known and reliable Edison dynamo, operated as a motor, with merely such changes as are necessary in reversing the direction of rotation of the armature. The differences between it and the incandescent dynamo of a similar size are scarcely discernible, and the windings are practically identical, except in the machines designed for special purposes. This is a sufficient commendation, for it is needless to say that the same characteristics that make a good and reliable dynamo are sufficient to ensure admirable performance in a motor. Fig. 340 shows the complete machine. The type and general appearance remain the same from the smallest motor manufactured, that of ¼ horse-

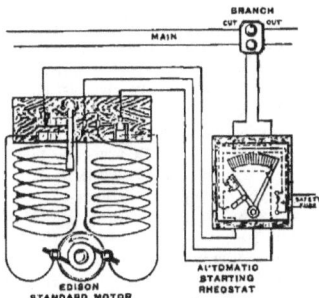

FIG. 341.—CONNECTIONS OF EDISON MOTOR.

power, up to the 150 horse-power motor, corresponding to the largest of Edison dynamos. Fig. 341 shows the diagram of connections, both of the motor itself and of the rheostat, while Fig. 342 gives a view of the self-oiling

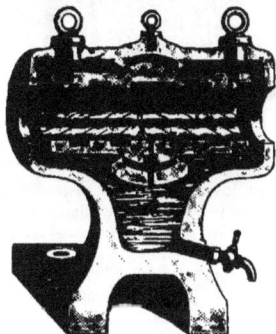

FIG. 342.—EDISON SELF OILING BEARING.

bearing that is one of the most desirable mechanical features of the machine. This device insures complete and excellent lubrication for long periods of running. The brushes usually employed with these motors are of the

regular Edison type and are of hard, straight, copper wires; occasionally they are replaced by carbon brushes.

The machines are made of all sizes, the smaller ones generally being intended for 110 volts, and the larger sizes for 220 or 500 volts. For these higher voltages of course the windings are special, but the dimensions and the arrangements of the machines remain practically the same. The speed of the motors is very nearly that of the corresponding sizes of dynamo of the same voltage, and ranges from 2,100 revolutions per minute in the ¼ and ½ horse-power motors to as low as 360 in the 150 horse-power machine, the largest one listed. The Edi-

FIG. 343.—DETAIL VIEW OF CROCKER-WHEELER MOTOR.

son Company makes at present only constant potential machines, using for constant current work mainly the excellent constant current motor manufactured by the Crocker-Wheeler Motor Co., whose machines are very widely known, and are of special interest as displaying a design and construction peculiarly their own and quite apart from ordinary dynamo types.

Fig. 343 gives an excellent detailed view of the standard Crocker-Wheeler motor, while Fig. 344 shows the general appearance of the 5 horse-power pattern. To begin with, the machine is of the inverted horseshoe type; each pole piece is continuous with its magnetic core, and is formed of the softest iron, drop-forged exactly to its finished shape. These forgings are fitted very carefully into recesses in the main casting of the motor that forms at once the magnet yoke and the support for the bear-

FIG. 344.—FIVE HORSE-POWER CROCKER-WHEELER MOTOR.

ings. The armature is relatively of very large diameter, and, compared to the field, quite powerful. The pole pieces are rather lean in figure, but their lack of cross-section is more than compensated by the excellent quality

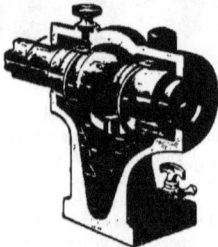

FIG. 345.—CROCKER-WHEELER SELF OILING BEARING.

of the iron. The result of this construction is a very powerful field obtained most economically. The armature is a Pacinotti ring with a comparatively small amount of wire wound upon it. The clearance of the armature is so

small that the magnetic resistance of the air gap is exceptionally low, and the coils, sunk flush with the surface of the armature, are subjected to a very powerful induction. This construction, too, gives almost complete immunity from burning out of the armature, as each section is isolated, and no two contiguous wires are subjected to any considerable difference of potential. Relatively large as the armatures are they are yet exquisitely balanced, and run almost without vibration, while the size gives a powerful torque and good efficiency at an unusually low speed, exceptionally low for a two-pole machine. The bearings, shown in detail in Fig. 345, have all the mechanical features of those employed in the largest machines, are self-oiling and self-centering. Both constant potential and constant current motors are made by the Crocker-Wheeler Company, and in general appearance closely resemble each other.

FIG. 346.—CROCKER-WHEELER FAN MOTOR.

It is the intention of the makers that the smallest machine turned out shall be as complete and perfect in all its details as the largest. A glance at the machine, Fig. 346, shows another excellent mechanical feature, for the supports of the bearings are fitted to the projecting base of the machine that carries them, not with a straight joint, but with a bearing that follows the arc of a circle, so that in taking out the armature and replacing it again there is no danger of getting it in the least out of line. Fig. 344, the 5 horse-power motor, is the largest size regularly made, and from that the motors run down to the little $\frac{1}{12}$ horse-power affair principally used in driving fans. All the sizes retain the same fundamental characteristics. The little fan motor just mentioned is shown in Fig. 346. It carries, usually, a 12-inch fan, and has come into very extensive use in offices, restaurants, and the like.

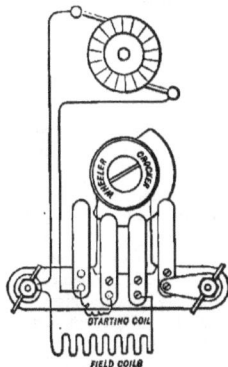

FIG. 347.—CROCKER-WHEELER STARTING SWITCH.

On its pole piece will be noticed a starting switch, which is supplied to all the small motors for starting and stopping, and in some cases for regulating. This switch when turned first charges the field, then starts the armature through a resistance wound on the machine, and finally cuts out the resistance and gives the full current to the armature. The details of this little device are well shown in Fig. 347. The constant potential motors are wound for almost every possible case, for voltages from 6 volts, for use with a battery, to 500 volts, to be employed on a railway circuit. The battery motors are usually, however, series wound. The sizes up to ½ horse-power may be provided with a starting device employing a two-speed switch for use if occasion requires, and motors of all the usual sizes and voltages are arranged with reversible switches, if desired. Regulating boxes similar to those used with the Edison motors furnish means for varying the speed of the larger sizes. Besides these constant potential machines, the Crocker-Wheeler Company makes a large number of

motors for constant current, to be employed where arc circuits only are available. A complete series of these machines is made, ranging from ½ horse-power up to 5 horse-power, and meet wide use where incandescent circuits are not within reach. They are in general appearance very similar to the constant potential machines. In the latest models regulation is accomplished by shifting the brushes. This movement in the larger motors is accomplished by a centrifugal governor acting upon the brush holders.

Fig. 348 shows one of the smaller constant current motors. They are wound regularly for 6½, 9½, or 18 ampère circuits. The arc motors below ½ horse-power are fitted with a simple hand-governor instead of the centrifugal arrangement, for they are generally used only on regular work, for which they may be set to run at any desired speed.

It will thus be seen that the Crocker-Wheeler machines are characteristically motors in construction, having an armature vastly more powerful than is usual in dynamos, and being intended to give a powerful torque and low speed. Nevertheless, the larger sizes of constant potential machines are not infrequently used for dynamos, and work admirably for this

FIG. 348.—SMALL CROCKER-WHEELER CONSTANT CURRENT MOTOR.

purpose. They may be said to represent, however, the motor type, just as the Edison motor is the typical reversible dynamo.

To the latter category belongs another modern motor that is mentioned here especially on account of the character of its magnetic circuit.

The Connecticut motor, which for the past two years has been manufactured at Plantsville, Conn., is almost unique from the fact that throughout all sizes of the machine the magnetic circuit is composed of a single casting without joints of any kind. It was designed especially with the idea of furnishing

FIG. 349.—THE CONNECTICUT MOTOR.

a simple, strong and reliable motor of fair efficiency and good mechanical qualities. There has been a certain tendency in electrical machine design to fly to one of two extremes, either to build a motor exceedingly efficient but somewhat thin-skinned and unable to resist severe strains, or, on the other hand, to neglect electrical efficiency too much. The present Connecticut motor is of a compromise pattern, and while remarkably simple in construction has a good efficiency and is mechanically well made. The form of the machine throughout is the inverted horseshoe. In the smaller sizes the armature is supported by brackets fastened to the pole pieces ; in the larger sizes by bearings carried on an extension of the base.

Fig. 349 shows the first pattern, which is adapted for machines of 2 horse-power and under. The magnetic circuit is short and its cross-section large ; it is a single casting of soft iron, so formed that the magnetizing coils can be wound on bobbins and dropped directly over the pole pieces. The armature is a drum about two diameters long, wound with a comparatively small number of turns, and with rather coarse wire. The air

gap is short and the magnetizing power required, owing to this fact and to the absence of any joints, is very small for a cast iron magnet. The smaller sizes have the shaft revolving in graphite bushings, while the larger ones are provided with self-oiling bearings. In all the journals are allowed a large bearing surface. The machines are manufactured for incandescent circuits only, and are shunt wound for 110, 220 and 500 volts. They are started through the medium of a switch box not unlike that employed in the Edison and several other systems. Generators of the same form are also manufactured, being either shunt or compound wound, as the occasion requires.

Another very well known electric motor is the Eddy, which like the Edison and Connecticut motors follows the usual lines of dynamo construction very closely. The magnetic circuit is of a modified horseshoe form, somewhat elliptical in shape, and of large cross-section, and it is usually mounted on a wooden base. The material is soft cast iron, and the motor is shunt wound with unusually fine wire. The armature is of the drum form, Siemens wound, as usual. It is wound with a comparatively small number of turns of rather coarse wire,

FIG. 350.—THE EDDY MOTOR.

giving a low armature resistance. All motors of above 7½ horse-power are wound with several wires in parallel for convenience and efficiency. The armature is supported by gun-metal yokes fastened rigidly to the pole pieces of the magnet by gun-metal studs. These yokes contain bearing sleeves of hard composition metal.

The bearings of all sizes are self-oiling and consequently require but little attention. The self-oiling device is a loose ring hanging on the shaft and dipping into an oil well, the form employed, with various modifications, on nearly every self-oiling bearing. After the oil works along the shaft it comes out at each end of the box, is caught in grooves turned for that purpose, and immediately returns to the well. The regulation, as might be expected from the design, is automatic, and the machines have given an excellent account of themselves in a wide variety of service. They are manufactured in all sizes and for all the usual voltages. Fig. 350 gives a good idea of the motor in question.

A very compact and well worked out design may be found in the motors recently brought out by the United States Electric Lighting Co. They present a radical departure from the usual shapes of magnetic circuit, the form presented requiring but a single magnetizing coil, and being virtually an inverted horseshoe in shape, with the coils wound around the yoke, as shown in Fig. 351. The magnetic circuit is cast in two pieces, the joint being in the center of the magnetizing coil, and the two portions being held together by the bolts shown in the cut. The mechanical construction is exceedingly simple, as the field magnets form their own base by projections cast solid with them, and similar projections form a support for the bearings of the armature shaft. The switch for controlling the motor is placed directly on top of the pole pieces. These motors are made in sizes from ¼ horse-power up to 20 horsepower, wound for all the usual potentials up

FIG. 351.—THE UNITED STATES MOTOR.

to 500 volts. The armature presents some interesting peculiarities; it is a drum of rather large diameter, and is of the toothed variety; the teeth are very numerous and small, so that no trouble is encountered from the heating that almost always follows the use of large projections in an armature. This construction accomplishes two ends in the most admirable fashion. In the first place it reduces the air gap to a very minute amount, inasmuch as the teeth run very close to the polar surfaces. In the second place it simplifies winding the armature immensely, for no special care need be taken in laying off the various sections as the armature is wound; it is simply necessary to take the size of wire used for that particular motor and fill the space between the teeth with it, thus forming an independent segment of the armature. The mechanical advantage secured by this construction is that all the armature wires and bands lie beneath the surface of the armature and are therefore completely protected from injury. The armature resistance is very low, and the field secured by the compact form of magnetic circuit is a very powerful one, so that the efficiency of the finished motor is high, and its speed moderate. By careful design of the pole pieces the non-sparking area is sufficiently increased for the motors to run without shifting the brushes, even under very violent changes of load. Sizes of 1 horse-power and above are supplied with a special base and belt-tightener when necessary. Like the other motors just mentioned, these are constant potential machines, and are frequently used as dynamos, making a very compact and efficient machine for generating purposes. When used as dynamos they are very often compound wound. For a motor to be used for potentials above 220 volts a special starting device is employed, enclosed in a glass-topped case, so that accidental contact with the terminals of the machine is impossible. For these high potential forms the brushes are held in hard rubber brush-holders lined with metal, so constructed as to avoid the danger of receiving shocks from either brush or holder.

A group of machines that possess some excellent features, and have served as a model for not a few imitations, are the Perret motors made by the Elektron Manufacturing Company of Brooklyn. Their distinctive feature is the lamination of the field magnet; instead of being cast or forged in one or more solid pieces, as usual, it is built up of thin plates of charcoal iron stamped into their finished form, and then clamped together by bolts. The advantage of such a construction is primarily the ready use of the best quality of soft iron for the magnetic circuit. The lamination further tends to check eddy currents in the pole pieces, and enables the toothed armature to be used without any special difficulties from the cause just mentioned. At first sight the laminated magnets might appear costly, but the simplicity of construction by stamping has in large part obviated this objection. The armatures are provided with teeth, and the coils wound in the spaces between them are completely beneath the surface of the armature, and thoroughly protected from injury. The toothed construction, as mentioned in describing the United States motor, enables the air gap to be reduced to a very small amount, consequently the magnetism is very economically obtained, and the armature wire is utilized to the best advantage. The Elektron Company does not confine itself to a single type of machine, but manufactures three distinct patterns. The first, used for the smallest sizes of motors, is of the erect horseshoe form, and is provided with a toothed drum armature. For somewhat larger motors, up to 2 horse-power, the consequent pole form of magnetic circuit with four magnetizing coils is employed; while for larger machines the two-pole con-

struction is abandoned and the Perret motors from 2 to 20 horse-power are six-pole machines with ring armatures. The special advantage to be gained by the multipolar construction is light weight and low speed for a given efficiency and output. The modern tendency in motor building is toward lowering the running speed as far as possible, because for most purposes for which small motors are used a speed of 1,500 to 2,000 revolutions per minute is far higher than is desirable in the machines that are to be driven, consequently the speed must be reduced by belting or gearing in a very considerable ratio; so much, in fact, as to cause the loss of considerable power through

FIG. 352.—PERRET MULTIPOLAR MOTOR.

countershafts. The more nearly the speed of any motor can be made to correspond with the driven machine the more economically it can be applied, so that there is a very great advantage in the use of the multipolar construction with its consequent reduction of speed. In these larger Perret motors the laminated field magnets are retained. Fig. 352 shows one of the latest multipolar motors of 5 horse-power. Fig. 353 exhibits the arrangement of the magnetic circuit. The six poles are furnished with three independent magnets arranged at equidistant points around the armature. There are but three magnetizing coils, each one energiz-

ing a single magnet. There are no joints at all
in the magnetic circuit except the air gap,
since the stampings that form the individual
magnets are of such shape that they can be
readily wound in a lathe. The armature is a
ring of relatively very large diameter, toothed,

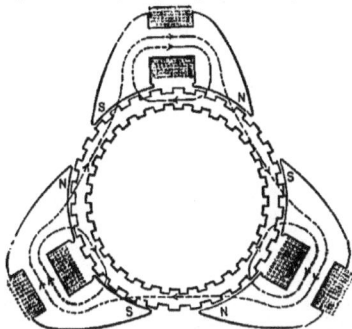

Fig. 353.—MAGNETIC CIRCUIT OF PERRET MULTI-
POLAR MOTOR.

and running very close to the pole pieces as in
the usual forms of the Perret motor. The resulting machine is a very compact, convenient
and efficient one, and its speed is singularly
low, only about half that employed in two-pole
motors of corresponding size. The very small

sizes have come into wide use for fans, sewing
machines and the like ; while the larger ones
recently introduced have made a good reputation for ordinary power purposes.

A large number of electric motors besides
those mentioned have been placed upon the
market within the past two years. Most of
them have no special distinguishing peculiarities that are not well shown in one or more of
the motors that have just been described. They
represent individual experience and fancy in
electrical design, and are most of them of good
quality. There is to-day no excuse for building a poor motor.

Of alternating current motors there is comparatively little to be said. A large number
of patents have been taken out, there has been
much interesting discussion as to details of
construction, but very few motors have been
built and fewer yet sold. None, at the time
of writing, have been used extensively enough
to enable a judgment to be formed as to their
practical qualities. The Tesla machines, mentioned in Chapter XV, are perhaps the best
known of any, but even they have not come
into any considerable use. Very little that is
radically new has been done in alternating current construction, and the present state of the
art is effectively covered by the chapter just
mentioned.

INDEX.

LIST OF BOOKS

—RECOMMENDED BY—

THE ✳ ELECTRICAL ✳ WORLD

FOR SPECIAL READING.

We are often asked by those who desire to inform themselves in regard to electrical matters to recommend to them a course of reading or books on particular subjects or in relation to certain special departments of electrical application. The following list will, we trust, meet the requirements of most of those who desire such information. With scarcely an exception, the books mentioned are substantially bound in cloth and copiously illustrated.

A. PRINCIPLES AND THEORY
of Electricity and Magnetism.

(1.) An Elementary Course.

PRICE.

Atkinson's Elements of Static Electricity, with a full description of the Holtz and Töpler Machines........ $1.50

Atkinson's Elements of Dynamic Electricity and Magnetism 2.00

Ayrton's Practical Electricity for First Year Students 2.50

Electrician Primers, vol. I, Theory, $1.00; vol. II., Practice 1.00

Fleming's Short Lectures to Electrical Artisans....... 1.50

Houston's Dictionary of Electrical Words, Terms and Phrases. Second edition, entirely re-written, containing about 5,000 distinct titles, 570 Illustrations and 502 double column pages. 8vo 5.00

Jenkin's Electricity and Magnetism, with an Appendix on the Telephone and Microphone............ .... 1.50

Kennelly & Wilkinson's Practical Notes for Electrical Students 2.50

Maycock's First Book of Electricity and Magnetism. 0.60

Thompson's Elementary Lessons in Electricity and Magnetism 1.25

Thompson's Lectures on the Electromagnet.......... 1.00

(2.) An Advanced Course.

Cumming's Introduction to the Theory of Electricity.. $2.25

Emtage's Introduction to the Mathematical Theory of Electricity and Magnetism 1.00

Ewing's Magnetism of Iron and other Metals (new) . 4.00

Fleming's Alternate Current Transformer in Theory and Practice. 2 vols., second vol. in press. 3.00

Faraday's Experimental Researches in Electricity. Three vols 20.00

Houston's Dictionary of Electrical Words, Terms and Phrases, second edition, entirely re-written, containing about 5,000 distinct titles, 570 Illustrations and 502 double column pages. 8vo...... .... 5.00

Lodge's Modern Views of Electricity........ .... 2.00

Mascart & Joubert's Treatise on Electricity and Magnetism. Two vols 15.00

Maxwell's Treatise on Electricity and Magnetism. Two vols. New edition. In press....

Thompson's Electromagnet and Electromagnetic Mechanisms 6.00

Watson & Burbury's Mathematical Theory of Electricity and Magnetism. Two vols 5.50

(B.) PRACTICAL APPLICATIONS
of Electricity and Magnetism.

(1.) General Treatises.

PRICE.

Electricity in Daily Life........................... .... $3.00

Hospitalier's Modern Applications of Electricity. Two vols 8.00

Houston's Dictionary of Electrical Words, Terms and Phrases, second edition, entirely re-written, containing about 5,000 distinct titles, 570 Illustrations and 502 double column pages. 8vo 5.00

Guillemin's Electricity and Magnetism..... .... 8.00

Slingo & Brooker's Electrical Engineering for Electric Light Artisans and Students.................. 3.50

Trevert's Electricity and Its Recent Applications ... 2.00

Wormell's Electricity in the Service of Man.... ..... 6.00

(a.) Electric Lighting.

Alglave & Boulard's Electric Light: Its History, Production and Application. $5.00

Atkinson's Elements of Electric Lighting 1.50

Day's Electric Light Arithmetic.................... ... 0.40

Desmond's Electricity for Engineers 2.50

Dredge's Electric Illumination. Vol. 1., $15., vol II . 7.50

Gordon's Decorative Electricity 3.75

Houston's Dictionary of Electrical Words, Terms and Phrases, second edition, entirely re-written, containing about 5,000 distinct titles, 570 Illustrations and 502 double column pages. 8vo............. .... 5.00

Latimer's Incandescent Electric Lighting...... .... 0.50

Russell's Electric Light Cables......................... 1.50

Urquhart's Electric Light: Its Production and Use..... $3.00

Urquhart's Electric Light Fitting 2.00

(b.) The Electric Motor.

Badt's Electric Transmission Hand-book $1.00

Bottone's Electro-motors; How Made and How Used.. .50

Crosby & Bell's Electric Railway in Theory and Practice 2.50

Houston's Dictionary of Electrical Words, Terms and Phrases, second edition, entirely re-written, containing about 5,000 distinct titles, 570 Illustrations and 502 double column pages. 8vo...................... 5.00

Kapp's Electric Transmission of Energy............ .... 3.00

Martin & Wetzler's Electric Motor and Its Applications; with an Appendix by Dr. Louis Bell.... ...... 3.00

Urquhart's Electro-motors 3.00

LIST OF BOOKS RECOMMENDED BY THE ELECTRICAL WORLD.

(c.) Telegraphy.

	PRICE.
Abernethy's Commercial and Railway Telegraphy ..	$2.00
Houston's Dictionary of Electrical Words, Terms and Phrases, second edition, entirely re-written, containing about 5,000 distinct titles, 570 illustrations and 562 double column pages. 8vo	5.00
Lockwood's Electricity, Magnetism, and Electric Telegraphy.............................	2.50
Mayer & Davis' Quadruplex, with Chapters on Telegraph Repeaters and the Wheatstone Automatic Telegraph	1.50
Pope's Modern Practice of the Electric Telegraph.....	1.50
Preece & Sivewright's Telegraphy...........	1.75
Prescott's Electricity and the Electric Telegraph. Two vols........	7.00
Plum's Military Telegraph During Our Civil War. Two vols	5.00
Reid's Telegraph in America..............	5.00

(d.) The Telephone.

Du Moncel's Telephone, The Microphone and the Phonograph......	$1.25
Houston's Dictionary of Electrical Words, Terms and Phrases, second edition, entirely re-written, containing about 5,000 distinct titles, 570 illustrations and 562 double column pages. 8vo........	5.00
Lockwood's Practical Information for Telephonists..	1.00
Poole's Practical Telephone Hand-book........... ...	1.25
Preece & Maier's Telephone......................	4.00
Prescott's Bell's Electric Speaking Telephone	6.00

(e.) Electro-Metallurgy.

Bonney's Electro-platers' Hand-book	$1.20
Gore's Art of Electrolytic Separation of Metals, etc....	3.50
Gore's Theory and Practice of Electro-deposition	0.80
Houston's Dictionary of Electrical Words, Terms and Phrases, second edition, entirely re-written, containing about 5,000 distinct titles, 570 illustrations and 562 double column pages. 8vo..............	5.00
Urquhart's Electrotyping......	2.00
Wahl's Galvanoplastic Manipulation...................	7.50
Watt's Electro-deposition..............................	3.50

(f.) Batteries.

Carhart's Primary Batteries.	$1.50
Gladstone & Tribe's Chemistry of Secondary Batteries of Plante and Faure......................	1.00
Houston's Dictionary of Electrical Words, Terms and Phrases, second edition, entirely re-written, containing about 5,000 distinct titles, 570 illustrations and 562 double column pages. 8vo..............	5.00
Niaudet's Elementary Treatise on Electric Batteries .	2.50
Niblett's Secondary Batteries....	1.50
Reynier's Voltaic Accumulator.......................	3.00
Salomons' Electric Light Installation and the Management of Accumulators............	2.00

(g.) The Dynamo.

Bad's Dynamo Tenders' Hand-book...............	$1.00
Bottone's Dynamo: How Made and How Used......	1.00
Croft's How to Make a Dynamo.....................	.80
Hering's Principles of Dynamo Electric Machines ..	2.50
Thompson's Dynamo Electric Machinery. New. Fourth edition. Revised. Re-written.....	6.00
Walker's Practical Dynamo Building for Amateurs..	.80

(h.) Alternating Currents.

	PRICE.
Blakesley's Papers on Alternating Currents of Electricity. Reprinting...	
Desmond's Electricity for Engineers	$2.50
Fleming's Alternate Current Transformer in Theory and Practice	3.00
Houston's Dictionary of Electrical Words, Terms and Phrases, second edition, entirely re-written, containing about 5,000 distinct titles, 570 illustrations and 562 double-column pages. 8vo..........	5.00

(C.) ELECTRICAL TESTING and Measurement.

Ayrton's Practical Electricity......	$2.50
Gray's Absolute Measurement in Electricity and Magnetism	1.25
Hering's Table of Equivalents of Units of Measurement	0.50
Houston's Dictionary of Electrical Words, Terms and Phrases, second edition, entirely re-written, containing about 5,000 distinct titles, 570 illustrations and 562 double-column pages. 8vo	5.00
Kempe's Hand book of Electrical Testing....	5.00
Lockwood's Electrical Measurement and the Galvanometer	1.50
Swinburne's Practical Electrical Measurement........	1.75
Webb's Testing of Insulated Wires and Cables	1.00

(D.) MISCELLANEOUS.

Allsop's Practical Electric Bell Fitting.................	$1.25
Atkinson's Elements of Static Electricity.............	1.50
Gray's Electrical Influence Machines	1.75
Houston's Dictionary of Electrical Words, Terms and Phrases, second edition, entirely re-written, containing about 5,000 distinct titles, 570 illustrations and 562 double-column pages. 8vo	5.00
Hering's Universal Wiring Computer....	1.00

(E.) BOOKS FOR THE NON-TECHNICAL READER.

Benjamin's Age of Electricity.....................	$2.00
Guillemin's Electricity and Magnetism...............	8.00
Hospitalier's Modern Applications of Electricity. Two vols......	8.00
Houston's Dictionary of Electrical Words, Terms and Phrases, second edition, entirely re-written, containing about 5,000 distinct titles, 570 illustrations and 562 double column pages. 8vo..............	5.00
Reid's Telegraph in America.........................	5.00
Wormell's Electricity in the Service of Man	6.00

(F.) HISTORICAL WORKS.

Alglave & Boulard's Electric Light	$5.00
Dredge's Electric Illumination. Vol. I, $15.00, vol. II	7.50
Fahie's History of Telegraphy to 1837	3.00
Martin & Wetzler's Electric Motor and Its Applications.....................................	
Pope's Evolution of the Electric Incandescent Lamp	1.00
Prescott's Telephone.............	6.00
Reid's Telegraph in America...........................	5.00
Thompson's Dynamo Electric Machinery. New. Fourth edition. Revised. Re-written............	6.00
Thompson's Philipp Reis, Inventor of the Telephone.	3.00

Copies of any of the books mentioned above, or any other electrical books published, will be mailed, POSTAGE PREPAID, to any address in the world on receipt of price. Remit by Post office order, express money order, draft or registered letter, and address all orders to

THE W. J. JOHNSTON COMPANY, LIMITED,

Publishers, Importers and Wholesale and Retail Dealers in Electrical Books,

167-176 TIMES BUILDING, NEW YORK.